Technology of Fluoropolymers

This third edition has been updated and expanded, providing industrial chemists, technologists, environmental scientists, and engineers with an accurate, compact, and practical source of information on fluoropolymers. Highlighting existing and new industrial, military, medical, and consumer goods applications, this edition adds more detailed information on equipment and processing conditions. It explores breakthroughs in understanding property-structure relationships, new polymerization techniques, and the chemistry underlying polymers, such as melt-processable fluoroplastics. It also expands on the important properties of fluoropolymers, including heat and radiation degradation, health effects, and recycling.

Features:

- Revised, updated, and expanded to continue to provide an accurate, compact, and practical source of information on fluoropolymers
- Explores the property-structure relationships, polymerization techniques, and the chemistry underlying polymers
- Fluoropolymers rank high on the specialty polymers group and, due to their unique properties, are naturally part of the solution to the industrial sustainability challenges of the twenty-first century
- Describes the technology of fluoropolymers, including thermoplastic and elastomeric products
- Expands upon the important characteristics of fluoropolymers and their recycling.

Technology of Fluoropolymers

A Concise Handbook

Third Edition

Jiri G. Drobny
Drobny Polymer Associates

Sina Ebnesajjad
FluoroConsultants Group

CRC Press
Taylor & Francis Group
Boca Raton London New York

CRC Press is an imprint of the
Taylor & Francis Group, an **informa** business

First edition published 2023
by CRC Press
6000 Broken Sound Parkway NW, Suite 300, Boca Raton, FL 33487-2742

and by CRC Press
4 Park Square, Milton Park, Abingdon, Oxon, OX14 4RN

© 2023 Jiri G. Drobny and Sina Ebnesajjad

CRC Press is an imprint of Taylor & Francis Group, LLC

ISBN: 978-1-032-01360-2 (hbk)
ISBN: 978-1-032-06886-2 (pbk)
ISBN: 978-1-003-20427-5 (ebk)

DOI: 10.1201/9781003204275

Typeset in Times
by SPi Technologies India Pvt Ltd (Straive)

Contents

Part I   Overview of Fluoropolymers

Part II   Thermoplastic Fluoropolymers

Part III   Fluoroelastomers

Part IV   Technology of Fluoropolymer Aqueous Systems

12 Characteristics and Properties of Fluoropolymer Aqueous Systems

Part VII   Safety and Sustainability

Abbreviations

APA	Advanced Polymer Architecture
APET	Amorphous polyethylene terephtalate
ASTM	American Society for Testing and Materials (now ASTM International)
BMI	Bismaleimide
CD	Cross-direction
CMD	Cross-machine direction
CSM	Cure site monomer
CTE	Coefficient of thermal expansion
CTFE	Chlorotrifluoroethylene
DIN	Deutsches Institut für Normung eV (German Institute for Standardization)
DMI	Bismaleimide
DTM	Differential thermal analysis
E	Ethylene
EB	Electron beam
ECTFE	Copolymer of ethylene and chlorotrifluoroethylene
EFEP	Ethylene-tetrafluoroethylene-hexafluoropropylene terpolymer
ETFE	Copolymer of ethylene and tetrafluoroethylene
FEP	Fluorinated ethylene-propylene (copolymer of tetrafluoroethylene and hexafluoropropylene)
FEPM	Copolymer of tetrafluoroethylene and propylene
FEVE	Fluorinated ethylene vinyl ether
FFKM	Perfluoroelastomer
FKM	Fluorocarbon elastomer
FMQ	Fluorosilicone
FPFPE	Functionalized perfluorinated polyether
FPM	ISO designation for fluorocarbon elastomer of the FKM type (ASTM)
FPU	Fluorinated polyurethane
FTIR	Fourier Transform Infrared Spectroscopy
FTPE	Fluorinated thermoplastic elastomer
FVE	Fluorovinyl ether
FVMQ	Fluoro-vinyl polysiloxane
FZ	Polyphosphazene elastomer
Gy	Grey (SI unit of radiation dose, larger, more widely used unit is kGy)
HDPE	High-density polyethylene
HFIB	Hexafluoroisobutylene
HFP	Hexafluoropropylene
HPFP	Hydropentafluoropropylene
IPN	Interpenetrating network
IR	Infrared
ISO	International Organization for Standardization
JIS	Japanese Industrial Standard
LAN	Local area network
LIM	Liquid injection molding
LLDPE	Linear low density polyethylene
LOI	Limiting oxygen index
MA	Maleic anhydride
MD	Machine direction

MDO	Machine direction orientation
MFA	Copolymer of tetrafluoroethylene and tetrafluoroethylene and perfluoromethyl vinyl ether
MVE	Methyl vinyl ether
M_n	Number average molecular weight
M_w	Weight average molecular weight
MQ	Silicone resins
MWD	Molecular weight distribution
NBS	National Bureau of Standards (in 1988 changed to NIST, see below)
NIST	National Institute of Standards and Technology
OI	Orientation ratio
P	Poise (unit of dynamic viscosity, superseded by SI unit of Pa•s)
P (or PP)	Polypropylene
PA	Polyamide
PAS	Polyarylsulfone
PAVE	Perfluoroalkyl vinyl ether
PCTFE	Polychlorotrifluoroethylene
PDD	Perfluoro-2,2-dimethyl dioxole
PE	Polyethylene
PES	Polyethylsulfone
PETG	Glycol-modified polyethylene terephtalate
PFA	Perfluoroalkoxy polymer (copolymer of tetrafluoroethylene and perfluoropropyl vinyl ether)
PFEVE	Poly(fluoroethylene vinyl ether)
PFOA	Perfluorooctanoic acid
PFOS	Perfluorooctanesulfonate
PI	Polyimide
PMTFPS	Polymethyltrifluoropropylsiloxane or poly[methyl(3,3,3-trifluoropropylsiloxane)
PMVE	Perfluoromethyl vinyl ether
P or PP	Polypropylene
PPVE	Perfluoropropyl vinyl ether
PTFE	Polytetrafluoroethylene
P, PUR	Polyurethane
PVC	Polyvinyl chloride
PVDF	Polyvinylidene fluoride
PVF	Polyvinyl fluoride
RF	Radiofrequency
RoHS	Restriction of hazardous substance
RR	Reduction ratio
SBS	Styrene-butadiene-styrene block copolymer
SEM	Scanning electron microscope
SI	International system of units
SKF	Fluorocarbon elastomer
SSG	Standard specific gravity,
T_g	Glass transition temperature
T_m	Crystalline melting point
TAC	Triallyl cyanaurate
TAIC	Triallyl isocyanurate
TD	Transverse direction (cross-machine direction)
TDO	Transverse direction orientation
TFE	Tetrafluoroethylene
TGA	Thermogravimetric analysis
THV	Terpolymer of tetrafluoroethylene, hexafluoropropylene, and vinylidene fluoride (THV Fluoroplastic, product of 3M Dyneon)

TMPTA	Trimethylolpropane triacrylate
TMPTMA	Trimethylolpropane trimethacrylate
UL	Underwriters Laboratory
UV	Ultraviolet
VDF	Vinylidene fluoride (also VF_2)
VF	Vinyl fluoride
VOC	Volatile organic compounds

Preface to the First Edition

The first major endeavor to review the growing field of fluoropolymers was the book *Fluorocarbons* by M. A. Rudner published by Reinhold Publishing Corporation in 1958 (second printing, 1964), covering the state of the art of fluoropolymer technology. The next major publication, which focused on the chemistry and physics of these materials, was *Fluoropolymers*, edited by L. A. Wall in 1972, published by Wiley Interscience. Without doubt, it has been and still is a valuable resource to scientists doing academic and basic research, but it placed relatively little emphasis on practical applications. Information applicable to industrial practice, whether development or manufacturing, has been available mostly in encyclopedias, such as the *Kirk-Othmer Encyclopedia of Chemical Technology* or the *Polymer Materials Encyclopedia*, and occasional magazine articles. The work *Modern Fluoropolymers: High Performance Fluoropolymers for Diverse Applications*, published in 1997 by John Wiley and Sons in the Wiley Series in Polymer Science and edited by J. Scheirs, covers the significant advancements in the field over the past decade or so. It is a collection of chapters written by a number of experts in their respective fields with an emphasis on structure/property behavior and diverse applications of the individual fluoropolymers.

Technology of Fluoropolymers has the goal of providing systematic fundamental information to professionals working in industrial practice. The main intended audience is chemists or chemical engineers new to fluoropolymer technology, whether the synthesis of a monomer, polymerization, or a process leading to a product. Another reader of this book may be a product or process designer looking for specific properties in a polymeric material. It can also be a useful resource for recent college and university graduates. Because of the breadth of the field and the wide variety of the polymeric materials involved, it does not go into details; this is left to publications of a much larger size. Rather, it covers the essentials and points the reader toward sources of more specific and/or detailed information.

With this in mind, this book is divided into nine separate sections, covering the chemistry of fluoropolymers, their properties, processing, and applications. A distinction is made between fluoroplastics and fluoroelastomers because of the differences in processing and in the final properties, as well as in applications. Technology, i.e., processing and applications, is combined into one chapter. Other topics include effects of heat, radiation, and weathering. Because processing of water-based systems is a distinct technology, it is covered in a separate chapter. Materials that have become commercially available during the past decade or so and some of their applications are included in Chapter 8. Chapter 9 covers recycling.

This book began as lectures and seminars given at the Plastics Engineering Department of the University of Massachusetts at Lowell and to varied professional groups and companies. It draws on the author's more than 40 years of experience as a research and development professional and more recently as an independent international consultant.

My thanks are due to the team from CRC Press particularly to Carole Gustafson, Gerald Papke, and Helena Redshaw for bringing this work to fruition and to my family for continuing support. Special thanks go to my daughter, Jirina, for meticulous typing and help in finishing the manuscript, and to Ms. Kimberly Riendeau for expert help with illustrations. Helpful comments and recommendations by Dr. T. L. Miller from DeWAL Industries are highly appreciated.

Jiri George Drobny
Merrimack, New Hampshire, January 2000

Preface to the Second Edition

The first edition of *Technology of Fluoropolymers* had the main goal to provide systematic fundamental information to professionals working in industrial practice. Since its publication in 2001 the industry has changed. Many technological developments have taken place, new applications were developed, and companies were sold and bought, reorganized, and/or renamed. New products were developed and commercialized and some already established ones were discontinued. Environmental issues such as toxicity of certain additives and of products of thermal decomposition of certain fluoropolymers etc. became quite important. Thus it was time to update the publication to include these changes and issues. As in the first edition, the second edition still stresses the practical aspects of fluoropolymers and their industrial application. Few illustrations were added and one of the major features is the addition of processing and engineering data of commercial products. In addition, the feedback from colleagues, students, clients, and attendants of various seminars and training sessions were helpful in preparing the manuscript of this updated and expanded edition.

My thanks are due to Lindsey Hofmeister, who was very helpful and encouraging in the initial stage of the preparation of the manuscript, Dr. Sina Ebnesajjad of Fluoroconsultants, for valuable advice and encouragement, to Steve Mariconti for valuable comments and recommendations in the text, to Corinne Gangloff from Freedonia Group, and to Ray Will from SRI Consulting for valuable capacity and market data. A special credit is due to the team from CRC Press, particularly to David Fusel, Hilary Rowe, and Sylvia Wood for bringing this work to fruition.

Jiri George Drobny
Merrimack, New Hampshire, and Prague,
Czech Republic, November 2007

Preface to the Third Edition

The previous two editions of *Technology of Fluoropolymers* had the main goal of providing systematic fundamental information to professionals who are entering industrial practice in the fluoropolymer industry and to those who are already working in this field. We also hope to address people in other industries, such as the automotive, electrical and electronic, aerospace, medical, construction, chemical processing, and food processing as well as other groups, such as educators and students of chemistry, chemical engineering, electrical and electronic engineering, and material science.

Since the publication of the second edition in 2009, the industry has grown greatly in size and changed in many ways. In order to meet the challenges of technology developments, new applications, and the entry of new practitioners, an industry professional and writer with decades of experience has joined the project. The goal was set to produce a new edition that would qualify as a *concise handbook* with a larger variety of topics and more depth than the previous two editions.

Consequently, the coverage of the book was expanded by adding new chapters to a total of 20, separated into seven parts. Each of these parts covers a specific topic, namely: an overview of fluoropolymers; thermoplastic fluoropolymers; fluoroelastomers; fluoropolymer aqueous systems; the effects of temperature and other variables on fluoropolymers; and safety and sustainability. New illustrations have been added and additional new engineering data have been included to aid the reader.

Some readers may require more in-depth knowledge in the future than this book covers. This serves as a gateway to finding more detailed references. We have sought to use and cite every trustworthy source of fluoropolymer information in the chapters of this book. Undoubtedly, we may have made inadvertent errors in the fulfilment of the ambitious goals of this book. A note indicating the specific error to the publisher, Taylor and Francis, for the purpose of correcting future editions would be much appreciated.

We are most grateful to the many who have contributed to this book. Institutions, companies, and researchers whose works have been referred to have been cited to the best of our abilities. In spite of our diligent efforts some names may have been missed. For that, we offer our sincere apologies.

J. G. Drobny
S. Ebnesajjad, PhD
November 2022

About the Authors

Jiri George Drobny is the president of Drobny Polymer Associates, an international consulting firm specializing in fluoropolymer science and technology, radiation processing, and elastomer technology. His career spans over 50 years in the polymer processing industry in Europe, the United States, and Canada, mainly in R&D with senior and executive responsibilities. Drobny is also active as an educator, author, and as a technical and scientific translator. He is a member of the Society of Plastic Engineers, the American Chemical Society (ACS), the Rubber Division of ACS, and RadTech International. He resides in New Hampshire in the United States.

Dr. Sina Ebnesajjad retired from DuPont fluoropolymers in 2005 after 23 years of service and founded the FluoroConsultants Group, LLC in 2006. Sina was the series editor of the Plastics Design Library published by Elsevier from 2010 to 2021. He is author, editor, and co-author of over a dozen technical and data books including five handbooks on fluoropolymer technologies and applications. He is the author of multiple volumes on the surface preparation and adhesion of materials. He has been engaged in technical writing and publishing since 1974. His experience includes thermoplastic fluoropolymer technologies, including polymerization, finishing, fabrication, and failure analysis.

Part I

Overview of Fluoropolymers

1

Introduction

Sina Ebnesajjad

People have become so accustomed to the use of plastics in their daily lives that one might think these materials have been around for centuries. Yet most key developments and scientific progress took place after the 1930s, when the word "polymer" joined the lexicon of technology and science terms. Nylon, Teflon®, Neoprene, Lucite®, Dacron®, Lexan®, Kevlar, Nomex®, and so on were new polymers born during those years and soon assumed vital roles in human life. Teflon® a trademark of the DuPont Co (now Chemours) referred to an unusual class of polymer generically called fluoropolymers.

Fluoropolymers represent a specialized group of polymeric materials. Their chemistry is derived from the fluorocarbon compounds first developed for us as refrigerants. In the 1920s, significant efforts were made to develop nontoxic, inert, low boiling, liquid refrigerants mainly for reasons of safety and toxicity. The fluorine-based refrigerants, based on compounds of carbon, fluorine, and chlorine, quickly became a commercial success. Since their adoption in the 1930s as refrigerants fluorocarbon compounds have found applications such as fire retardants, foaming agents, aerosol propellants, and anesthesia agents.

The serendipitous discovery of polytetrafluoroethylene (PTFE) by Roy Plunkett [1] in 1938 in the laboratories of E. I. du Pont de Nemours & Co. during the ongoing refrigerant research was the dawn of a new era. Until the end of the Second World War PTFE was used exclusively in the Manhattan Project for the construction of the atomic bomb because of its properties including resistance to nearly all chemicals. Wartime needs rescued Roy Plunkett's discovery from oblivion. Thousands of applications have since been developed for fluoropolymers, impacting every facet of human life. PTFE was commercialized by the DuPont Company under the trademark of Teflon® in 1950.

PTFE is basically a fully fluorinated analog of linear polyethylene on paper (Figure 1.1).

Since the inception of PTFE, a large number of perfluorinated and partially fluorinated thermoplastic and elastomeric fluorine-containing polymers have been developed. Some are derivatives of the original PTFE; some contain other elements, such as chlorine, silicon, or nitrogen, and represent a sizable group of materials with a formidable industrial utility as seen from its growth rate since the 1940s (Figure 1.2). The factor determining the unique properties of fluoropolymers are the strong bonds between carbon and fluorine which shield the carbon backbone by fluorine atoms.

Monomers for commercially important large-volume fluoropolymers are shown in Table 1.1. Those monomers can be combined to yield homopolymers, copolymers, and terpolymers. The resulting products range from rigid resins to elastomers with unique properties not achievable by any other polymeric materials. Several fluoropolymers have very high melting points, notably PTFE and perfluoroalkoxy (PFA) and fluorinated ethylene propylene (FEP) resins; perfluorinated polymers have excellent dielectrics, low coefficients of friction, stain resistance, and high resistance to common solvents and aggressive chemicals. In general the properties of fluoropolymers are a function of the fluorine content of their molecules (Table 1.2).

Commercial fluoropolymers with the exception of PTFE and polyvinyl fluoride (PVF) are melt processible into films, sheets, profiles, and moldings using conventional manufacturing methods. PTFE is processed in alternative ways to fabricate all types of shapes and parts. PVF is processed by a plastisol method which involves the addition of a solvent to suppress its melting point. Fluoropolymers are widely used: in the chemical, automotive, electrical, and electronic industries; in aircraft and aerospace; in communications, construction, medical devices, special packaging, protective garments, and a variety of other

DOI: 10.1201/9781003204275-2

H   H   H   H H   H H   H
| | | | | | | |
– C – C – C – C – C – C – C – C -   Polyethylene
| | | | | | | |
H   H   H   H   H   H   H   H

F   F   F   F   F   F   F   F
| | | | | | | |
–C—C—C—C—C—C—C—C–   Polytetrafluoroethylene
| | | | | | | |
F   F   F   F   F   F   F   F

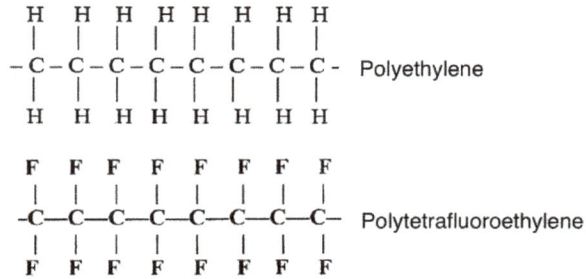

FIGURE 1.1   Molecular structure of polyethylene and polytetrafluoroethylene.

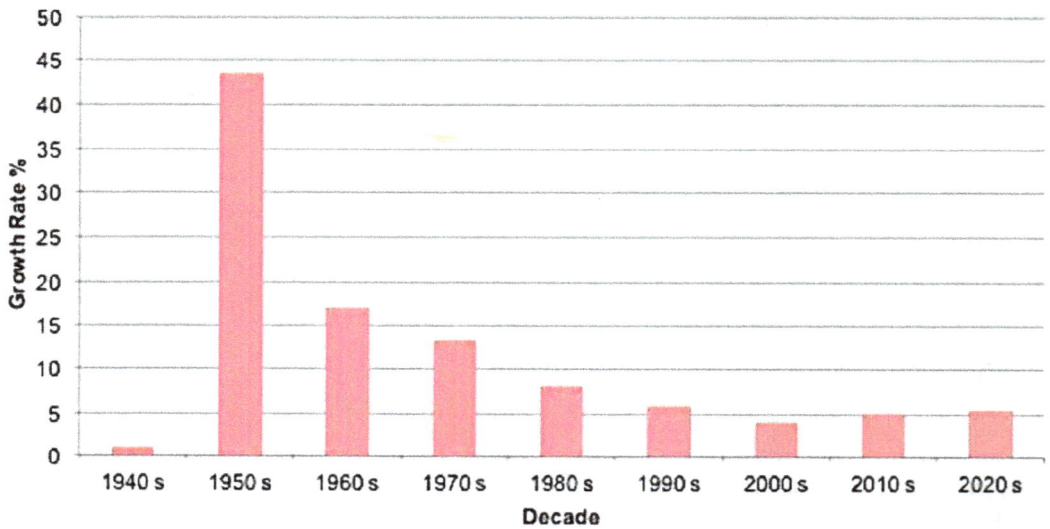

FIGURE 1.2   Growth rate of fluoropolymers during the decades since the 1940s (%).

TABLE 1.1

Monomers Used in Commercial Fluoropolymers

Compound	Formula
Ethylene	$CH_2 = CH_2$
Tetrafluoroethylene	$CF_2 = CF_2$
Chlorotrifluoroethylene	$CF_2 = CClF$
Vinylidene fluoride	$CH_2 = CF_2$
Vinyl fluoride	$CFH = CH_2$
Propene	$CH_3CH = CH_2$
Hexafluoropropene	$CF_3CF = CF_2$
Perfluoromethylvinyl ether	$CF_3OCF = CF_2$
Perfluoropropylvinylether	$CF_3CF_2CF_2OCF = CF_2$

TABLE 1.2

Effect of Increase in the Fluorine Content of Fluoropolymers

Property	Impact
Chemical resistance	Increases
Melting point	Increases
Coefficient of friction	Decreases
Thermal stability	Increases
Dielectric constant	Decreases
Dissipation factor	Decreases
Volume and surface resistivity	Increase
Mechanical properties	Decrease
Flame resistance	Increases
Resistance to weathering	Increases
Surface energy	Decreases

TABLE 1.3

Fluoropolymer Categories (LUMIFLON, CYTOP, and AFLAS are trademarks of AGC Chemicals)

Phase	Resin Type	Partially Fluorinated	Perfluorinated
Crystalline	Thermoplastic resin	ETFE	PTFE
		PVDF	PFA
		PVF	FEP
		PCTFE	
Amorphous	Resin	LUMIFLON (FEVE)	CYTOP
		Copolymer of TFE and dioxole	
	Elastomer	FKM	FFKM
		AFLAS (FEPM)	

Note: ETFE, copolymer of ethylene and tetrafluoroethylene; ECTFE, copolymer of ethylene chlorotrifluoroethylene; FEVE, fluorinated ethylene vinyl ether; MFA, copolymer of perfluoromethylvinylether and tetrafluorethylene; PFA, copolymer of perfluoropropylvinylether and tetrafluoroethylene; FEP, fluorinated ethylene-propylene copolymer; PVDF, poly(vinylidene fluoride).

industrial and consumer products. Classification of commercial fluoropolymers based on crystallinity is listed in Table 1.3. Figure 1.3 shows the chronological development of perfluorinated and partially fluorinated fluoropolymers. A list of current manufacturers of fluoropolymers can be found at the end of this chapter.

The global fluoropolymers market is projected to grow from $7.23 billion in 2021 to $10–12 billion in 2028 at a compounded annual growth rate (CAGR) of 5.2–6.5% in the forecast period, 2021–2028. The global market experienced a 6% decline in 2020 from the average annual growth during the prior three years [2, 3]. Asia Pacific is expected to undergo the highest rate of usage growth of fluoropolymers during the rest of the 2020s (Figure 1.4). The future growth of fluoropolymers may be affected by the safety, health, and environmental issues associated with these products. A longer term factor is the development work focused on finding non-fluorinated alternatives to fluoropolymers.

PTFE is the oldest product among the fluoropolymers, albeit after polychlorotrifluoroethylene which is a specialty resin because of its relatively inferior properties and high cost. Yet PTFE, over seven decades after its commercialization, has the highest consumption quantity (>50% globally) among all thermoplastic fluoropolymers (Figure 1.5). PVDF and FEP have the second and third highest consumptions. Both polymers especially PVDF have grown more rapidly than others in recent decades. PFA, in spite of its PTFE-like properties, has not grown rapidly because of its high price stemming from comonomer perfluoroalkyl vinyl ethers. The "Other" category in Figure 1.4 includes fluoroelastomers and specialty

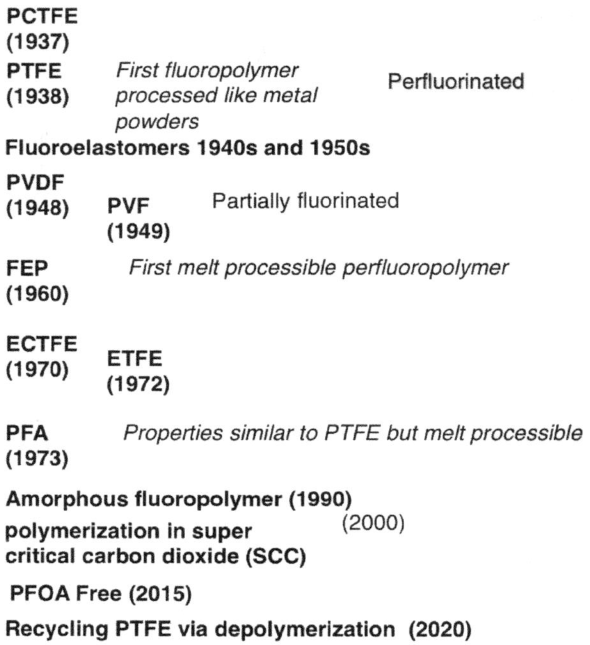

PCTFE
(1937)
PTFE
(1938)
First fluoropolymer
processed like metal
powders
Perfluorinated
Fluoroelastomers 1940s and 1950s
PVDF
(1948)
PVF
(1949)
Partially fluorinated
FEP
(1960)
First melt processible perfluoropolymer
ECTFE
(1970)
ETFE
(1972)
PFA
(1973)
Properties similar to PTFE but melt processible
Amorphous fluoropolymer (1990)
polymerization in super
critical carbon dioxide (SCC)
(2000)
PFOA Free (2015)
Recycling PTFE via depolymerization (2020)

FIGURE 1.3 Chronological development of thermoplastic fluoropolymers and fluoroelastomers.

Total Value, US$ Billion
4
3.5
3
2.5
2
1.5
1
0.5
0
2018
2020
2022
2024
2026
2028
2030
Year

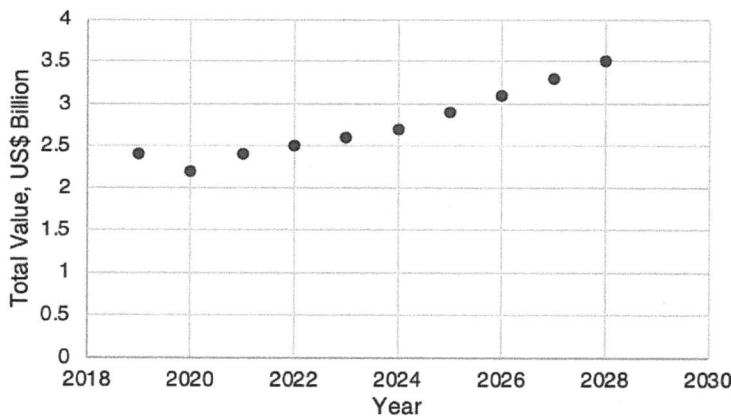

FIGURE 1.4 Projected growth of fluoropolymers in Asia Pacific ($ billions) [2].

fluorinated polymers. Little change has occurred over time in the distribution of the product usage because of the maturity of fluoropolymers.

The main applications of thermoplastic fluoropolymers by consumption are shown in Figure 1.6. Electrical/electronics and automotive/aviation round out the top three consumers of fluoropolymers. A surprisingly high quantity of fluoropolymers is used in medical applications, some of which are permanent implants in human bodies. Industrial/chemical processing industries are the leading consumer of PTFE and other fluoropolymers because of the chemical resistance and high temperature properties of this group of resins.

An amazing aspect of fluoropolymers is their longevity of use in industries. Over time some applications of these plastics have disappeared by the normal attrition process of technological and product

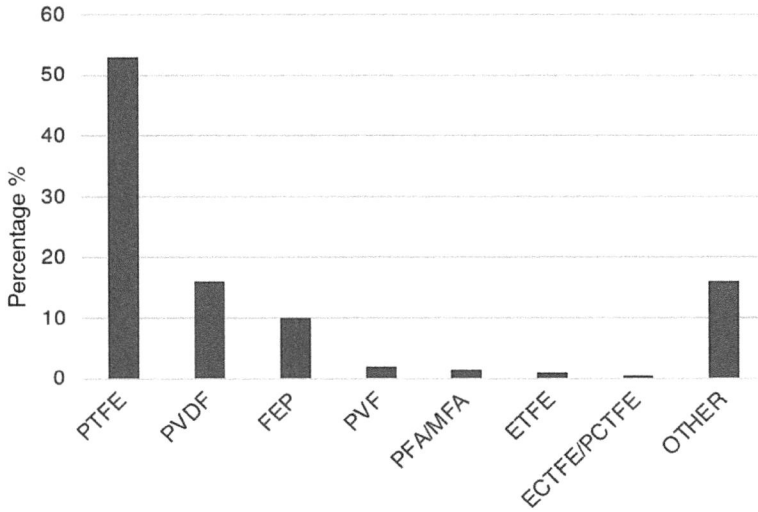

FIGURE 1.5 World consumption of fluoropolymers by type [4].

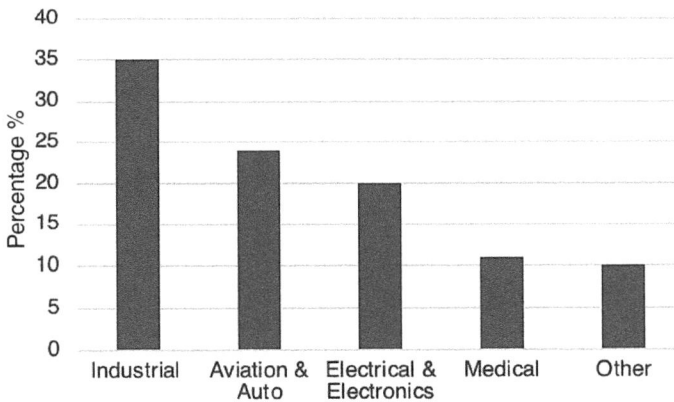

FIGURE 1.6 The global high performance fluoropolymer market share by end use industry [5].

evolution. The unique characteristics and versatility of fluoropolymers have led to the development of new uses and parts. The latest growth of fluoropolymers has come in the areas of sustainability and clean energy. Solar cells and electric vehicles are two examples (more in Chapter 2) of new end uses for fluoropolymers. In the end it all comes down to the performance characteristics for which there are no viable alternatives:

- Resistance to a wide variety of organic and inorganic chemicals.
- Does not support flame under atmospheric conditions.
- Weather resistance.
- Withstands elevated temperatures; the continuous use temperature of PTFE and PFA is 260°C.
- Low surface energy, thus non-wetting properties.
- Non-sticking and release properties.
- High-performance dielectric properties.
- Surfaces may be modified to allow adhesive bonding.

Evolution fluoropolymers and their value to human society are discussed in Chapter 2. Basic chemistry, monomer synthesis methods, and polymerization/finishing of fluoropolymers are covered in Chapters 3 and 4. The properties of commercial products and fabrication processes of fluoropolymers are described in Chapters 5–7. The applications of commercial products are discussed in Chapter 8. Chapters 9–11 discuss commercial fluorocarbon-based elastomers. Fluoropolymer dispersions and their applications are the subjects of Chapters 12 and 13. Amorphous fluoropolymers, fluorinated polyurethanes, fluorinated ionomers, and fluoropolymer additives and their applications are covered in Chapters 14 and 15. The effects of temperature, environmental elements, and radiation are discussed in Chapters 16–18. Safety and sustainability topics are covered in Chapters 19 and 20.

Current Major Manufacturers of Fluoropolymers

Manufacturer	Products
Arkema (www.arkemagroup.com)	PCTFE, PVDF
Asahi Glass Co. (www.agc.co.jp)	ETFE, FEP, PFA, FEVE, PVDF fluoroelastomers, PTFE, amorphous fluoropolymers, PTFE micropowders
Central Glass Co., Ltd. (www.cgco.co.jp)	TPE
Daikin Industries Ltd. (www.daikin.com)	EFEP, ETFE, FEP, PCTFE, PFA, PTFE, PTFE micropowders, fluorocarbon elastomers (FKM, fluorinated TPE), FKM latex
Dow Corning (www.dowcorning.com)	Fluorosilicones
Chemours Co (www.Chemours.com)	ETFE, FEP, PFA, PTFE, amorphous fluoropolymer (AF), PTFE micropowders, PVF, fluorinated ionomers
Dyneon LLC (www.dyneon.com)	ETFE, FEP, HTE fluoroplastic, THV fluoroplastic, PTFE, PTFE micropowders, PFA, PVDF, fluorocarbon elastomers
Honeywell (www.honeywell.com)	PCTFE
JSC Halogen (www.halogen.com)	PTFE, ETFE, FEP, PVDF, PTFE, modified PTFE fluorocarbon elastomers (FKM)
Kureha Chemical Industry Co. (www.kureha.co.jp)	PVDF
Momentive Performance Materials (www.momentive.com)	Fluorosilicones
Shandong Dongyue Chemical Co. (www.dongyuechem.com)	PTFE
Shin-Etsu Chemical Co., Ltd. (www.shinetsu.co.jp)	Fluorosilicones
Solvay Solexis S.p.A. (www.solvay.com)	ECTFE, MFA, PFA, PTFE, PTFE micropowders, PVDF, fluorocarbon elastomers (FKM, FFKM), FKM latex, ionomers

REFERENCES

1. Plunkett, R. J., U.S., Patent 2,230,654 to Kinetic Chemicals Inc. (1941).
2. Data from Fluoropolymers Market, Size, Share and Covid Impact Analysis, and Regional Forecast, 2021–2028, www.fortunebusinessinsights.com, (2022).
3. Fluoropolymer Market by Product Type, Application and End Use Industry, 2020–2027, Allied Market Research, www.alliedmarketresearch.com, (2022).
4. Data from Fluoropolymers, Chemical Economics Handbook, pub by HIS Markit, https://ihsmarkit.com, (2019).
5. Global High Performance Fluoropolymer Market information, MRFR/CnM/2885-HCR, pub by Market Research Future®, www.marketresearchfuture.com, (2021).

2

Societal Benefits of Fluoropolymers

Sina Ebnesajjad

The title of this chapter may sound silly to anyone who works with materials or has knowledge of them or designs parts and devices in just about any industry. Yet the question is certain to be raised by many who may not know the extent of the role of this plastic family in 21st-century living standards. The title of this chapter could have easily been: Do the Fluoropolymers Still Matter in the 21st Century? The answer is a resounding yes as will become clear by a few examples other than everyone's favorite of non-stick pots and pans. The most important applications of this plastic family are hidden from view.

The descriptions of fluoropolymers' characteristics and properties offered in this chapter are selective with no attempt at completeness. These plastics have been described to the extent required to answer the question posed by the title of this chapter. Readers can readily find detailed additional information by referring to the existing body of literature published about fluoropolymers [1–10].

2.1 Basic Fluoropolymer Properties

Fluoropolymers possess a combination of properties for most of which no realistic alternatives have been found. Therefore, this polymer family finds use in new applications well into the 21st century. The key properties of fluoropolymers include:

- High and low temperature resistance.
- Chemical resistance.
- Non-wetting properties.
- Non-sticking properties.
- High-performance dielectric properties.
- Flame resistance – limiting oxygen index (LOI) is 95%.
- Ultraviolet radiation resistance.
- Resistance to weather.

2.2 Examples of Fluoropolymer Applications

The 2019 edition of *Fluoropolymers: Chemical Economics Handbook* [11] states: "Fluoropolymers are among the most useful modern materials, providing nonstick surfaces for cookware and industrial products, waterproofing surface treatments for clothing and other substrates, stain barriers for textiles, high-purity fluid handling" plumbing, "medical applications (e.g., vascular grafts), wire and cable insulation jackets, high-performance coatings for harsh environments, mar-free coatings for touch screen electronic devices, architectural and marine coating additives, back sheets for photovoltaic panels, films and membranes for technical, waterproof clothing, and industrial applications."

DOI: 10.1201/9781003204275-3

A few common examples of applications containing fluoropolymers with their specific applications are:

- Wire and cable insulation for the electrical/electronic, automotive, and aerospace fields.
- Aerospace, automotive, and industrial hoses.
- Medical devices and instruments.
- Whole fluoropolymer and fluoropolymer-lined vessels, pipes, valves, pumps, and fittings for fluid processing equipment for chemical, petroleum, environmental, and semiconductor applications.
- Polytetrafluoroethylene (PTFE) plumber's tapes for water, liquid, and gas thread sealing.
- Breathable and water repellent textile microporous expanded PTFE membranes.
- Industrial fabrics and fibers for clothing, dental floss, and industrial/environmental applications, including laminates for industrial and automotive filtration.
- Lubricants for printing inks, including lubricity materials for mechanical joints and contact points in mechanisms.
- Cookware coatings.
- Mechanical, coating, and lubrication applications for vehicles, building construction, industrial machines, and appliances.

2.3 Automotive Applications

Uses of fluoropolymers in internal combustion automobiles rely on several basic properties including high temperatures and chemical resistance, low-friction, and durability. Fluoropolymers have routinely been used in automobile fuel systems, power trains, brake systems, electrical systems, heat, ventilation, and air conditioning (HVAC) systems, chassis, and interiors.

The ever-rising under-hood temperatures and requirements to prevent atmospheric emissions of fuel and other fluids can be accomplished quite efficiently using fluoropolymers. PTFE is used to insulate wires connected to oxygen sensors which are mounted in vehicle exhaust manifolds. PTFE is quite suitable for this hot environment by providing reliable dielectric protection for the wiring to ensure adequate control exhaust emissions.

Ethylene tetrafluoroethylene copolymer (ETFE) is used to insulate wiring for other high-temperature locations near auto engines, and wiring exposed to hot hydraulic fluid within automatic transmissions. Resistance of ETFE to the permeation of fuel components is critical in applications in fuel tanks, filler necks, and fuel and vapor management hoses. These components allow compliance with strict emission regulations. One example is fuel hose components constructed from ETFE lined elastomers.

In recent decades electric vehicles (EVs) have begun to replace internal combustion vehicles. Fluoropolymers have found applications in EVs for some of the reasons they have been used in conventional automobiles. EVs have a complex design and require special materials and parts. Even though they do not contain motor oil they require special lubricants because of the many moving parts in those vehicles.

Fluoropolymers and fluorinated materials find use both in the drive train and interior of EVs. Polyvinylidene fluoride copolymers are used as the binder in lithium-ion batteries because of the corrosive internal environment of the battery. Battery gaskets are made from perfluoroalkoxy polymer (PFA) because it offers the ultimate thermal and chemical resistance in addition to low permeation [12].

2.4 Aerospace Wire and Cable

Fluoropolymers are typically used in general hook-up wire, military, or high-temperature applications. Durability is the most desirable factor of these cables as they are able to survive in temperatures reaching up to 400°C, intermittently. Generally, fluoropolymers are used in applications that require heat

TABLE 2.1

A Partial List of Aircraft Open Wiring Constructions [13]

Document	Voltage Rating (Maximum)	Rated Wire Temperature (°C)	Insulation Type	Conductor Type
MIL-W-22759/1	600	200	Fluoropolymer insulated TFE and TFE coated glass	Silver coated copper
MIL-W-22759/2	600	260	Fluoropolymer insulated TFE and TFE coated glass	Nickel coated copper
MIL-W-22759/3	600	260	Fluoropolymer insulated TFE-glass-TFE	Nickel coated copper
MIL-W-22759/4	600	200	Fluoropolymer insulated TFE-glass-FEP	Silver coated copper
MIL-W-22759/5	600	200	Fluoropolymer insulated extruded TFE	Silver coated copper
MIL-W-22759/6	600	260	Fluoropolymer insulated extruded TFE	Nickel coated copper
MIL-W-22759/7	600	200	Fluoropolymer insulated extruded TFE	Silver coated copper
MIL-W-22759/8	600	260	Fluoropolymer insulated extruded TFE	Nickel coated copper
MIL-W-22759/9	1,000	200	Fluoropolymer insulated extruded TFE	Silver coated copper
MIL-W-22759/10	1,000	260	Fluoropolymer insulated extruded TFE	Nickel coated copper
MIL-W-22759/13	600	135	Fluoropolymer insulated FEP PVF2	Tin coated copper
MIL-W-22759/16	600	150	Fluoropolymer insulated extruded ETFE	Tin coated copper
MIL-W-22759/17	600	150	Fluoropolymer insulated extruded ETFE	Silver coated high strength copper alloy
MIL-W-22759/20	1,000	200	Fluoropolymer insulated extruded TFE	Silver coated high strength copper alloy
MIL-W-22759/21	1,000	260	Fluoropolymer insulated extruded TFE	Nickel coated high strength copper alloy
MIL-W-22759/34	600	150	Fluoropolymer insulated crosslinked modified ETFE	Tin coated copper

TFE = PTFE FEP = perfluorinated ethylene propylene copolymer.
PVF2 = polyvinylidene fluoride.
ETFE = ethylene tetrafluoroethylene copolymer.

resistance, corrosion resistance, and friction or wear resistance. Aerospace is considered a critical end use for obvious reasons. Different types of wire and cable connect a variety of systems, some critical, throughout the fuselage of civilian and military aircraft.

Maximum operating temperature ratings vary anywhere from 150 to 260°C with short duration peaks approaching 300°C. An aircraft may experience low temperatures approaching −65°C resulting in wide range thermal cycling. Aircraft contain corrosive fluids such as hydraulic oils and jet fuel, thus requiring chemical resistant materials. The insulation material must also be flame resistant as required of other materials installed in airframes and which do not produce toxic fumes if ignited. This is why the installation of polyvinyl chloride insulation in aerospace applications has been banned since the 1990s, although there exist a few exceptions outside passenger planes. Tables 2.1 and 2.2 show lists of insulation materials for aerospace wire and cable in which the prevalence of fluoropolymers is clear.

TABLE 2.2

A Partial List of Protected Aircraft Wiring Constructions [13]

Document	Voltage Rating (Maximum)	Rated Wire Temperature (°C)	Insulation Type	Conductor Type
MIL-W-22759/11	600	200	Fluoropolymer insulated extruded TFE	Silver coated copper
MIL-W-22759/12	600	260	Fluoropolymer insulated extruded TFE	Nickel coated copper
MIL-W-22759/14	600	135	Fluoropolymer insulated FEP-PVF2	Tin coated copper
MIL-W-22759/15	600	135	Fluoropolymer insulated FEP-PVF2	Silver plated high strength copper alloy
MIL-W-22759/18	600	150	Fluoropolymer insulated extruded ETFE	Tin coated copper
MIL-W-22759/19	600	150	Fluoropolymer insulated extruded ETFE	Silver coated high strength copper alloy
MIL-W-22759/22	600	200	Fluoropolymer insulated extruded TFE	Silver coated high strength copper alloy
MIL-W-22759/23	600	260	Fluoropolymer insulated extruded TFE	Nickel coated high strength copper alloy
MIL-W-22759/32	600	150	Fluoropolymer insulated crosslinked modified ETFE	Tin coated copper
MIL-W-22759/33	600	200	Fluoropolymer insulated crosslinked modified ETFE	Silver coated high strength copper alloy
MIL-W-22759/44	600	200	Fluoropolymer insulated crosslinked modified ETFE	Silver coated copper
MIL-W-22759/45	600	200	Fluoropolymer insulated crosslinked modified ETFE	Nickel coated copper
MIL-W-22759/46	600	200	Fluoropolymer insulated crosslinked modified ETFE	Nickel coated high strength copper alloy
MIL-W-81044/12	600	150	Crosslinked polyalkene - PVF2	Tin coated copper
MIL-W-81044/13	600	150	Crosslinked polyalkene - PVF2	Silver coated high strength copper alloy
MIL-W-81381/17	600	200	Fluorocarbon polyimide	Silver coated copper
MIL-W-81381/18	600	200	Fluorocarbon polyimide	Nickel coated copper

TFE = PTFE FEP = perfluorinated ethylene propylene copolymer.
PVF2 = polyvinylidene fluoride.
ETFE = ethylene tetrafluoroethylene copolymer.

2.5 Aircraft Fuel Hoses

All powered aircraft, civilian and military, require fuel on board to operate the engines. A fuel system consisting of storage tanks, pumps, filters, valves, fuel lines, metering devices, and monitoring devices is designed and certified under the strict Title 14 of the Code of Federal Regulations (14 CFR) guidelines. Each system must provide an uninterrupted flow of contaminant-free fuel regardless of the aircraft's altitude. One of the components of the fuel system is flexible lines that consist of steel or polyaramid (e.g., Kevlar® by DuPont Co.) braided fluoropolymer-lined hoses. Flexible hoses, often rubber-lined, are used in areas where vibration exists between components, such as between the engine and the aircraft structure.

The fire resistant PTFE lined hose in Figure 2.1 is unaffected by any known fuel, petroleum, or synthetic base oils, alcohol, coolants, or solvents commonly used in aircraft. PTFE hose has the distinct advantages of practically unlimited storage time, a greater operating temperature range, and broad usage (hydraulics, fuel, oil, coolant, water, alcohol, and pneumatic systems). Medium-pressure PTFE hose

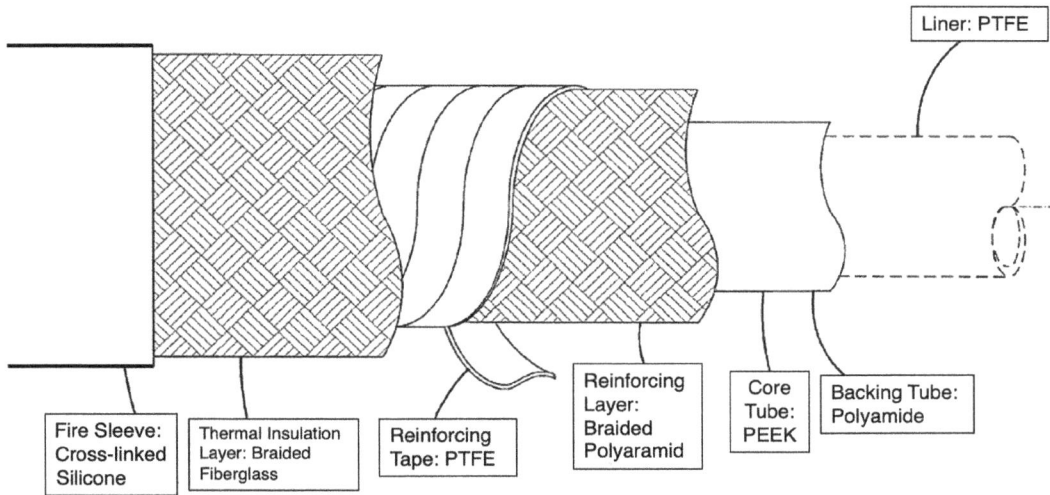

| Liner: PTFE |

| Fire Sleeve: Cross-linked Silicone | Thermal Insulation Layer: Braided Fiberglass | Reinforcing Tape: PTFE | Reinforcing Layer: Braided Polyaramid | Core Tube: PEEK | Backing Tube: Polyamide |

FIGURE 2.1 Example of the design of a fire resistant PTFE-lined hose for aerospace applications [15].

assemblies are sometimes preformed to clear obstructions and to make connections using the shortest possible hose length. Since preforming permits tighter bends that eliminate the need for special elbows, preformed hose assemblies save space and weight [14].

2.6 Heart Rhythm Management: The Implantable Cardioverter Defibrillator

Figures 2.2 shows an example of an implantable cardioverter defibrillator (ICD), an important medical device. This is a small, battery-powered device placed under the skin that holds a tiny computer. If an abnormal rapid heart rhythm is detected the device will deliver an electric shock to restore a normal heartbeat. It is implanted in the body to watch for and treat abnormal heart rhythms. Newer-generation ICDs may have a dual function which also includes the ability to serve as a pacemaker. The pacemaker feature stimulates the heart to beat faster if the heart rate is detected to be too slow. It can detect and treat both fast and slow heart rhythms. In other words, an ICD acts as both a pace maker and a defibrillator. Typically, the battery lasts up to ten years during which time an average human heart beats about 400 million times.

ICDs have been very useful in preventing sudden death in patients with known, sustained, ventricular tachycardia or fibrillation. Studies have shown that they may have a role in preventing cardiac arrest in high-risk patients who have not had, but are at risk of, life-threatening ventricular arrhythmias.

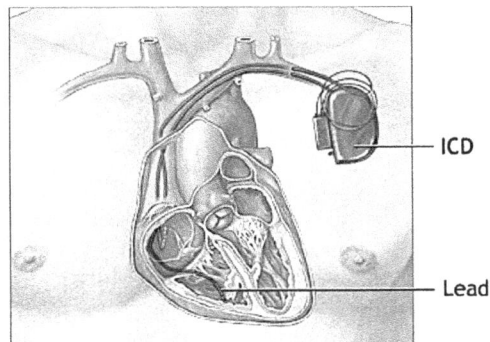

FIGURE 2.2 An example of an implantable cardioverter device in the human body [16].

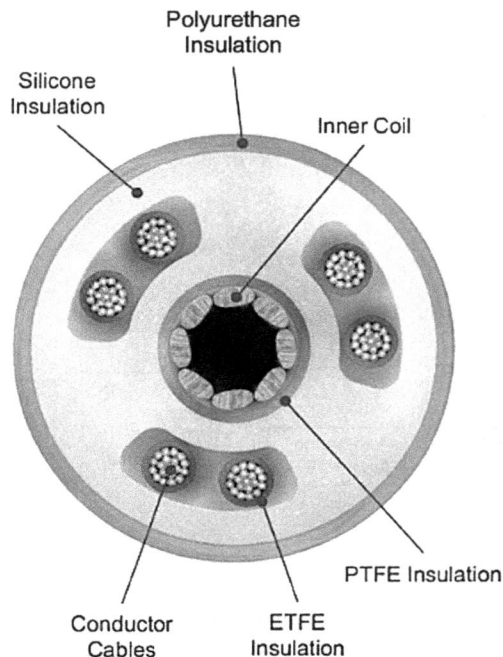

FIGURE 2.3   Cross-section of the lead of a typical implantable cardioverter defibrillator.

An important part of ICD is the *lead body*. This must also withstand the human body's *in vivo* chemistry, which is surprisingly harsh. It must also withstand flex action generated by the heart beat motion, requiring the insulation to possess significant flex fatigue resistance. This operation of the ICD subjects the insulation to significant voltage (Figure 2.3). ETFE and PTFE are both biocompatible and provide adequate dielectric strength for the lead to continue to function normally throughout the life of the ICD. Moreover, ETFE is abrasion resistant and has proven resilient to allow high voltage therapy (as high as 900 V).

Another important medical device is the guide catheter in which PTFE and other fluoropolymers play an important role. The lubricity of PTFE allows ease of movement of the guidewire. This is a catheter that makes it easier to enter a vessel with other devices or instruments. Guide catheters are used to facilitate the placement of lasers, stents, and balloons for angioplasty [17]. Catheters allow easy, atraumatic access to areas of the vasculature that are otherwise difficult to access.

Steerable guide catheters are constructed with components that are selected to provide optimal navigability, torque transfer, and push-ability for a variety of typical percutaneous access routes. The catheter wall thickness in the deflecting segment of the guide catheter is about ⅓ mm or less, and includes a slotted deflection tube. This construction allows a very tight turning radius (Figure 2.4).

2.7 Pediatric Heart Repair

Some children are born with congenital heart defects which must be repaired by corrective surgical fixes. A baby born with one or more heart defects has a congenital heart disease(s). Surgery is often needed when the defect could harm the child's long-term health [19].

For example, coarctation of the aorta occurs when a part of the aorta has a very narrow section (Figure 2.5). The shape looks like an hourglass timer. The narrowing makes it difficult for blood to get through to the lower extremities. This leads to very high blood pressure before the point of coarctation and low blood pressure beyond the point of coarctation. The most common way to repair it is to cut the narrow section and make it bigger by using a patch made of expanded PTFE which is a synthetic material [9].

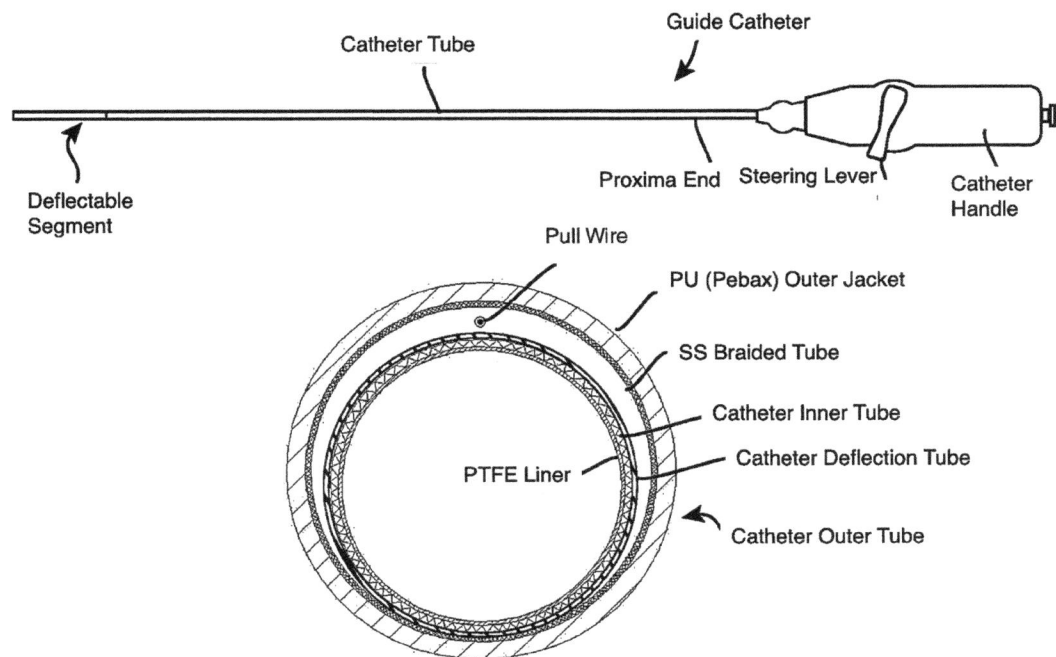

FIGURE 2.4   Construction design of a steerable guide catheter [18].

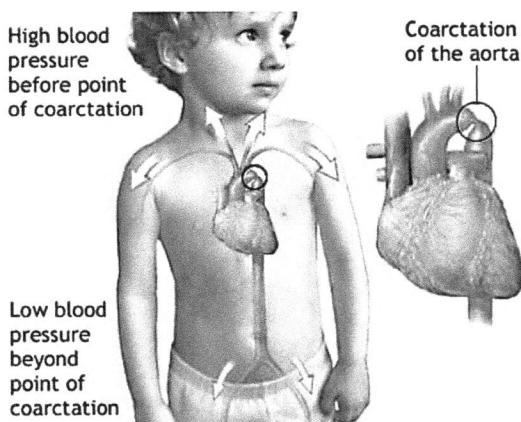

FIGURE 2.5   Depiction of pediatric coarctation of the aorta [20].

2.8 Thread Sealant

An everyday example of the usefulness of fluoropolymers is ordinary thread sealant tape (Figure 2.6). Tape for thread sealant utilizes the outstanding properties of PTFE to ensure long-term, highly reliable service plus easy disassembly. The tape provides permanently sealed joints, which will outlast the pipe under practically all service conditions. The tape can be used with all types of pipe including steel, iron, galvanized, aluminum, brass, exotic metals, Monel®, plastic, and even rubber.

The advantage of PTFE tape is the ease of its use and safety. It does not contain any chemicals that may cause exposure as compared to pipe dopes used for thread sealing in the past. It is also easy to remove to reseal a pipe. In the case of pipe dope or plumber's putty, cleaning of the seals is required before resealing.

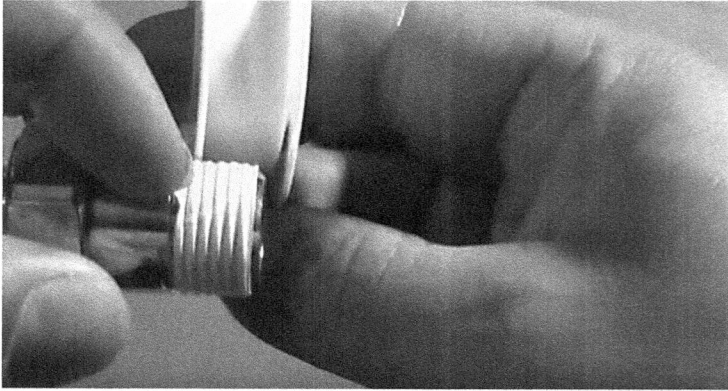

FIGURE 2.6   PTFE thread sealant tape as applied to a pipe thread.

2.9 Chemical Processing Industry-lined Pipes, Fittings, and Vessels

In the chemical process industry, manufacturing is usually carried out with chemically aggressive fluids. A major capital cost item in those factories includes the expenses of pipes, fittings, and vessels (Figures 2.7–2.9). If the chemicals are aggressive to the point of corroding carbon steel, then more expensive metals including stainless steel and exotic metals must be considered. The cost of moving to stainless steel and higher alloys is prohibitively high thus requiring a more economical solution.

Contamination of process streams by corrosion by-products or ions from metallic equipment is detrimental to most processes and products. Increasing the resistance of equipment to corrosion is highly

FIGURE 2.7   Examples of fluoropolymer-lined fittings and linings.

FIGURE 2.8 A fluoropolymer-lined pipe.

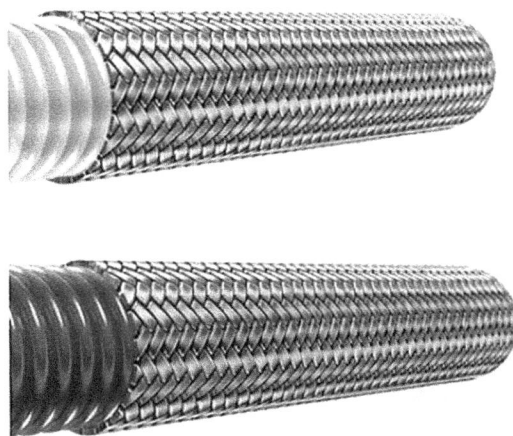

FIGURE 2.9 Convoluted and smooth-bore PTFE-lined braided hoses for fluid transfer.

beneficial because it extends service life and cuts unscheduled downtime. Modified PTFE, PFA and, perfluoromethylvinyl ether and tetrafluoroethylene copolymer (MFA) are quite resistant to permeation, and form smooth surfaces thus preventing surface build up and release of particles.

Fluoropolymers have found major applications in chemical processing equipment as liners as well as parts to prevent corrosion and maintain purity. Even though fluoropolymers may increase the initial equipment cost, lifetime costs can be reduced, allowing manufacturers to be more competitive.

Severe services such as hydrogen fluoride processing have been made possible by lining carbon steel components with PTFE, modified PTFE, or perfluoroalkoxy polymers (PFA and MFA).

2.10 Semiconductor Chip Fabrication

Our lives have been, and continue to be, revolutionized by electronics that impact the way we work, treat patients, communicate, shop, travel, bank, and learn. The driving force of these changes has been the semiconductor manufacturing industry (semicon). It has renewed itself rapidly by improving the state of technology and reducing the cost of its products. Historically, nearly every two years (Moore's Law) there has been a step change toward smaller, faster, and less costly semiconductor devices (chips) [3].

From the beginning, the semiconductor industry has relied on polymers for the containment and supply of ultrapure water (UPW), acids, bases, oxidizers, aqueous salt solutions, and solvents. Growth of the semicon industry has relied on fluoropolymers as the material of construction and liner for critical components (Figure 2.10). Examples of fluoropolymer uses include bulk chemical systems, fluid transport and distribution systems, chemical mechanical polishing equipment, wet processing and etching equipment, tanks and containers, piping, and wafer handling and carrying tools. Suppliers of high-purity chemicals to semicon fabrication factories (fabs) also use fluoropolymer components for process surfaces in manufacturing equipment.

The growth in the semicon industry has relied on the advancement of the manufacturing of silicon chip fabs, each of which can easily exceed $2 billion in capital investment. Precision of assembly, contamination control (purity), automation, and speed are some of the key contributing factors for the improved productivity of fabs. Consequently, over the decades every year more computation power has been packed into smaller processors, shrinking device sizes and lowering costs to consumers [3]. Fluoropolymers have had an outsized role in the success of the semicon industry. Some of the benefits of fluoropolymer use in the semicon industry and the places in which fluoropolymers are used are:

Benefits:

- Maximizing yields by reducing contamination.
- Maximizing design freedom.
- Minimizing cost of ownership.

Use of fluoropolymers in fabs:

- Bulk chemical systems.
- Cleaning and etching.

FIGURE 2.10   High-purity fluid handling valves made from PFA and other fluoropolymers.

- Chemical mechanical planarization.
- Analyses of process fluid composition.
- High-purity chemical manufacturing.

2.11 Biomedical Applications

PTFE and other fluoropolymers have found use in medical applications for nearly a half-century. PTFE is the dominant fluoropolymer in biomedical devices and implant applications. Medical devices utilize these polymers in enabling parts, temporary implants, and permanent grafts. In spite of its inertness, the PTFE surface may be modified to change it from hydrophobic to hydrophilic.

The versatility of PTFE has allowed the development of microporous membranes (ePTFE) with innumerable industrial and biomedical applications [9]. Chemical functional groups can be grafted to a PTFE surface to achieve characteristics such as hydrophilicity and non-thrombogenicity to prevent blood clot formation on the surface where grafts are exposed to blood. Proteins have complex chemical structures and can interact with a PTFE surface hydrophobically which is beneficial in many applications. For applications where cell growth in the pores of ePTFE membranes is beneficial, technology has been developed to modify its surface architecture and chemistry to enhance growth.

2.12 PTFE Micropowders

PTFE micropowders, also referred to as fluoroadditives, are homopolymer grades of PTFE with a considerably lower molecular weight than standard PTFE [21]. They are produced either by controlled suspension or dispersion polymerization to a lower molecular weight [2, 22] or by degradation of PTFE (often scrap) by thermal cracking (pyrolysis) or by irradiation by a high-energy electron beam (EB). The EB process is the most widely used commercial method. Micropowders are white, free-flowing powders with very small particle size (typically in the range 2–20 µm). They have different particle shapes and morphology from those of granular and fine powder grades of PTFE. The molecular weight of micropowders is in the range 10^4 to 10^5 compared with that of standard PTFE, which is typically in the range 10^6 to 10^7. The melt viscosity of micropowders ranges from 10^2 to 10^5 poise, considerably lower than the typical values of standard PTFE of 10^9 to 10^{11}. They are used predominantly as additives to lubricants to improve their performance and to plastics and rubber to reduce their coefficient of friction, and as additives to printing inks and coatings to reduce their nonstick properties [2].

2.13 Applications of Fluoroelastomers in Transportation

Fluorocarbon elastomers, commonly known as FKMs, are a key component in industries that are exposed to harsh chemical conditions, ozone attacks, and intense temperatures [10]. FKMs can handle environments from as low as −40°C to as high as 250°C – or higher for short periods of time. FKMs contain a high ratio of fluorine to hydrogen content which gives them an extraordinarily strong resistance to a wide range of industrial chemicals including acids, steam, methanol, petroleum-based and silicone oils, diesel fuels, and other highly polar fluids.

Advancements in gas turbine engines are pushing fluoroelastomers to their thermal limits. As vehicles, aircraft, and ships become more powerful and energy efficient they require durable components that are more reliable, operate safely, and last longer. FKM polymers provide a solution to both the aerospace and automobile industries and are also used to build high-performance machinery that require premium quality parts to provide stability in high endurance conditions. FKM compounds typically range from 55 to 90 durometers [23].

Typical uses for FKMs in the aeronautical industry include O-rings, gaskets, shafts, fuel hoses, joints, and other electrical connector components that are subjected to intense temperatures and pressure changes during flights. In the automotive industry, FKM synthetic rubbers help power high-performance engines that combine oil and chemicals with high temperatures.

2.14 Properties of Thermoplastic Fluoropolymers

Some of the industrially useful properties of fluoropolymers include a high melting point, high thermal stability at extremely low and high temperatures, insolubility in solvents, chemical inertness, a low coefficient of friction, a low dielectric constant/dissipation factor, low water absorptivity, excellent weatherability and ultraviolet light resistance, and combustion resistance. The vast majority of the properties of fluoropolymers are superior to those of the hydrocarbon-based polymers. Through the decades a large family of fluorinated polymers has been developed leading to a revolution in human life in the 20th and 21st centuries.

Chemical resistant pipes and tubes, transportation parts, comfortable apparel, semiconductor fabs, stadium roofs, vascular grafts, space lab parts, ultra chemical resistant rubbers, and many others have been made possible by fluoropolymers. Fluoropolymer plastics have replaced untold quantities of metals.

2.15 Delving Deeper into Properties

Table 2.3 shows some of the observed changes in the properties of hydrocarbon-based polymers as a result of increased F substation. Fully fluorinated polymers such as PTFE, PFA, and fluorinated ethylene propylene copolymer (FEP) possess the ultimate properties of fluoropolymers. The impact of complete fluorine substitution for hydrogen atoms, on the properties of polyethylene, is considerable (Table 2.4).

Some of the conclusions that can be derived from the comparison of polyethylene and PTFE properties in Table 2.4 are:

- PTFE is one of the lowest surface energy polymers.
- PTFE is the most chemically resistant polymer.
- PTFE is one of the most thermally stable polymers.
- The melting point and specific gravity of PTFE are more than double those of polyethylene.

To summarize, fluoropolymer resins are essentially chemically inert. In the case of perfluorinated polymers resistance lasts to the upper-use temperature (260°C for PTFE and PFA). Very few chemicals are known to chemically react with those resins, that is, molten alkali metals and gaseous fluorine; and a few

TABLE 2.3

Typical Effect of Increase in the Fluorine Content of Polymers [2]

Property	Impact
Chemical resistance	Increases
Melt temperature	Increases
Coefficient of friction	Decreases
Thermal stability	Increases
Dielectric constant	Decreases
Dissipation factor	Decreases
Volume and surface resistivity	Increases
Mechanical property	Decreases
Flame resistance	Increases
Resistance to weathering	Increases
Surface energy	Decreases

TABLE 2.4

A Comparison of PTFE and Polyethylene (PE) Properties [24–26]

Property	Polyethylene	PTFE
Specific gravity	>2	~1
Melt temperature (^{O}C)	First: 342; second: 327	
Dynamic coefficient of friction	0.33	0.04
Critical surface energy (dynes/cm)	33	18
Refractive index	1.51	1.35
Propensity to chain branching	Yes	No
Decomposition temperature (^{O}C)	330	450

TABLE 2.5

Comparison of Select Properties of Commercial Fluoropolymers

Property	PTFE	FEP	ETFE	PFA	PVDF	THV
Specific gravity	2.15	2.15	1.70	2.15	1.77	1.97
Melting point (2nd) (°C)	327	260	270	310	162	115–180
Tensile strength (MPa)	20	20	40	28	50	23
Elongation (%)	300	300	200	300	150	500
Flexural modulus (MPa)	560	650	1,100	650	2,000	80–210
Temperature rating (°C)	260	200	150	260	120	Various
Dielectric constant	2.1	2.1	2.6	2.1	6.0	4.0
Coefficient of friction	0.1	0.2	0.4	0.2	—	—
Cut-through (kg)	4.5	4.5	18	4.5	—	—
Chemical resistance	Excellent	Excellent	Very good	Excellent	Good	Good

FEP = perfluorinated ethylene propylene copolymer.
ETFE = ethylene tetrafluoroethylene copolymer.
PFA = perfluoroalkoxy polymer.
PVDF = polyvinylidene fluoride.
THV = tetrafluoroethylene hexafluoropropylene vinylidene fluoride terpolymer.

fluorochemicals, such as chlorine trifluoride, ClF_3, or oxygen difluoride, OF_2, which readily liberate free fluorine at elevated temperatures [27] (Table 2.5).

The unique degree of the inertness of fluoropolymers reflects their chemical structure. Molecules of perfluorinated resins are formed simply from strong carbon–carbon and super-strong carbon–fluorine bonds; moreover, the fluorine atoms form a protective sheath around the carbon core of each molecule. This structure also produces other special properties, such as insolubility and low-surface adherability and friction. To a minor degree fluoropolymer resins may absorb halogenated organic chemicals. This will cause a very small weight change and in some cases slight swelling. If absorption is very high, it usually indicates a fabricated part of high porosity.

Partially fluorinated fluoropolymers retain some or most of the properties of perfluorinated resins mostly depending on fluorine. These polymers present advantages over the perfluorinated resins including higher mechanical properties such as cut-through resistance in wire and cable, lower melting points, and ease of processability into complex shapes.

Even though they are invisible to the eyes, because of playing functional roles as opposed to decorative ones, fluoropolymers play a critical part in the lives of people in the 21st century. It is surprising for an industry that has been in existence since the late 1940s that it continues to generate large numbers of new applications. There are, however, trends in the new products of the 21st century.

Traditional uses of fluoropolymers such as lined pipe and vessel lining consume large quantities of resin. These applications continue in basic industries. Resin and part manufacturing continue their migration to developing economies, especially China, India, Mexico, and Vietnam.

New applications typically consume smaller quantities of resins. They are often material for the construction of a component in complex designs such as electronic and medical devices. Another trend is the desire to apply fluoropolymer coatings to surfaces or impart fluoropolymer-like properties on the surfaces of parts. The savings are significant when surface functionality is sufficient to meet the end use.

2.16 Forces Affecting the Fluoropolymer Industries

In spite of the incredible value fluoropolymers bring to raising everyday living standards they are subject to global market forces. To be sure we are not speaking of the industrial impact of the pandemic caused by the novel coronavirus SARS-CoV-2, commonly known as "COVID-19." Some of the major factors bringing about changes in the fluoropolymer industries include: (1) environmental and health concerns; (2) the need to improve the sustainability of products; (3) globalization of businesses including manufacturing and consumption; and (4) the transition of legacy companies out of traditional fluoropolymers and the entry of new players. These drivers have shaken the relatively old fluoropolymer industries. The long term fate and shape of those industries remain to be defined over the next decades.

There is no escaping from the realities of global warming, the risks presented by emissions to people and ecosystems, and the increasing regulations of plastics, including fluoropolymers. Another factor is the exhaustion of key ingredients of fluoropolymers that could result in cost increases in addition to commercial trade frictions among countries. These forces are but certain to compel the fluoropolymer industries to reconsider their business models and rethink the entire value chain. To cite an example, a generation ago biobased polymers mostly existed on drawing boards, in journal articles, and in books. In recent years they have begun to become a reality by starting to replace petroleum-based polymers in many applications.

REFERENCES

1. Scheirs, J., *Modern Fluoropolymers: High Performance Polymers for Diverse Applications*, Wiley, New York, (1997).
2. Ebnesajjad, S., *Fluoroplastics Handbook: Non-melt Processible Fluoropolymers*, vol *1*, 2nd ed, Elsevier, Oxford, UK, (2015).
3. Ebnesajjad, S., *Fluoroplastics Handbook: Melt Processible Fluoropolymers*, vol *2*, 2nd ed, Elsevier, Oxford, UK, (2015).
4. Ebnesajjad, S., *Polyvinyl Fluoride Technology and Applications of PVF*. 1st ed. Elsevier, Oxford, UK, (2012).
5. Ebnesajjad, S. *Introduction to Fluoropolymers - Materials, Technology and Applications*. 2nd ed. Elsevier, Oxford, UK, (2020).
6. Drobny, J. G. J. G. *Technology of Fluoropolymers*, 2nd ed., CRC Press, Boca Raton Fl, (2009).
7. Smith, D. W., Iacono, S. T., Iyer, S. S., *Handbook of Fluoropolymer Science and Technology*, Wiley, New York, (2014).
8. Banerjee, S., *Handbook of Specialty Fluorinated Polymers, Preparation, Properties, and Applications*, 1st ed, Elsevier, Oxford, UK.
9. Ebnesajjad, S., *Expanded PTFE (ePTFE) Applications Handbook, Technology, Manufacturing and Applications*, 1st ed, Elsevier, Oxford, UK, (2016).
10. Drobny, J. G., *Fluoroelastomers Handbook*, 2 nd ed, Elsevier, Oxford, UK, (2016).
11. Fluoropolymers, Chemical Economics Handbook, pub by HIS Markit, https://ihsmarkit.com, April (2019).
12. Daikin Industries, www.daikinchemicals.com/solutions/industries/automotive.html, December 15, (2021).
13. Advisory Circular AC 43.13-1B, CHAPTER 11, Aircraft Electrical Systems, United States Federal Aviation Administration, September 8, (1998).
14. Fluid Lines and Fittings, Chapter 7, Aviation Maintenance Technician Handbook – General, United States Federal Aviation Administration, www.faa.gov, (2008).

15. US Patent Application US20120125470, Nanney, S., Ramaswamy, N, Stroempl, P. J., assigned to Parker Corp, May 24, (2012).
16. Implantable cardioverter-defibrillator, https://medlineplus.gov/ency/article/007370.htm, MedlinePlus, National Library of Medicine, Dep Health and Human Services, USA, May (2022).
17. Medical Dictionary, https://medical-dictionary.thefreedictionary.com, May (2022).
18. US Patent 8,939,960, Rosenman, D., Kayser, D., et al, assigned to Bio Cardia, Jan. 27, (2015).
19. Congenital heart defect - corrective surgery, https://medlineplus.gov/ency/article/002948.htm, MedlinePlus, National Library of Medicine, Dep Health and Human Services, USA, (May 2022).
20. Congenital heart defect, Coarctation of Aorta, https://medlineplus.gov/ency/article/000191.htm, MedlinePlus, National Library of Medicine, Dep Health and Human Services, USA, May (2022).
21. Ebnesajjad, S. and Morgan, R., *Fluoropolymer Additives*, 2nd Ed, Elsevier, Oxford, UK, (2019).
22. U.S. Patent 5,641,571, Mayer, L., Tonnes, K.-U., Bladel, H., Hoechst, A.G., June 14, (1997).
23. Northern Engineering Sheffield, UK, www.nes-ips.com, December (2021).
24. Chambers, R. D. *Fluorine in Organic Chemistry*, 1st ed. New York: John Wiley and Sons; (1973).
25. Hudlicky, M., *Chemistry of Fluorine Compounds*. 1st ed. New York: The McMillan Company; (1962).
26. Van Krevelen, D. W.. *Properties of Polymers: Their Estimation and Correlation with Chemical Structure*. 2nd ed. Amsterdam: Elsevier; (1976).
27. Properties Handbook No. H-37051-3. Teflon® PTFE fluoropolymers resin Published by DuPont; July (1996).

Part II

Thermoplastic Fluoropolymers

3

Synthesis and Properties of Monomers of Thermoplastic Fluoropolymers

Sina Ebnesajjad

Today it is hard to imagine a world in which automobiles did not have reliable air conditioning systems. There was once such a world before the 1930s. The growing use of automobiles requires air conditioning to make travel bearable in hot weather especially in hot and humid climates. The existing refrigerants such as ammonia were toxic, flammable, and inefficient. In 1928, Freon®, a chlorofluorocarbon, was invented. In 1930, General Motors and DuPont formed the Kinetic Chemical Company to produce Freon.

Roy Plunket discovered polytetrafluoroethylene (PTFE) while working on tetrafluoroethylene for the development of new fluorocarbon refrigerants. The story of the discovery has been told and retold often [1, 2]. The most fortunate and indeed miraculous event was tetrafluoroethylene which, in spite of being highly flammable and explosive, did not cause a disaster. Thus, tetrafluoroethylene became the first monomer for manufacturing fluoropolymers.

Commercially significant thermoplastic fluoropolymers include PTFE, perfluoroalkoxy polymer (PFA), fluorinated ethylene propylene polymer (FEP), ethylene-tetrafluoroethylene copolymer (ETFE), ethylene-chlorotrifluoroethylene copolymer (ECTFE), polychlorotrifluoroethylene (PCTFE), polyvinylidene fluoride (PVDF), and polyvinyl fluoride (PVF). This chapter describes the important preparation methods and properties of monomers used in the synthesis of thermoplastic fluoropolymers.

3.1 Preparation of Tetrafluoroethylene

There is some controversy over the first synthesis of tetrafluoroethylene (TFE; CAS # 116-14-3, $CF_2=CF_2$). Articles published in the last decade of the 19th century describe a number of efforts to prepare TFE via the direct fluorination of carbon and chloromethanes. Various papers reported the fluorination of tetrachloroethylene using silver fluoride [3–5]. Humiston reported the first documented preparation of TFE in 1919 but his invention has been disputed because of the errors in the TFE property data [6]. The first complete and definitive preparation of TFE was reported by Ruff and Bretachneider in 1933 [7].

Commercially, TFE is prepared by the pyrolysis of chlorodifluoromethane ($CHClF_2$), also known as HCFC-22, by which a molecule of HCl is removed, and a reaction of the degradation products [8]. Two CF_2 free radicals produced by dehydrochlorination of HCFC-22 combine, yielding C_2F_4 molecules. The synthesis process begins with the reaction of the fluorine ore fluorspar with an acid, usually sulfuric acid. The reaction yields highly reactive hydrofluoric acid (HF). HF is reacted with halogenated aliphatic hydrocarbons where halogen exchange in favor of fluorine results in hydrochlorofluorocarbons. Dehydrohalogenation allows the formation of fluorocarbons.

The entire chain of reactions in a fully integrated commercial TFE manufacturing operation is shown here. Fluorspar (CaF_2), sulfuric acid, and chloroform are the starting ingredients for the production of chlorodifluoromethane, also called difluorochloromethane, for the introduction of fluorine into an organic reaction sequence [9–15]. The conventional reaction scheme for the synthesis of tetrafluoroethylene from fluorspar is:

HF formation:

$$CaF_2 + H_2SO_4 \rightarrow 2HF + CaSO_4 \tag{3.1}$$

Chloroform formation:

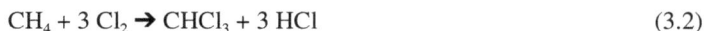

$$CH_4 + 3\ Cl_2 \rightarrow CHCl_3 + 3\ HCl \tag{3.2}$$

Chlorodifluoromethane (HCFC-22) preparation:

$$CHCl_3 + 2\ HF \rightarrow CHClF_2 + 2\ HCl \tag{3.3}$$

(SbF$_3$ catalyst)
TFE synthesis:

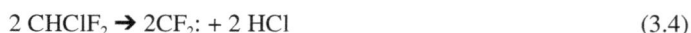

$$2\ CHClF_2 \rightarrow 2CF_2: +\ 2\ HCl \tag{3.4}$$

(pyrolysis)

$$2CF_2: \rightarrow CF_2{=}CF_2 \text{ (tetrafluoroethylene)} \tag{3.5}$$

The overall pyrolysis reaction is:

$$2\ CHClF_2 \rightarrow CF_2{=}CF_2 + 2\ HCl \text{ (an equilibrium reaction)} \tag{3.6}$$

Some by-products are also produced during the pyrolysis of HCFC-22. They include hexafluoropropylene (HFP), perfluorocyclobutane and octa-fluoroisobutylene, 1-chloro-1,1,2,2-tetrafluoroethane, 2-chloro-1,1,1,2,3,3-hexafluoropropane, and a small amount of highly toxic perfluoroisobutylene. The type and the amount of by-products, also called high boilers, depend on the reaction conditions because both Equations (3.4 and 3.5) are equilibrium reactions.

Steam has been used to minimize the HFP by-product. By holding the steam to CHClF$_2$ in the range of 7: 1 to 10:1 yields approaching 95% can be obtained at an 80% chlorodifluoromethane conversion. The TFE kinetics of the pyrolysis of chlorodifluoromethane have been studied by Chinoy and Sunavala. They concluded that operating the pyrolysis reaction at above 850–900°C is effective by maximizing the conversion of chlorodifluoromethane, minimizing the hexafluoropropylene content, and maximizing the TFE yield [16].

All traces of telogenic hydrogen or chlorine-bearing impurities must be removed from the reaction mixture because the polymerization of TFE to high molecular weights requires extreme purity. Commonly TFE is purified to 99.99995%. Trifluoroethylene is a powerful chain transfer agent and must be reduced to less than 0.2–0.3 parts per million of TFE. Separation of TFE begins with the cooling of pyrolysis products, followed by scrubbing with a dilute basic solution to remove HCl, then dried. The gas is then compressed and distilled to recover the unreacted CHClF$_2$ for recycling and recovering high purity TFE [17].

The production of TFE requires the generation of a reactive fluorinated carbon fragment, difluorocarbene (:CF$_2$). Upon quenching, difluorocarbenes may dimerize to produce TFE (C$_2$F$_4$). Hence, it is important for the processes that generate :CF$_2$ to operate in good yields at comparatively low temperatures to reduce energy consumption as well as reduce TFE production costs. To increase the yield of TFE new processes have been developed that lead to the formation of :CF$_2$, the TFE precursor, in good yields [18].

The largest consumption of TFE is for polymerization into a variety of PTFE homopolymers, modified PTFE, and micropowders, and as a comonomer in copolymerization with hexafluoropropylene, ethylene, perfluorinated ether, and other monomers, and also as a comonomer in a variety of terpolymers. Other uses of TFE are to prepare low-molecular-weight polyfluorocarbons and carbonyl fluoride oils as well as to form PTFE *in situ* on metal surfaces [19] and in the synthesis of hexafluoropropylene, perfluorinated ethers, and other oligomers [20].

3.2 Properties of Tetrafluoroethylene

TFE is a colorless, tasteless, odorless, and nontoxic gas that is highly flammable in addition to being explosive in the absence and presence of oxygen [21]. Some of the important properties of TFE are listed in Table 3.1. It is stored as a liquid (its vapor pressure at −20°C is 1 MPa) and is polymerized usually above its critical temperature and critical pressure. A TFE polymerization reaction is highly exothermic, requiring the removal of heat from the reaction medium to ensure a safe and controlled process.

TFE is a highly reactive compound that appears as a gas at room temperature (20°C). Under sufficient pressure it can be liquified in a container allowing availability of large quantities of this monomer for polymerization. TFE is stored or transported as a liquid in the presence of an inhibitor such as d-limonene. The inhibitor prevents auto-polymerization of TFE which can take place in the presence of small amounts of oxygen or other agents.

Oxygen can result in the auto-polymerization of TFE, which can in turn cause sufficient localized heating within the container to cause the TFE vapor above the liquid within the container to explode [24]. The explosivity of TFE may not depend on the presence of air, however, since air will normally be excluded (purged) from the container of liquid TFE. Even in the absence of air, exposure of the TFE vapor within the container to a spark, such as caused by the discharge of static electricity, a hot metal surface, such as is caused by metal surfaces rubbing together, or an external fire, can cause the TFE vapor to explode. Thus, for example, a saturated TFE vapor can explode at a temperature of −16°C or greater when it is under a pressure of at least 1.032 MPa. In contrast an unsaturated TFE vapor can explode at 25°C and 0.79 MPa.

A mixture of air and TFE are flammable when exposed to an ignition source such as a hot surface (240°C or higher) even under atmospheric pressure. The most dangerous characteristic of TFE, however, is its ability to degrade explosively in the absence of any oxidizing agent. Typically, TFE is most susceptible to degradation while vaporizing from a pressurized liquid state. TFE decomposition called "deflagration" yields tetrafluoromethane (CF_4) and carbon. The susceptibility to explosion increases with increasing pressure and temperature. Deflagration is highly exothermic, generating 57–62 kcal/mole at 25°C and 1 atm [25], about the amount of heat released by the explosion of black gun powder. Typically, because of transportation concerns, TFE manufacturing and polymerization are conducted at the same location.

The problem of TFE explosivity during transportation has been solved by adding an inhibitor such as hydrochloric acid and terpenes, α -pinene, terpene B, and d-limonene, which appear to scavenge oxygen, a polymerization initiator [25]. Commercially, HCl is added to the liquid TFE within the pressurized

TABLE 3.1

Properties of Tetrafluoroethylene [22, 23]

Property	Value
CAS No.	116-14-3
Molecular weight	100.02
Boiling point at 101.3 kPa, °C	−76.3
Boiling point, °C	−142.5
Critical temperature, °C	33.3
Critical pressure, MPa	39.2
Critical density, g/ml	0.58
Heat of formation for ideal gas at 25°C, ΔH, kJ/mol	−635.5
Heat of polymerization to solid polymer at 25°C, ΔH, kJ/mol	−172.0
Flammability limits in air at 101.3 kPa, volume%	14–43

container, the proportion of TFE to HCl being about 33 mol % TFE and 67 mol % of HCl. This mixture forms an azeotrope, so as TFE vaporizes, so does HCl in about the same proportion. The presence of HCl with the TFE in the vapor state renders the TFE non-ignitable, and therefore non-explosive, in the absence of air.

HCl resolves the issue of the explosion hazard of TFE. However, it introduces a new problem of the separation and disposal of large quantities of HCl present, with TFE required for the polymerization of polytetrafluoroethylene and other polymers. HCl toxicity presents a hazard in the event it is released.

3.3 Preparation of Hexafluoropropylene

Hexafluoropropylene is produced as a by-product of the pyrolysis of chlorodifluoromethane to produce TFE. The basic reaction for the formation is:

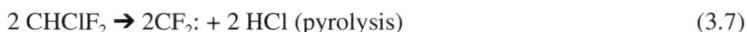

$$2 \ CHClF_2 \rightarrow 2CF_2 : + 2 \ HCl \ (pyrolysis) \tag{3.7}$$

$$2CF_2 : \rightarrow CF_2 = CF_2 \tag{3.8}$$

$$CF_2 : + CF_2 = CF_2 = CF_2 \rightarrow CF_3CF = CF_2 \ (hexafluoropropylene) \tag{3.9}$$

Hexafluoropropylene (HFP; $CF_3CF=CF_2$, CAS number 116-15-4) was first synthesized by Downing et al.. [26] by a pyrolysis reaction. Henne conducted the complete synthesis and identification of HFP [27]. Henne used a six-step reaction, starting with the fluorination of 1, 2, 3-trichloropropane, also known as allyl trichloride, and glycerol trichlorohydrin. The reaction product was 2,2-dichloro-1,1,1,3,3,3-hexafluoropropane. To obtain HFP he dehalogenated the latter with metallic zinc in a medium of boiling ethanol. HFP is produced as in the preparation of TFE. HFP yields may be improved by variation of the reaction process conditions at the expense of a lower TFE yield. Basically, the pyrolysis temperature is reduced and a steam diluent is added to the reaction medium [28, 29].

HFP may also be prepared by a multistep fluorination hexachloropropylene which is the chlorine analog of HFP [30]. In the follow-up steps the fluorination products are converted to CF_3-$CFCl$-CF_3, which is finally dechlorinated to HFP. Other methods describe the synthesis of HFP from the mixtures of a number of linear and cyclic tricarbon hydrocarbons with a partially halogenated three-carbon acyclic hydrocarbon.

TFP and HFP are co-produced by pyrolyzing compounds such as fluoroform, chlorodifluoromethane, chlorotetrafluoroethane, mixtures of chlorodifluoromethane and chlorotetrafluoroethane, or mixtures of perfluorocyclobutane and chlorodifluoromethane [31]. The reaction products consist of fluorinated olefins including TFE and HFP. The reaction reported by Gelblum et al. was conducted at temperatures between 600 and 1,000°C, using a gold-coated tube reactor because of the corrosivity of the reaction compounds and products.

3.4 Properties of Hexafluoropropylene

HFP is a non-flammable, colorless, odorless, tasteless gas with relatively low-toxicity (Table 3.2). It boils at −29.4°C and freezes at −156.2°C. HFP has relatively low reactivity compared to TFE. No commercial homopolymers of HFP have been developed. HFP ($CF_3CF=CF_2$) is used as a comonomer in a number of fluoropolymers along with TFE and vinylidene fluoride. It is one of the monomers used to "modify" the properties of homo-fluoropolymers, primarily to reduce the extent and rate of recrystallization.

HFP is extremely stable with respect to auto-polymerization, in contrast to TFE. It can be stored in a liquid state in the absence of a reaction inhibitor. HFP is thermally stable up to 400–500°C. At about 600°C, under vacuum, HFP decomposes and produces octafluoro-2-butene ($CF_3CF=CFCF_3$) and octafluoroisobutylene [33].

TABLE 3.2

Properties of Hexafluoropropylene [32]

Property	Value
CAS No.	116-15-4
Molecular weight	150.02
Boiling point at 101.3 kPa, °C	−29.4
Freezing point, °C	−156.2
Critical temperature, °C	85
Critical pressure, MPa	3254
Critical density, g/ml	0.60
Heat of formation for ideal gas at 25°C, ΔH, kJ/mol	−1,078.6
Heat of polymerization to solid polymer at 25°C, ΔH, kJ/mol	879
Flammability limits in air at 101.3 kPa, volume%	Nonflammable at all mixing ratios

HFP readily reacts with a number of elements, namely hydrogen, chlorine, and bromine by an addition mechanism analogous to other olefins [34–36]. Hydrogen halides such as hydrofluoric, hydrochloric, and hydrobromic acids react with HFP. The reaction of HFP with ammonia, alcohols, and mercaptans produces tetrafluoropropionitrile (CF_3CFHCN), hexafluoro ethers (CF_3CFHCF_2OR), and hexafluoro sulfides (CF_3CFHCF_2SR). The reaction of HFP with butadiene and cyclopentadiene yields Diels–Alder adducts. Alkali metal halide catalysis of HFP in N,N-dimethylacetamide can generate dimers and trimers of HFP [37].

3.5 Synthesis of Perfluoroalkylvinylethers

Perfluoroalkylvinylethers (PAVEs), such as perfluoroethylvinylether and perfluoropropylvinylether (PPVE) (CF_2=CF-O-C_3F_7, CAS number 1623-05-8), are synthesized by three reaction steps.

PPVE is prepared according to the following steps:

1. Oxidation of PPVE to hexafluoropropylene oxide (HFPO) by reacting it with oxygen in a diluent inert gas. Alternatively, oxygen can be replaced with hydrogen peroxide reacted at temperatures between 50 and 250°C.
2. HFPO is reacted with a perfluorinated acyl fluoride to obtain perfluoro-2-alkoxy-propionyl fluoride.
3. Perfluoro-2-alkoxy-propionyl fluoride is reacted with an alkali or alkaline earth metal salt containing oxygen at elevated temperatures to produce PPVE.

Alternative techniques including electrochemical methods are used for the production of perfluoro-2-alkoxy-propionyl fluoride. There are also electrochemical processes for the production of perfluoro-2-alkoxy-propionyl fluoride [38, 39].

A process has been reported for the preparation of perfluoroalkylvinylethers by fluorination using elemental fluorine of selected novel partially fluorinated dichloroethyl ethers, followed by dehalogenation to the corresponding perfluoroalkylvinylether. Perfluoroalkylvinylethers have been found to be useful as monomers for thermoplastic resins and elastomers [40].

An efficient process with a high reaction rate has been reported for the production of perfluoro (alkyl vinyl ethers). This process consists of dehalogenation, preferably chlorine, from carbon atoms α and β to an ether oxygen in an α,β-dihaloperfluoro ether in which the α,β-dihaloperfluoro ether reacts with zero valent zinc in pyrrolidinone. More specifically the process results in the production of CF_2=CFOCF$_2$CF$_2$SO$_2$F from CF$_2$ClCFClOCF$_2$CF$_2$SO$_2$F at relatively high yields in which CF$_2$ClCFClOCF$_2$CF$_2$SO$_2$F reacts with the zero valent zinc in the presence of N-methyl-2-pyrrolidinone [38].

TABLE 3.3

Properties of PPVE [41, 42]

Property	Value
CAS no.	1623-05-8
Molecular weight	266.04
Boiling point (at 101.3 kPa), °C	35–36
Melting point, °C	−70
Flash point, °C	−20
Density, g/cm^3	1.53
Critical temperature, °C	150.58
Critical pressure, MPa	1.91
Flammability limits in air at 101.3 kPa, volume%	1

3.6 Properties of Perfluoroalkylvinylethers

PAVEs are a useful class of monomers in the polymerization of fluoropolymers. They are effective monomers for the "modification" of the properties of fluorinated homopolymers such as PTFE. Modification here refers to the incorporation of less than 1% by weight of PAVEs in the fluoropolymers which may be called "minor copolymers". The small amount of incorporation has no impact on important properties of PTFE including thermal stability or chemical resistance. Conversely, HFP reduces the thermal stability of its copolymers with TFE.

PAVEs are also used in more extensive quantities as comonomers of TFE. Important commercial examples include perfluoropropyl vinyl ether (PPVE, C_3F_7-O-CF=CF_2), perfluoroethyl vinyl ether (PEVE, C_2F_5-O-CF=CF_2), and perfluoromethyl vinyl ether (PMVE, CF_3-O-CF=CF_2), all of which are used to produce melt processible copolymers of TFE.

PEVE is an odorless, colorless to yellow liquid at room temperature. It is less toxic than HFP. It is highly flammable, burning with a colorless flame. Information about the properties of PAVEs can be found in a number of publications [41, 42]. Some of the important properties of PPVE are listed in Table 3.3.

3.7 Synthesis of Chlorotrifluoroethylene (CTFE)

Chlorotrifluoroethylene (CTFE) is easier to prepare than the other perfluorinated monomers. The original commercial method starts with hexachloroethane and hydrofluoric acid. A mixture of these two compounds in the presence of chloro antimony fluoride produces R-113 [1,1,2-trichloro-1,2,2-trifluoroethane (TCTFE)]. Chlorine is removed from TCTFE by pyrolysis at 500–600°C in the presence of steam. Zinc can also be used to dechlorinate TCTFE at <100°C in methanol. By-products of zinc dechlorination such as chlorodifluoroethylene, trifluoroethylene, dichlorotrifluoroethane, and others are removed by distillation. Other by-products are removed by contact with sulfuric acid. In a final step moisture and hydrochloric acid are removed by adsorption in a dryer column such as alumina followed by purging of the vapor phase from the liquid CTFE.

Methods using R-113 are not favored because of its ozone depleting effect. Other methods have been developed to produce CTFE without using R-113. For example, a process for preparing CTFE consists of reacting tetrafluoroethylene with hydrogen chloride in the presence of a metallic catalyst [43]. Typically, the reaction is conducted at 330°C (range: 150–350°C) and at a pressure of 98 kPa (<9,800 kPa) in the presence of chromic oxide (Cr_2O_3). The residence time is related to the reaction temperature and ranges from 1 to 120 seconds. Detailed information on reaction conditions and products are presented in Table 3.4.

TABLE 3.4

Reaction Conditions and Product Composition for the Reaction of Hydrochloric Acid and TFE [43]

Temperature (°C.)	Flow Rate (ml/min)		TFE Conversion (%)	Selectivity (%)					
	TFE	HCl		CTFE	R-125	R-115	R-124	R-114	R-1112
350	45	45	52	53	2	12	6	5	17
310	45	45	37	51	8	7	9	5	12
300	90	30	24(8)	58(78)	6(0)	12(3)	8(4)	0(4)	11(5)
300	120	30	16(8)	71(79)	2(1)	7(5)	5(5)	1(2)	9(7)
250	45	45	33(17)	58(70)	6(4)	4(2)	9(6)	2(2)	15(12)

R-125 = pentafluoroethane; R-115 = chloropentafluoroethane; R-124 = 2-chloro 1,1,1,2-tetrafluoroethane; R-114 = 1,2-dichlorotetrafluoroethane.

TABLE 3.5

Properties of CTFE

Property	Value
Cas No.	79-38-9
Molecular weight	116.47
Appearance	Colorless gas
Odor	Faint etheral color
Density	1.54 g/cm^3 at –60°C
Melting point	–158.2°C
Boiling point	–27.8°C
Solubility in water	4.01 g/100 ml
Solubility	In benzene and chloroform
Magnetic susceptibility	–49.1×10^{-6} cm^3/mol
Refractive index	1.38 at 0°C
Critical temperature, °C	107
Critical pressure, MPa	3.03

3.8 Properties of Chlorotrifluoroethylene

CTFE is a colorless gas at room temperature and atmospheric pressure. The monomer will not autopolymerize at ambient temperatures; therefore, it can be transported without an inhibitor. Like all fully or partially fluorinated ethylenes, CTFE can undergo a disproportionation reaction and thus must be handled carefully. It forms high-molecular peroxides in reaction with oxygen, and these can precipitate from a solution. Thus, the oxygen concentration of commercial CTFE is maintained below 50 ppm [44]. CTFE hydrolyzes slowly in water containing oxygen, whereas it is completely stable in degassed water. It is used for the preparation of polychlorotrifluoroethylene (PCTFE) and as a comonomer to a variety of copolymers.

3.9 Synthesis of Vinylidene Fluoride

The main monomer for manufacturing polyvinylidene fluoride polymers is vinylidene fluoride (VDF). In addition to homopolymers, copolymers of VDF are produced using monomers such as CTFE, tetrafluoroethylene, and hexafluoropropylene. A variety of methods have been developed for the preparation of VDF. A few of these rely on the dehydrohalogenation of halogenated hydrocarbons. In these reactions a hydrogen halide (HX) is abstracted from the halo-hydrocarbons [45, 46].

The original method for VDF manufacturer was developed by Pennsalt Corp [45]. The process began by passing 1,1,1-trifluoroethane through a platinum-lined Inconel tube, which was heated to 1,200°C.

TABLE 3.6

Effect of Contact Time and Temperature on VDF Yield [45]

Variable	Case 1	Case 2
Temperature, °C	1,200	800
Contact time, sec.	0.01	4.4
Space velocity, l/hr	9,700	200
Total conversion, mole %	75.4	76
Conversion to VDF, mole %	74	66
VDF yield, %	98.1	86.5
By-product yield, %	1.9	13.5

Contact time was about 0.01 seconds. The exit gases were passed through a sodium fluoride bed to remove the hydrofluoric acid and were collected in a liquid nitrogen trap. VDF (boiling point: −84°C) was separated by low temperature distillation. Unreacted trifluoroethane was removed at −47.5°C and was recycled. Generally, a high temperature process improves VDF yield (Table 3.6).

$$CH_3\text{-}CF_3 \rightarrow CH_2 = CF_2 + HF$$

Commercially, VDF is produced by the dehydrochlorination of 1-chloro-1,1-difluoroethane (HCFC-142b) [75-68-3]. Many patents exist for the preparative routes based upon dehydrohalogenation of various chlorofluorohydrocarbons or related compounds. New research work on the manufacture of VDF was conducted after the phase out of HCFC production was announced in the mid-2000s. The industry had to devise alternative technology, such as producing VDF from 1,1-difluoroethane (HFC-152a).

Another method begins with the hydrofluorination of acetylene and subsequent chlorination [47], or by the hydrofluorination of trichloroethane [48], or by the hydrofluorination of vinylidene chloride [49]. In each case the final product, 1-chloro-1,1-difluoroethane, is stripped of a molecule of hydrochloric acid to yield VDF. The following one-step reaction scheme exhibits the use of vinylidene chloride as a starting compound:

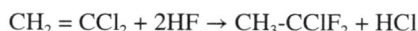

$$CH_2 = CCl_2 + 2HF \rightarrow CH_3\text{-}CClF_2 + HCl$$

$$CH_3\text{-}CClF_2 \rightarrow CH_2 = CF_2 + HCl$$

The mixture of vinylidene chloride and hydrofluoric acid is passed through a heated catalyst bed. The catalyst is prepared by heating $CrCl_3\bullet6H2O$ under vacuum to 300°C until it changes color from dark green to a solid violet throughout the porous mass. In this operation, crystallized water is removed (35% weight loss). The cooled mass is comminuted and screened into particles of 2–5 mm in diameter and which are loaded into a cylindrical reactor and heated to the reaction temperature (250–350°C). The resulting gases are condensed and vinylidene fluoride (boiling point: −84°C) is separated by low temperature distillation.

In another one-step process a mixture of vinylidene chloride and hydrofluoric acid is heated to 400–700°C in the presence of oxygen and a catalyst [50].

In yet another method [51] pyrolysis of 1,2-dichloro-2,2-difluoroethane in the presence of hydrogen is accomplished in the absence of a catalyst in an essentially empty reactor at temperatures exceeding 400°C. The term "absence of catalyst" refers to the lack of a conventional catalyst. In this method, the construction material of the reactor is a metal, such as iron, titanium, chromium, molybdenum, cobalt, or gold, or alloys thereof. The metal is selected to inhibit corrosion and/or catalytic effect.

Important manufacturers of VDF include key PVDF manufacturers such as Arkema (France), Solvay S.A. (Belgium), Daikin Industries Ltd (Japan), Dyneon GmbH (Germany), Kureha Corporation (Japan), Shanghai 3F New Materials Company Limited (China), Shanghai Ofluorine Chemical Technology Co. Ltd. (China), Zhejiang Fotech International Co. Ltd. (China), and Zhuzhou Hongda Polymer Materials Co. Ltd. (China).

TABLE 3.7

Properties of VDF [52]

Cas No.	75-38-7
Molecular weight	64.03
Appearance	Colorless gas
Odor	Faint ether-like odor
Density (liquid)	0.617 g/cc at 24°C
Melting point	−85.7°C
Boiling point	−144°C
Solubility in water	0.018 g/100 g at 25°C at 1 atm
Solubility	Slightly in dimethylacetamide
Refractive index	1.38 at 0°C
Flammability range, %	5.5–21%
Critical temperature, °C	30.1
Critical pressure, MPa	4.34

3.10 Properties of Vinylidene Fluoride

VDF ($CH_2=CF_2$, HFC-1131a) is a colorless gas at room temperature and atmospheric pressure. It is toxic by inhalation and contact. It has a faint ether-like odor and is flammable in the concentration range of 5.5 to 21.0 percent by volume. Table 3.7 presents some of its important physical properties. When polymerized it generates a great deal of heat which must be removed from the reaction vessel. VDF may be stored or transported in gas cylinders or high-pressure trailers without the addition of polymerization inhibitors. However, terpene and quinone inhibitors may be added to VDF to inhibit autopolymerization.

3.11 Synthesis of Vinyl Fluoride

There are a variety of reaction schemes to synthesize vinyl fluoride (VF) [53–56]. One of the methods described in the literature is a two-step method [57]. The first step is the reaction of hydrogen fluoride with acetylene in the presence of a suitable catalyst to yield ethylidene fluoride, which is subsequently pyrolyzed:

$$CH \equiv CH + HF \rightarrow CH_3CHF_2 \xrightarrow{pyrolysis} CH_2 = CHF$$

The commercial process for the synthesis of VF has not been published for proprietary reasons. The addition of HF to acetylene and the fluorination of vinyl chloride are, however, the most direct industrial route to the production of VF. As with TFE, it is essential that the VF monomer be purified prior to polymerization.

VF is stabilized with compounds like D-Limonene to prevent auto-polymerization. Transportation and storage of VF requires stabilization to avoid autopolymerization. Stabilizers are removed from VF prior to its polymerization.

3.12 Properties of Vinyl Fluoride

Table 3.8 shows a summary of the properties of VF. It is a colorless gas in atmospheric conditions. VF is highly flammable and may be ignited in the concentration range of 2.6 and 22% by volume. The ignition temperature for VF and air mixtures is >400°C. To prevent spontaneous polymerization of VF, adding a trace amount (<0.2%) of terpenes is effective. The US Department of Transportation has classified inhibited VF as a flammable gas [53].

TABLE 3.8

Properties of VF [58]

Cas No.	75-02-5
Molecular weight	46.04 g/mol
Appearance	Colorless gas
Odor	Faint ether-like odor
Density (liquid)	0.636 g/cc at 21°C
Melting point	−160.5°C
Boiling point	−72.2°C
Solubility in water	Slight
Refractive index	1.34 at 0°C
Flammability range, %	2.6–21.7%
Critical temperature, °C	54.8
Critical pressure, MPa	5.24

REFERENCES

1. Ebnesajjad, S., *Introduction to Fluoropolymers: Materials, Technology, and Applications*, 2nd ed, Elsevier, New York, (2021).
2. Ebnesajjad, S., *Fluoroplastics Vol 1 – Non-melt Processible Fluoropolymers*, 2nd ed, Elsevier, Oxford, (2015).
3. Chabrie, C., *Compt. Rend. de l'Academ des Sc*, *110*:279, 1202, (1890).
4. Moissan, H., *Compt. Rend.*, *110*:276–279, (1890).
5. Moissan, H., *Compt. Rend.*, *110*:951–954, (1890).
6. Humiston, B., *J. Phys. Chem.*, *23*:572–577, (1919).
7. Ruff, O., Bretschneider, O., *Z. Anorg. Che.*, *210*:73, (1933).
8. Downing, F. B., Benning, A. F., McHarness, R. C., US Patent 2,384,821, assigned to DuPont, (1945).
9. Park, J. D., Benning, A. F., Downing, F. B., Laucius, J. F., McHarness, R. C. (1947) Synthesis of tetra-fluoroethylene. *Ind. Eng. Chem.* 39, 354–358.
10. Hamilton, J. M., in Stacey, M., Tatlow, J. C., Sharpe, A. G., eds., *Advances in Fluorine Chemistry*, Vol. *3*, 12. Butterworth & Co., Ltd., Kent, U. K., p. 117, (1963).
11. Edwards, J. W., Small, P. A., *Nature*, *202*:1329, (1964).
12. Gozzo, F., Patrick, C. R., *Nature*, *202*:80, (1964).
13. Scherer, O., et al., US Patent 2,994,723, assigned to Farbewerke Hoechst, (1961).
14. Edwards, J. W., Sherratt, S., Small, P. A., Brit. Patent 960,309, assigned to ICI, (1964).
15. Ukahashi, H., Hisasne, M., US Patent 3,459,818, assigned to Asahi Glass Co., (1969).
16. Chinoy, Percy B., Sunavala, Pharokh D., *Ind. Eng. Chem. Res.*,*26*, pp 1340–1344, (1987).
17. Sherratt, S., in: Kirk-Othmer Encyclopedia of Chemical Technology 2nd ed., (A. Standen. ed.,), 9: 805–831, Interscience Publishers, Div. of John Wiley and Sons, New York, (1966).
18. Hintzer, K., Streiter, A. et al, US Patent US 9,139,496, assigned to 3M Innovative Properties Co, (2015).
19. Toy, M. S., Tiner, N. A., U.S. Patent 3,567,521 (1971) to McDonnell Douglas Co.
20. Gangal, S. V., in Mark, H. F., Kroschwitz, J. I., (Eds.), *Encyclopedia of Polymer Science and Technology*, vol. 16, John Wiley & Sons, New York, (2010).
21. Sheratt, S. *Encyclopedia of Chemical Technology*, Vol. *9*, John Wiley & Sons, New York, p. 808, (1966).
22. Ref Renfrew, M. M., Lewis, E. E. *Ind Eng Chem*, (1946), No. *38*, pp 870–7.
23. Perfluorinated Polymers, Polytetrafluoroethylene, Subhash V. Gangal, Paul D. Brothers, June 15, (2010), Wiley-Online Library, https://doi.org/10.1002/0471440264.pst233.pub2
24. US Patent 5,345,013, Van Bramer, D. J., Shiflett, M. B., Yokozeki, A., assigned to DuPont Company, (1994).
25. Dietrich, M. A., Joyce, R. M., U.S. Patent 2,407, 405 (1946), to E.I. du Pont de Nemours & Co.
26. Downing, F. B., Benning, A. F., McHarness, R. C., US Patent 2,384,821, assigned to DuPont, Sep. 18, (1945).

27. Henne, A. L., Woalkes, T. P., *J. Am. Chem. Soc.*, *68*: 496, (1946).
28. Chinoy, P. B., Sunavala, P. D., Thermodynamics and kinetics for the manufacture of tetrafluoroethylene by the pyrolysis of chlorodifluoromethane, *Ind. Eng. Chem. Res.*, *26*: 1340–1344, (1987).
29. Brayer, E., Bekker, A. Y., Ritter, A. R., Kinetics of the pyrolysis of chlorodifluoromethane, *Ind. Eng. Chem. Res. 27*: 211, (1988).
30. Webster, J. L., Trofimenko, S., Resnick, P. R. et al, US Patent 5,068,472, assigned to DuPont, (1991).
31. Gelblum, P. G., Herron, N., Noelke, C. J., Rao, V. N. M., US Patent 7,271,301, assigned to DuPont, (2007).
32. Perfluorinated Polymers, Perfluorinated Ethylene Propylene Cppolymers, Subhash V. Gangal, Paul D. Brothers, (2010), Wiley-Online Library, https://doi.org/10.1002/0471440264.pst233.pub2.
33. Gibbs, H. H., Warnell, J. J., British Patent 931,587, assigned to DuPont, (1963).
34. Knunyants, I. L., Mysov, E. I., Krasuskaya, M. P., Izvezt. Akad. Nauk S. S. S. R., Otdel. Khim. Nauk, pp. 906–907, (1958).
35. Haszeldine, R. N., Steele, B. R., *J. Chem. Soc.*, pp. 1592–1600, (1953).
36. Miller, W. T., Jr., Bergman, E., Fainberg, A. H., *J. Am. Chem. Soc.*, *79*: 4159–4164, (1957).
37. Coffman, D. D., Cramer, R., Rigby, G. W., *J. Am. Chem. Soc.*, *71*: 979–980, (1949).
38. Resnick, P. R., US Patent 6,388,139, assigned to DuPont, May 14, (2002).
39. Brice, T. J., Pearlson, W. H., US Patent 2,713,593, assigned to 3M Co., (1955).
40. Hung, M. H., Rozen, S., US Patent 5,350,497, assigned to DuPont, (1994).
41. Gangal, S. V., Brothers, P. D., Perfluorinated polymers, tetrafluoroethylene–perfluorovinyl ether copolymers, in: *Encyclopedia of Polymer Science and Technology*, online edition, John Wiley and Sons, (2010).
42. Technical Information PPVE Perfluoropropylvinyl Ether, DuPont FluoroIntermediates, publication No. H-88804-2, (2007).
43. US Patent 5,243,104, F. Yamaguchi, T. Otsuka and K. Nakagawa, assigned to Daikin Industries, (1993).
44. Stanitis, G. *Modern Fluoropolymers* (Scheirs, J., Ed.), John Wiley & Sons, (1997), p. 526.
45. Hauptschein, A., Feinberg, A. H., US Patent 3,188,356, assigned to Pennsalt Chemicals Corp., (1965).
46. Trager, F. C., Mansell, J. D., Wimer, W. E., US Patent 4,818,513, assigned to PPG Industries, Inc., (1989).
47. Schultz, N., Martens, P., Vahlensieck, H. J., German Patent 2,659,712, assigned to Dynamit Nobel AG, (1976).
48. McBee, E. T., et al., *Industrial Eng. Chem.*, *39*(3), pp 409–412, (1947).
49. Kaess, F., Michaud, H., US Patent 3,600,450, assigned to Sueddeutsche Kalkstickstoff-Werke AG, (1971).
50. Elsheikh, M. Y., US Patent 4,827,055, assigned to Pennwalt Corp., (1989).
51. US 8,350,101, Sylvain Perdrieux, Serge Hub, assigned to Arkema, France, (2013).
52. PubChem, National Cen for Biotech Inf, NIH, https://pubchem.ncbi.nlm.nih.gov/compound/Vinylidene-fluoride, (2021).
53. Ebnesajjad, S., *Fluoroplastics–Melt Processible Fluoropolymers*, Vol 2, 2nd ed, Elsevier, (2015).
54. Coffman, D. D., Cramer, R., Rigby, G. W., *J. Am. Chem. Soc.*, *71*: 979–980, (1949).
55. Coffman, D. D., Raasch, M. I., Rigby, G. W., Barrich, P. L., Hanford, W. E., *J. Org. Chem.*, *14*: 747–753, (1949).
56. Pajaczkowski, A., Spoors, J. W., *Chem. Ind., London*, *16*: 659, (1964).
57. US Patent US2,471,525, Hillyer, J. C., Wilson, J. F., assigned to Phillips Petroleum, (1949).
58. PubChem, National Cen for Biotech Inf, NIH, https://pubchem.ncbi.nlm.nih.gov/compound/Vinyl-fluoride, (2021).

4

Polymerization of Commercial Thermoplastic Fluoropolymers

Sina Ebnesajjad

The first industrial fluorinated polymer was polychlorotrifluoroethylene (PCTFE). Its preparation was first reported in 1937 by the notorious IG Farben Werk memorialized by the British Patent number 465520. It had impressive chemical resistance but degraded significantly at elevated temperatures even at those lower than its melting point. Today one of the measures of PCTFE's molecular weight is zero strength time (ZST), measured in seconds, when a specimen is held under a modest load at PCTFE's melting point.

The era of fluoropolymers began in April 1938 earnestly, though rather serendipitously. Roy Plunket of DuPont discovered a powdery product in a tetrafluoroethylene (TFE) cylinder. The white powder was later found to be a polymer of TFE, formed by autopolymerization. TFE gas was intended for the synthesis of new refrigerants for the up and coming automobile market. Ostensibly, the PTFE polymer was a laboratory nuisance rather than a success. Plunket's curiosity in identifying the powdery substance became the initial step towards the inception of fluoropolymer industries.

The story of Roy Plunkett's discovery has been told and retold many times by Plunket himself and other authors:

- Plunkett RJ. US Patent 2,230,654, assigned to DuPont Co. February 4, 1941.
- Plunkett RJ. The history of polytetrafluoroethylene: discovery and development. In: Seymour RB, Kirshenbaum GS, editors. High performance polymers: their origin and development, Proceed. of symp. on the hist. of high perf. polymers at the ACS meeting in New York. New York: Elsevier; April 1986. 1987.
- Kinnane A, editor. From the banks of the Brandywine to miracles of science. Published by DuPont Company, 200th Anniversary DuPont; 2002.
- Bahadur P, Sastry NV. Principles of polymer science. Pub Alpha Science Int'l Ltd.; 2005.
- Ebnesajjad, S., *Fluoroplastics, Vol.1, Non-Melt Processible Fluoroplastics*, 2nd Edition, Elsevier, Oxford, UK, 2015.

4.1 Polymerization of Tetrafluoroethylene

TFE, which is currently the most widely used monomer in the fluoropolymer industries, was first synthesized by Ruff and Brettschneider in 1933 by the pyrolysis of tetrafluoromethane in an electric furnace [1].

The largest consumer of TFE is its polymerization into a variety of PTFE homopolymers, modified PTFE, micropowders, and as a comonomer in copolymerization with hexafluoropropylene, ethylene, perfluorinated ether, and other monomers, and as a comonomer in a variety of terpolymers. Other uses of TFE are to prepare low-molecular-weight polyfluorocarbons and carbonyl fluoride oils as well as to form PTFE *in situ* on metal surfaces and in the synthesis of hexafluoropropylene, perfluorinated ethers, and other oligomers [2].

DOI: 10.1201/9781003204275-6

TFE is polymerized via a free radical addition mechanism in an aqueous medium using a water-soluble, free radical initiator, such as peroxydisulfate, organic peroxide, or a reduction-activation system [3]. The additives must be selected carefully to ensure they do not interfere with TFE polymerization detrimentally. The types of undesirable interference include inhibition of the polymerization of the reaction or chain transfer leading to the formation of undesirable low molecular species. During aqueous dispersion polymerization, highly halogenated surfactants, such as fully fluorinated acids [4], are added. TFE polymerizes readily at moderate temperatures (40 to 80°C) and moderate pressures (0.7 to 2.8 MPa). The reaction is extremely exothermic (the heat of polymerization is 41 kcal/mol), requiring the removal of heat from the polymerization vessel.

In principle, there are two distinct methods of polymerization of tetrafluoroethylene: *suspension* and *dispersion*. When little or no dispersing agent is used and the reaction mixture (i.e., suspension) is agitated vigorously, TFE polymerizes in the gas phase and a polymer is precipitated, known as *granular* resin. In the presence of sufficient amounts of dispersants in the reaction mixture and with a mild agitation maintained, polymerization proceeds by the dispersion regime. The dispersion polymer consists of small, negatively charged, round, colloidal particles (the longer dimension is less than 0.5 μm). The two PTFE products have unique and differing properties even though both have high-molecular-weight TFE polymers. Those differences require the two products to be fabricated by vastly different processes.

The aqueous dispersion recovered in the second process is used to produce fine powders or to be further concentrated/finished into paint-like products used to produce *dispersion* grade PTFE fabricated by direct dipping and other coating application methods. The schematics of the processes for TFE polymerization and the finishing steps are shown in Figures 4.1 and 4.2. The details of the manufacturing of the three different PTFE products are described in the following sections. The last section in this chapter discusses the surfactants used in polymerization.

4.1.1 Granular Resins

Granular PTFE resins are produced by polymerizing TFE alone or in the presence of trace amounts of comonomers [5, 6] using an initiator and sometimes in the presence of an alkaline buffer in an aqueous suspension medium. Other additives in the reaction mixture modify PTFE's properties. The polymerization

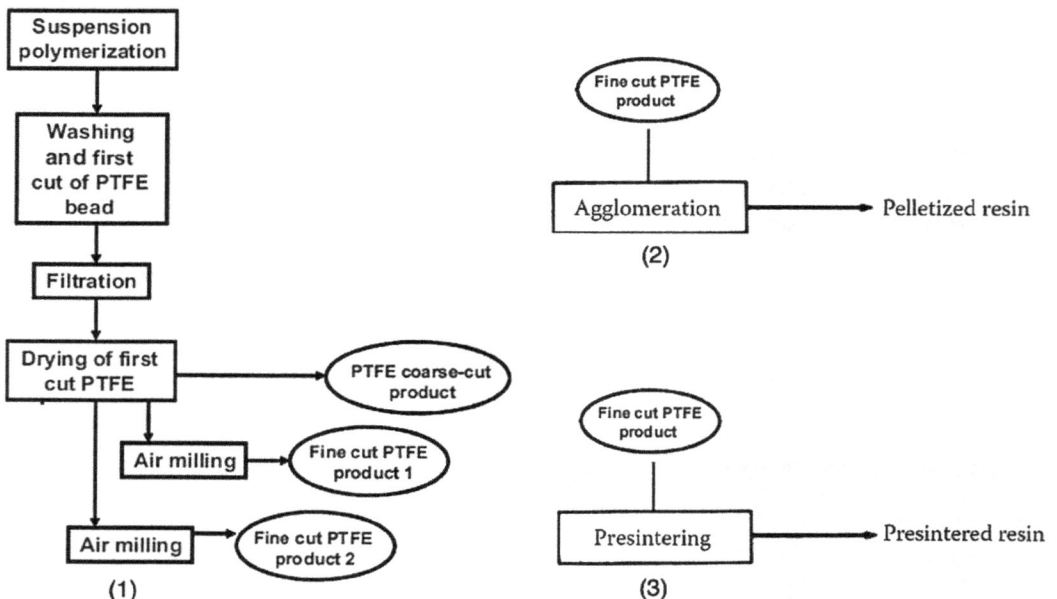

FIGURE 4.1 Manufacturing schematic of granular PTFE products.

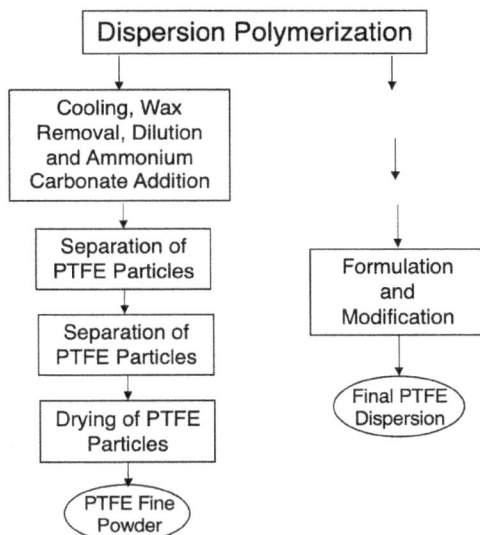

FIGURE 4.2   Schematic of the manufacturing steps of fine powder and dispersion PTFE products.

product from the autoclave consists of a mixture of water with PTFE particles resembling coconut shreds. There are different approaches to the comminution of the recovered polymerization particles.

Most frequently the polymer recovered from the reactor (Figure 4.3) is cut into smaller particles (Figure 4.4). After the water is removed from the mixture, the polymer is dried. To obtain moldable resins, known as "fine cut resins", the dry product is disintegrated by hammer milling. An alternative method

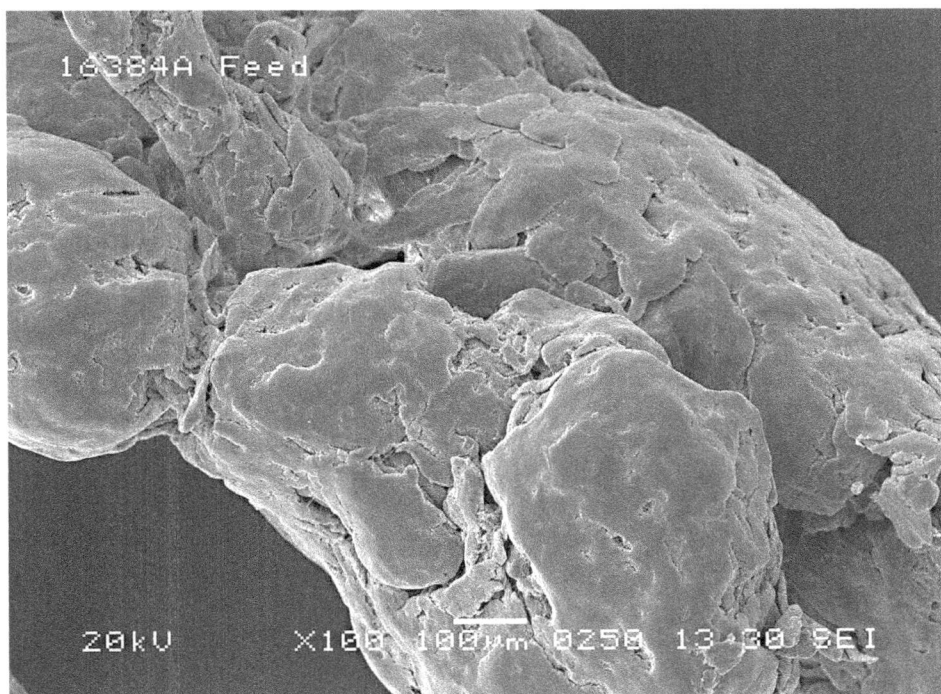

FIGURE 4.3   Optical micrograph of a PTFE granular particle (100X magnification).

FIGURE 4.4   Optical micrograph of cut PTFE granular particles (500X magnification).

is to cut the PTFE particles to the final size before water removal and drying. Finished powders usually have a mean particle size from <10 to 700 μm and an apparent density from 200s to >700 g/l, depending on the grade [7].

Fine-cut granular resins produced by size reduction of the suspension polymer have a typical average particle size from 20 to 40 μm. The small particle size of fine-cut PTFE imparts the highest possible mechanical properties to articles made from granular resins. Fine-cut resins (powders) have poor "flow" and low apparent (bulk) density (less than 500 g/l). Their consistency resembles wheat flour with poor flow characteristics. Granular powders with a smaller particle size (and lower density) are prepared by pulverizing the ordinary dry raw suspension using air or hammer mills equipped with a bladed rotor rotating at a speed of 3000 m/min in a vortex of air or another gas. This process results in powders that are difficult to handle during processing because of poor flow and low apparent densities (<300 g/liter).

The advantage of "lighter" powders is improved mechanical properties of parts made from them. The disadvantage of fine-cut resins is poor flow and low bulk density. These characteristics render them unsuitable for use in automatic and isostatic molding techniques. Fabricating low density PTFE requires significantly taller molds than higher bulk density powders so as to obtain useful length parts. To address the poor flow of PTFE powders they are pelletized, which renders them useful for automatic and isostatic molding methods that require free flow powder. The free flow allows proper filling of the mold cavities quickly.

Pelletized granular resins are obtained by the agglomeration of fine-cut resins. The agglomeration process increases the powder flow and bulk density. The goal of this process is to make the small PTFE particles adhere. Essentially, there are two processes of agglomeration: namely *wet* and *dry* techniques. The wet process involves forming a suspension of PTFE powder in a liquid phase (aqueous or organic) usually in the presence of a surfactant. The suspension is warmed and stirred which gives rise to the formation of round particles consisting of small PTFE particles that adhere. The particles are filtered out and dried subsequently. The dry method consists of adding a small amount of a surfactant or another suitable liquid followed by tumbling. Particles are formed by the tumbling action which are dried to remove the liquid [8].

Presintered resins are prepared by sintering PTFE particles, cooling the melt, and disintegrating the slabs of PTFE back into small particles. The average particle size of these resins is several hundred microns, and their melting point is reduced from 342 to 327°C. The powder flow is improved by this process considerably, and the presintered products are particularly suitable for the ram extrusion of thin-walled tubes and thin solid rods. Ram extrusion is the only semicontinuous process for granular PTFE. It allows extrusion of long tubes, rods, and other profiles from PTFE.

4.1.2 Fine Powder Resins

Fine powder resins are prepared by *dispersion* polymerization of TFE in an aqueous colloidal *dispersion* in the presence of an initiator, a surfactant, and low melting wax to prevent coagulation when the solids content increases. Although the polymerization mechanism is not a typical emulsion regime, some of the principles of emulsion polymerization apply here. Both the process and the ingredients have significant impact on the properties of the final product. The products of dispersion are round particles in the diameter range of 0.1–0.3 µm. The solids content of such dispersions can be as high as 45% by weight (approximately 22% by volume) because of the high density of PTFE. The dispersion must be sufficiently stable (i.e., contain sufficient surfactant and a wax) throughout the polymerization so that it does not coagulate in the autoclave. Gentle stirring ensures the stability of the dispersion and the uniformity of the particle shape and size.

The finished dispersion is then diluted to a solids content of about 10% by weight to destabilize it, followed by coagulation by controlled stirring and the addition of an electrolyte. The low concentration dispersion first thickens to give a gelatinous mass; then the viscosity decreases again. The coagulum changes to air-containing, water-repellent agglomerates (diameter 300–700 µm) that float on the aqueous medium [3].

The wet agglomerates are dried by placing them in trays in an oven or continuously using a conveyor belt. Shearing of the particles must be avoided, especially after particles are dried, because of a phenomenon called "fibrillation" that occurs at temperatures above its ambient transition point of 19°C (Figure 4.5). The fibrillation of fine powder is the removal of groups of chains of PTFE from the crystalline phase. Fibrillation is required during the processing of fine powder to impart strength to them. Its occurrence prior to processing leads to the formation of defects in fabricated parts. Fine powder PTFE is stored and transported at temperatures <19°C to avoid untimely fibrillation.

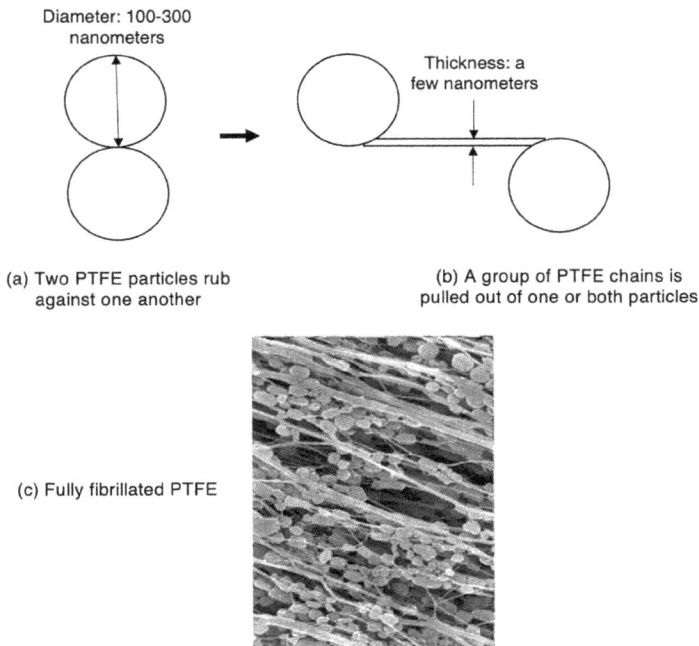

Diameter: 100-300 nanometers

Thickness: a few nanometers

(a) Two PTFE particles rub against one another

(b) A group of PTFE chains is pulled out of one or both particles

(c) Fully fibrillated PTFE

FIGURE 4.5 Mechanism of PTFE fibrillation.

Core-shell polymers consist of particles with a composite structure. The inner portion of the particle (core) has a different composition to that of the outer portion (shell). Typically, they are prepared by introducing a comonomer later during polymerization under specific conditions, thus altering the composition of the particle exteriors from the cores [9]. An example is a composition with a core constituting 65 to 75% of the total weight of the particle. The remaining 25 to 35% of the polymer forms the shell with a lower comonomer content than the core. Such resins exhibit improved paste extrudability, especially in the production of tubing or wire insulation, and a reduced number of flaws in the final product.

4.1.3 Aqueous Dispersions

PTFE aqueous dispersions are made by the polymerization process used to make fine powders. Raw dispersions are polymerized to different particle sizes [10]. The optimum particle size for most applications is about 0.2 µm. The dispersion from the autoclave is stabilized by the addition of nonionic or anionic surfactants, followed by concentration into a solids content of 60 to 65% by electrodecantation, evaporation, or thermal concentration. After further formulation (modification) with chemical additives as required by the end or the fabrication process, the commercial product is sold with a polymer content of about 60% by weight, various viscosities, and a specific gravity around 1.5.

The fabrication processing characteristics of the dispersion depend on the conditions of the polymerization and the type and amounts of the chemical additives contained in it. Stability is a key requirement of the final product. Typically, a PTFE aqueous dispersion must have a shelf life of several months to one year. It should withstand transportation and handling during processing. The shear rate during processing must be low enough as to not cause the agglomeration of the particles. Ideally, the temperature during transportation, storage, and processing should be below 19°C, to prevent fibrillation, as is the case for fine powders.

4.1.4 Filled Compounds

To alter or improve the properties of the raw PTFE (to enhance wear resistance, creep resistance, and thermal and electrical conductivity), various fillers, such as glass fibers, powdered metals, and graphite, are combined with all three types of PTFE polymers, mostly by intimate mixing. Filled fine powders are produced by adding fillers during lubricant addition or when adding into a dispersion and then coagulating the mixture. Aqueous dispersion can also be modified by the addition of certain fillers, pigments, heat resistant dyes, carbon blacks, mica, and powdered metals.

4.1.5 Modified PTFE

Modified PTFE represents a relatively new product line designed to overcome some of the limitations of conventional PTFE, namely poor creep resistance (i.e., a tendency to cold flow), difficult welding, and high level of microvoids [11]. These changes are accomplished by a reduction in the molecular weight of PTFE. To prevent excessive recrystallization, a small amount of a special comonomer is incorporated in amounts less than 0.5% by weight. The most effective modifier has been perfluoropropyl vinyl ether (PPVE). The *long* pendent group C_3F_7- (seen in Figure 4.6) prevents excessive recrystallization of the

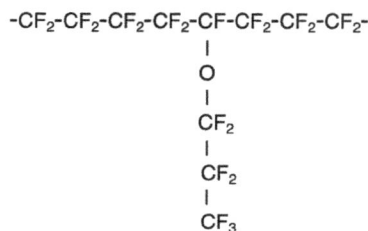

$$-CF_2-CF_2-CF_2-CF_2-CF-CF_2-CF_2-CF_2-$$
$$|$$
$$O$$
$$|$$
$$CF_2$$
$$|$$
$$CF_2$$
$$|$$
$$CF_3$$

FIGURE 4.6 Molecular structure of modified PTFE.

PTFE thus allowing parts to possess useful mechanical properties. The ether link structure of modified PTFE is stable and does not detract from the continuous use temperature of PTFE (260°C).

Copolymerization is carried out in an aqueous suspension for granular polymer and in a dispersion for fine powder PTFE. Polymerization conditions are like those of the homopolymerization process. TFE pressure is in the range 5 to 20 bar and a temperature range from 35 to 90°C [12].

4.2 Fluorinated Ethylene Propylene

Fluorinated ethylene propylene (FEP) is a copolymer of TFE and hexafluoropropylene (13.5% by weight incorporation) and has a branched structure containing monomer units of –CF$_2$–CF$_2$– (TFE) and –CF$_2$–CF (CF$_3$)– (HFP). The pendent methyl fluoride group (-CF$_3$) disrupts the crystalline structure of the copolymer resulting in a significantly lower recrystallization extent. FEP retains most of the favorable properties of PTFE, but its melt viscosity is sufficiently lower than PTFE thus allowing conventional melt processing. The introduction of HFP allows the reduction of the molecular weight of FEP which reduces its melting point from 327°C for PTFE to 260°C (nominally) [13].

4.2.1 Industrial Process for the Production of FEP

There are several methods for the copolymerization of hexafluoropropylene and tetrafluoroethylene using different catalysts at different temperatures [14–22]. Both aqueous and nonaqueous dispersion polymerization regimes have been used for the commercial manufacturing of perfluoroalkoxy (PFA). In recent decades, solvent-based polymerization has been discontinued to eliminate atmospheric emissions of organic solvents.

The conditions for this type of process are like those of the dispersion homopolymerization of TFE except for the use of potassium peroxides for initiation. FEP is a random copolymer; that is, HFP units add to the growing chain at random intervals. The optimal composition and molecular weight of the copolymer are such that the mechanical properties of FEP are in the usable range as required by applications. Commercial FEP is available at various medium and high viscosity grades, for injection molding, extrusion, and other melt processing methods. Aqueous dispersions are available at concentrations as high as 55% solids by weight.

4.3 Perfluoroalkoxy Resin

PFA resins are prepared by the copolymerization of TFE and perfluoroalkyl vinyl ether (PAVE) monomers in either aqueous or nonaqueous media [23–25]. Perfluoropropyl vinyl ether (PPVE) and perfluoroethyl vinyl ether (PEVE) are the two most common among the PAVE comonomers. Mixtures of (PEVE) and PPVE are also used in some grades of PFA. The mechanism of the action of PAVEs is similar to their impact in modifying PTFE, as described in Section 4.1.5. The difference is that the molecular weight of the polymer must be reduced more extensively than the modified PTFE to allow processing by customary melt methods. Larger amounts of PPVE must be incorporated to prevent the extensive recrystallization of PFA thus allowing useful mechanical properties in fabricated parts.

4.3.1 Industrial Process for the Production of Perfluoroalkoxy Resins

Presently, PFA resins are produced by aqueous copolymerization. The process has similar reaction conditions to the *dispersion* polymerization of TFE. Inorganic peroxy compounds (e.g., ammonium persulfate) are used as an initiator. A fluorinated surfactant (e.g., ammonium perfluorooctanoate and its replacements) is added to the reaction medium to allow dispersion polymerization [26]. Since the end of 2015 ammonium perfluorooctanoate is no longer used in the USA, EU, and Japan. It has been replaced by less toxic, less bio-accumulative, and more easily degradable compounds [13].

In a nonaqueous copolymerization, fluorinated acyl peroxides are added that are soluble in the medium [27]. A chain transfer agent may be added to control the molecular weight of the resin. The polymer is separated from the medium and converted into useful forms such as melt-extruded cubes (pellets) for melt processing (e.g., extrusion, injection molding). PFA resin is also available as an aqueous dispersion (60% solids), molding powders, and fine powders for powder coating [28, 29].

There is a variant of the PFA polymer popularly called "MFA" to distinguish it from PFA made from the polymerization of PPVE and TFE. MFA is made by the copolymerization of TFE and perfluorome-thyl vinyl ether (PMVE). MFA is offered commercially only by the Solvay Corporation. The maximum continuous use temperature of MFA is in the range 200–260°C; the intermediate to continuous use temperature of FEP is 200°C; and that of PFA is 260°C [30]. MFA is fully melt processible and is used where superior properties to FEP, but inferior ones to those of PFA, are required.

4.4 Polychlorotrifluoroethylene

4.4.1 Industrial Process for the Production of Polychlorotrifluoroethylene

Polychlorotrifluoroethylene (PCTFE) is prepared by different polymerization techniques including bulk, suspension, and emulsion techniques [31]. The commercial process for the production of PCTFE is essentially polymerization initiated by free radicals at moderate temperatures and pressures in an aque-ous system at low temperatures and moderate pressures. Inorganic peroxy catalysts initiate the reaction in the presence of halogenated alkyl acid salt surfactants. Even though it is possible to polymerize CTFE by several techniques, the emulsion system produces the most thermally stable polymer [32]. Emulsion polymerization produces a polymer with a normal molecular weight distribution and a molecular weight–melt viscosity relationship resembling that of a bulk-polymerized polymer. The tendency of PCTFE to become brittle during use, because of increased crystallinity, can be reduced by incorporating a small amount (less than 5%) of vinylidene fluoride (VDF) during the polymerization process [33].

4.5 Polyvinylidene Fluoride

4.5.1 Industrial Process for the Production of Polyvinylidene Fluoride

The most common methods of producing homopolymers and copolymers of VDF are the emulsion and suspension polymerizations, although other methods have also been used [8].

Emulsion polymerization requires the use of free radical initiators, fluorinated surfactants, and often chain transfer agents. The polymer isolated from the reaction vessel consists of agglomerated spherical particles ranging in diameter from 0.2 to 0.5 pm [34]. It is then dried and supplied as a free-flowing pow-der or as pellets, depending on the intended use. If very pure polyvinylidene fluoride (PVDF) is required, the polymer is rinsed before the final drying to eliminate any impurities such as a residual initiator and surfactants.

Aqueous suspension polymerization requires the usual additives, such as free radical initiators, colloidal dispersants (though not always), and chain transfer agents to control molecular weight. After the process is completed, the suspension contains spherical particles approximately 100 pm in diameter. Suspension polymers are available as free-flowing powder or in pellet form for extrusion or injection molding [34].

The powder polymers from emulsion or suspension polymerizations intended to be used for sol-vent-based coatings are often milled into a finer particle size with a higher surface area for easier disso-lution when used as coatings for metals and other substrates.

Small amounts of comonomers (typically less than 6%) are often added to improve specific perfor-mance characteristics in cases where homopolymer performance is deficient. Higher levels of comono-mers (e.g., HFP) result in products with elastomeric characteristics [34].

Commercial products based on PVDF contain various amounts of comonomers such as HFP, CTFE, and TFE that are added at the start of polymerization to impart chemical resistance and obtain products with different degrees of crystallinity. Products based on such copolymers exhibit higher flexibility and elongation, higher chemical resistance or solubility, impact resistance, optical clarity, and thermal stability during processing. However, the incorporated monomers lower the melting points, increase permeation, decrease tensile strength, and produce higher creep than the PVDF homopolymer [34]. VDF also copolymerizes with other monomers, such as acrylic compounds, famously used to produce Kynar® 500 series coatings offered by the Arkema Corp [35]. Barium and strontium salts have been added to PVDF to improve its thermal stability [36].

4.6 Polyvinyl Fluoride

4.6.1 Industrial Process for the Production of Polyvinyl Fluoride

Vinyl fluoride is polymerized by free radical processes as with other commercial fluoropolymers. It is, however, more difficult to polymerize than TFE or VDF [37]. The successful polymerization of vinyl fluoride to high molecular polymers requires significantly higher pressures [38]. The temperature range for the polymerization in aqueous media is reported to be in the range 50–150°C. Commercial polymerization pressures as high as 50 MPa have been reported. Catalysts for the polymerization are peroxides and azo-compounds [33]. A continuous aqueous process has been reported at a temperature of 100°C and a pressure of 27.5 MPa [39]. The use of perfluoroalkylpropyl amine salts as surfactants in aqueous polymerization enhances the polymerization rate and yield and produces a polymer with an excellent color [40]. The polymerization temperature has an influence on the crystallinity and the melting point of the resulting polymer. Higher temperatures increase branching. Polyvinyl fluoride (PVF) characterization as a resin has not been published. DuPont produces films (Tedlar®) that are specified by typical film properties.

4.7 Ethylene Chlorotrifluoroethylene Copolymer

4.7.1 Industrial Process for the Production of Ethylene Chlorotrifluoroethylene

The copolymerization of ethylene and chlorotrifluoroethylene takes place as a free radical suspension process in aqueous media at low temperatures. A commercial polymer with an overall CTFE-to-ethylene ratio of 1:1 contains ethylene blocks and CTFE blocks in proportions lower than 10 mol% each [41]. The reaction pressure is adjusted to obtain the desired copolymer ratio [42]. Typical pressures during the process are of the order of 3.5 MPa. Small amounts of comonomers are added to the reaction mixture to reduce the extent of recrystallization and thus enhance the stress crack resistance of the ethylene chlorotrifluoroethylene (ECTFE) copolymer. The modified products have lower melting points in addition to lower crystallinity [43].

During copolymerization the polymer product precipitates as a fine powder, with particles typically less than 20 μm in the major dimension. These particles eventually agglomerate into roughly spherical beads, and the reactor product is a mixture of beads and powder. The product is then dewatered and dried. It is pelletized by extrusion for melt processing (e.g., extrusion, injection molding, blow molding) or ground and screened into powder coating grades [44]. Additional methods for the copolymerization of ethylene and CTFE are discussed elsewhere [13].

4.8 Ethylene Tetrafluoroethylene Copolymer

4.8.1 Industrial Process for the Production of Ethylene Tetrafluoroethylene

Commercial products based on the copolymers of ethylene and TFE are made by addition copolymerization initiated by free radicals [45]. Small amounts (1 to 10 mol%) of modifying comonomers are added to improve stress cracking and the rapid embrittlement of the product at exposure to elevated temperatures. Examples of the modifying comonomers are perfluorobutylethylene, hexafluoropropylene, perfluorovinyl ether, and perfluorobutyl ethylene. Additional information on the methods of preparation of ethylene tetrafluoroethylene (ETFE) copolymers can be found elsewhere [13]. ETFE resins are essentially alternating copolymers and in the molecular formula they are isomeric with PVDF with a head-to-head, tail-to-tail structure. However, in many important physical properties, the modified ETFE copolymers are superior to PVDF with the exception of the latter's remarkable piezoelectric and pyroelectric characteristics.

4.9 Terpolymers of Tetrafluoroethylene, Hexafluoropropylene, and Vinylidene Fluoride (THV Fluoroplastic)

Tetrafluoroethylene, hexafluoropropylene, and vinylidene (THV) fluoroplastic is prepared by emulsion copolymerization. The resulting dispersion may be used directly or may be concentrated with the addition of a surfactant. If it is coagulated, washed, and dried, the final products are either powders (after grinding) or pellets (after extrusion and pelletizing). No additives are used in the polymer since the product is inherently very stable and easy to process. At the time of writing, Dyneon Company is producing two types of commercial grades of THV fluoroplastic, namely, dry products and an aqueous dispersion, differing in monomer ratio, which affects the melting points, chemical resistance, and flexibility. Because they contain a VDF monomeric unit, they are cross-linkable by an electron beam; one grade is soluble in common solvents [13].

4.10 Terpolymers and Quarterpolymers of Hexafluoropropylene, Tetrafluoroethylene, and Ethylene

These polymers are prepared by the emulsion polymerization of ethylene, TFE, and HFP [46]. They may have, for example, ethylene monomeric units in a range of from at least about 2, 10, or 20% by weight up to 30, 40, or even 50%, and HFP monomeric units in a range of from at least about 5, 10, or 15% by weight up to about 20, 25, or even 30%, with the remainder of the weight of the polymer being TFE monomeric units [47]. Quarterpolymers of ETFE have also been reported in which two comonomers were added to TFE and ethylene. For example one such polymer was composed of 30 to 55% TFE, 40 to 60% ethylene, 1.5 to 10.0% HFP, and 0.05 to 2.50% of a vinyl monomer [48].

REFERENCES

1. Ruff, O., Brettschneider, O., The Preparation of Hexafluoroethane and Tetrafluoroethene from Tetrafluoromethane. *Z. Anorg. Allgem. Chem. 210*, p. 173, (1933).
2. Gangal, S. V. *Encyclopedia of Polymer Science and Technology*, Vol. *16* (Mark, H. F. and Kroschwitz, J. I., Eds.), John Wiley & Sons, New York, p. 579, (1989).
3. Sheratt, S. *Encyclopedia of Chemical Technology*, Vol. *9*, John Wiley & Sons, New York, p. 812, (1966).
4. Berry, K. L., U.S. Patent 2,559,752, assigned to DuPont Co., (1951).
5. Doughty, T. R., Russell, T., Sperati, C. A., Ho-Wei Un, H., U.S. Patent 3,855,191, assigned to DuPont Co., (1974).
6. Mueller, M. B., Salatello, P. P., Kaufman, H.S. U.S. Patent 3,655,611, assigned to Allied Chemicals, April 11, (1972).
7. Taylor Stiles High Speed Rotary, published by LittleFord Day Company, www.Littleford.com; (1976).
8. Ebnesajjad, S., *Fluoroplastics, Vol.1, Non-Melt Processible Fluoroplastics*, 2nd Edition, Elsevier, Oxford, UK, (2015).
9. Chinese Patent Application No. CN112409528B, assigned to Zhejiang Juhua Technology Center Co, (2020).
10. Gangal, S. V., U.S. Patent 4,342,675, assigned to DuPont Co., (1982).
11. Holmes, D. A., Fasig, E. W.. US Patent 3,819,594, assigned to DuPont; (1974).

12. Hintzer, K., Löhr, G. *Modern Fluoropolymers: High Performance Polymers for Diverse Applications* (Scheirs, J., Ed.), John Wiley & Sons Ltd., Chichester, UK, (1997).

13. Ebnesajjad, S., *Fluoroplastics, Vol.2, Melt Processible Fluoroplastics*, Elsevier, Oxford, UK, (2015).

14. Miller, W. T.. US Patent 2598283, assigned to US Atomic Energy Commission; (1952).

15. British Patent No. 781532, assigned to Dupont Co.; (1954).

16. Bro, M. I., Sandt, B. W.. US Patent 2946763, assigned to Dupont Co.; (1960).

17. Couture, M. J., Schindler, D. L., Weiser, R. B.. US Patent 3132124, assigned to DuPont Co.; (1964).

18. Couture, M. J., Schindler, D. L., Weiser, R. B.. US Patent 3132124, assigned to DuPont Co.; (1964).

19. Khan, A. S., Morgan, R. A.. US Patent 4380618, assigned to DuPont Co.; (1983).

20. Buckmaster, M. D., Morgan, R. A.. US Patent 4675380, assigned toDuPontCo.; (1987).

21. McDermott, D. E., Piekarski, S.. US statutory invention registration H130, assigned to DuPont Co.; (1986).

22. Anolick, C., Petrov, V. A., Smart, B. E., Stewart, C. W., Wheland, R. C., Farnham, W. B., Feiring, A. E., Qiu, W., US Patent 5637663, assigned to DuPont Co.; (1997).

23. Harris, Jr., J. F., McCrane, D. I., U.S. Patent 3,132,123, to Du Pont, (1964).

24. Carlson, D. P., U.S. Patent 3,536,733, to Du Pont, (1997).

25. Gresham, W. F., Vogelpohl, A. F., U.S. Patent 3,635,926, to DuPont, (1972).

26. Hartwimmer, R., Kuhls, J., U.S. Patent 4,262,101, assigned to Hoechst, (1981).

27. Hintzer, K., Löhr, G. *Modern Fluoropolymers: High Performance Polymers for Diverse Applications* (Scheirs, J., Ed.), John Wiley & Sons, Ltd, Chichester, UK, (1997).

28. Sperati, C. A. *Handbook of Plastics Materials and Technology* (Rubin, I. I., Ed.) John Wiley & Sons, New York, (1990).

29. Gangal, S. V. *Encyclopedia of Polymer Science and Technology*, Vol. *16* (Mark, H. F. and Kroschwitz, J. I., Eds.), John Wiley & Sons, New York, (1989).

30. Solvay Corp, www.Solvay.com, (2022).

31. Miller, W. A., Chlorotrifluoroethylene-Ethylene Copolymers, in: *Encyclopedia of Polymer Science and Engineering*, (Miller, W. A., Ed.), 2nd ed., Vol. *3*, pp. 480–491, John Wiley and Sons, New York, (1989).

32. Booth, H. S., Burchfield, P. E., *J. Am. Chem. Soc.* 55, (1933).

33. Ref Carlson, D. P., Schmiegel, W. *Ullmann's Encyclopedia of Industrial Chemistry*, (Gerhartz, W., Ed.), Vol. *A 11*, VCH Publishers, Weinheim, West Germany, (1988).

34. Dohany, J. E., Humphrey, J. S. in *Encyclopedia of Polymer Science and Engineering*, Vol. *17* (Mark, H. F. and Kroschwitz, J. I., Eds.), John Wiley & Sons, New York, (1989).

35. Arkema Corp, www.Kynar500.com, (2022).

36. Carlson, D. P., U.S. Patent 3,536,733, assigned to Du Pont Co, (1997).

37. Ebnesajjad, S. *Polyvinyl Fluoride Technology and Applications of PVF*. 1st ed. Oxford, UK, Elsevier, (2012).

38. Brasure, D., Ebnesajjad, S. *Encyclopedia of Polymer Science and Engineering*, Vol. *17* (Mark, H. F. and Kroschwitz, J. I., Eds.), New York, John Wiley & Sons, p. 471, (1989).

39. James, V. E., US Patent 3129207, assigned to DuPont Co., April 14, (1964).

40. Uschold, R. E., U.S. Patent 5,229,480, assigned to DuPont Co, July 20, (1993).

41. Reimschuessel, H. K., Marti, J., Murthy, N. S., *J. Polym. Sci., Part A: Polym. Chem* 26:43, (1988).

42. Schulze, S., U.S. Patent 4,053,445, assigned to DuPont Co, (1977).

43. Reimschuessel, H. K., Marti, J., Murthy, N. S., *J. Polym. Sci., Part A: Polym. Chem*, 26:43, (1988).

44. Stanitis, G. *Modern Fluoropolymers* (Scheirs, J., Ed.), John Wiley & Sons, Ltd., Chichester, UK, (1997).

45. Ref Kerbow, D. L. *Modern Fluoropolymers* (Scheirs, J., Ed.), John Wiley & Sons, Ltd., Chichester, U.K., (1997).

46. Grootaert, W. M., U.S. Patent 5,285,002, assigned to Minnesota Mining and Manufacturing Company, (1994.)

47. Jing, N., Hofmann, G. R. A., U.S. Patent 6,986,947, assigned to 3M Innovative Properties Company, (2006).

48. Sulzbach, R. A.. US Patent 4381387, assigned to Hoechst Aktiengesellschaft; (1983).

5

Properties of Commercial Thermoplastic Fluoropolymers

Jiri G. Drobny

5.1 Properties as Related to the Structure of Fluoropolymers

In general, fluoropolymers represent a group of macromolecules offering a variety of unique properties, in particular, a good-to-outstanding chemical resistance and stability at elevated temperatures. Because of these, they have been used increasingly in applications where most hydrocarbon-based polymers would fail, such as chemical processing, oil wells, motor vehicle engines, nuclear reactors, and space applications. On the other hand, they exhibit some deficiencies when compared with most engineering polymers. They typically have poorer mechanical properties, higher permeabilities, and often considerably higher cost. Knowing the advantages and disadvantages and understanding how properties, performance, and structure are correlated is very important for a proper selection of the processing technology and suitability for specific practical applications.

In general, the unique properties of fluoropolymers are the result of the very strong bond between carbon and fluorine [1] and shielding the carbon backbone by large fluorine atoms and of the fact that they are fully saturated macromolecules. Fluorine itself is a highly reactive element with the highest electronegativity of all elements. The atomic structure of fluorine gives rise to some of the strongest chemical bonds known [1]. The change in the properties of compounds where fluorine has replaced hydrogen can be contributed by the differences between C–F and C–H bonds. These differences are shown in Table 5.1 [1].

As pointed out in Chapter 1, there are several categories of fluoropolymers. For the purpose of this book, we will predominantly discuss thermoplastic fluoropolymers (also known as fluoroplastics) in Chapters 5 to 8, fluoroelastomers in Chapters 9 to 11, and other fluoropolymers in Chapters 14 and 15.

5.1.1 Fluoroplastics

Polytetrafluoroethylene (PTFE) has a conformation of a twisting helix comprising 13 CF_2 groups every 180° turn. This configuration is thermodynamically favored over planar zigzag (typical for polyethylene), because of the mutual repulsion of the adjacent fluorine atoms and their relatively large size. The helix forms an almost perfect cylinder comprising an outer sheath of fluorine atoms enveloping a carbon-based core (Figure 5.1) [2]. This morphology is conducive for PTFE molecules to pack like parallel rods. However, individual cylinders can slip past one another. This contributes to a relatively strong tendency of PTFE to cold flow. The mutual repulsion of fluorine atoms tends to inhibit the bending of the chain backbone. Therefore, the PTFE chain is very stiff.

The outer sheath of fluorine atoms protects the carbon backbone, thus providing chemical inertness and stability. It also lowers the surface energy, giving PTFE a low coefficient of friction [1] and nonstick properties. The extremely high molecular weight of the PTFE polymer results in a melt viscosity that is

DOI: 10.1201/9781003204275-7

TABLE 5.1

Carbon Bond Energies

Bond	Bond Energy (kcal/mole)
C–F	116
C–H	99
C–O	84
C–C	83
C–Cl	78
C–Br	66
C–I	57

Source: Adopted from [1].

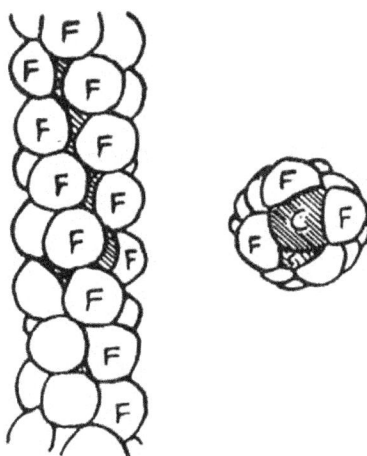

FIGURE 5.1   Schematic representation of the PTFE helix. Modified from [13].

about six orders of magnitude higher than that of most common thermoplastic polymers, namely, 10^{10} to 10^{12} P (10^9 to 10^{11} Pa.s). Such an extremely high viscosity even suppresses normal crystal growth. Thus, the virgin polymer has a degree of crystallinity in excess of 90% and a melting point of approximately 340°C (644°F). After being melted, even after slow cooling of the melt, the degree of crystallinity rarely reaches 70% and the melting point is reduced to about 328°C (622°F).

The exceptional chemical resistance, resistance to ultraviolet (UV) radiation, and thermal stability can be further explained by the fact that the C-F and C-C bonds in fluorocarbons are among the strongest known in organic compounds [1].

Fluorinated ethylene propylene (FEP), a copolymer of tetrafluoroethylene (TFE) and hexafluoropropylene (HFP), is essentially PTFE with an occasional methyl side group attached. The methyl groups have the effect of defects in crystallites and therefore reduce the melting point. These side groups also impede the slipping of the polymer chains past each other, thus reducing the cold flow.

Perfluoroalkoxy resin (PFA) is a copolymer of TFE and perfluoropropylvinyl ether (PPVE) in a mole ratio of approximately 100:1. Even such a small amount of comonomer is sufficient to produce a copolymer with a greatly reduced crystallinity. The relatively long side chains also markedly reduce the cold flow. Methylfluoroalkoxy resin (MFA), a copolymer of TFE and perfluoromethylvinyl ether (PMVE), has similar properties with a somewhat lower melting point.

Ethylene tetrafluoroethylene (ETFE), a copolymer of TFE and ethylene, has a higher tensile strength than PTFE, FEP, and PFA because its molecular chains adopt a planar zigzag configuration [3]. A strong electronic interaction between the bulky CF_2 groups of one chain and the smaller CH_2 groups of an adjacent chain causes an extremely low creep [4].

Polyvinylidene fluoride (PVDF) comprises alternating CH_2 and CF_2 groups. These alternating units can crystallize with larger CF_2 groups adjacent to smaller CH_2 units on an adjacent chain [4]. This interpenetration gives rise to a high modulus. In fact, PVDF has the highest flexural modulus of all fluoropolymers (see Figure 5.4). The aforementioned alternating groups create a dipole that renders the polymer soluble in highly polar solvents, such as dimethylformamide, tetrahydrofuran, acetone, and esters. Other consequences of this structure are a high dielectric constant, high dielectric loss factor, and piezoelectric behavior under certain conditions. The shielding effect of the fluorine atoms adjacent to the CH_2 groups provides the polymer with a good chemical resistance and thermal stability.

Polychlorotrifluoroethylene (PCTFE) has better mechanical properties than PTFE because the presence of the chlorine atom in the molecule promotes the attractive forces between molecular chains. It also exhibits greater hardness and tensile strength, and considerably higher resistance to cold flow than PTFE. Since the chlorine atom has a greater atomic radius than fluorine, it hinders the close packing possible in PTFE, which results in a lower melting point and reduced propensity of the polymer to crystallize [5]. The chlorine atom present in ethylene chlorotrifluoroethylene (ECTFE), a copolymer of ethylene and chlorotrifluoroethylene (CTFE), has a similar effect on the properties of the polymer.

5.1.1.1 Mechanical Properties

The mechanical properties of fluoroplastics can be ranked into two categories based on whether the polymers are fully fluorinated or contain hydrogen atoms in their structures. Generally, the fluoroplastics with hydrogen in their structure have about 1.5 times the strength of fully fluorinated polymers and are about twice as stiff. Fully fluorinated polymers, on the other hand, exhibit a higher maximum service temperature and greater elongation (Figures 5.2 and 5.3) [4, p. 5 and 6].

PVDF has the highest flexural modulus among the known commercial fluoropolymers (Figure 5.4) [6, p. 6]. Its high modulus can be intentionally reduced by copolymerization with HFP (typically less than 15%). Such lower-modulus copolymers have increased impact strength and elongation. ECTFE and ETFE also possess relatively high moduli due to interchain attractive forces. PTFE, FEP, and PFA display low stiffness (despite the rigidity of their molecular chains) because of their very low intermolecular attractive forces [6, p. 6].

5.1.1.2 Optical Properties

FEP and PFA, despite being melt processable, are crystalline (between 50 and 70%). The crystallinity results in poor optical properties (low clarity) and a very poor solubility in organic solvents. The latter

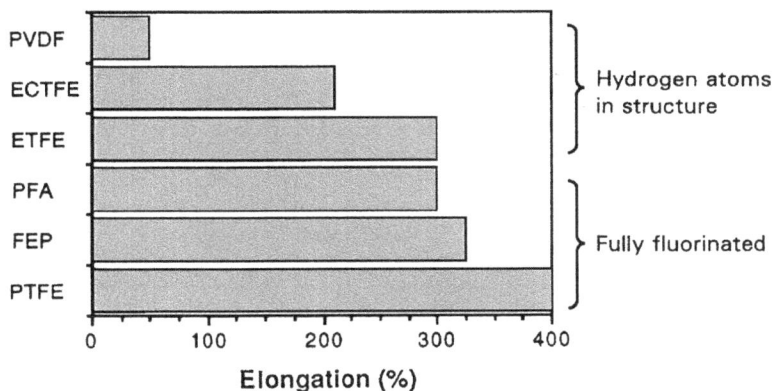

FIGURE 5.2 Elongation values of commercial fluoropolymers. Data from Imbalzano, J. F., *Chemical Engineering Progress*, April 1991, p. 69).

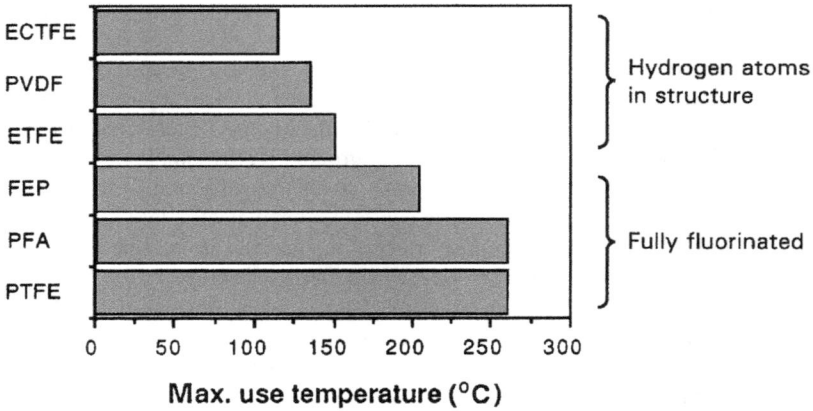

FIGURE 5.3 Maximum service temperatures of commercial fluoropolymers. (Data from Imbalzano, J. F., *Chemical Engineering Progress*, April 1991, p. 69).

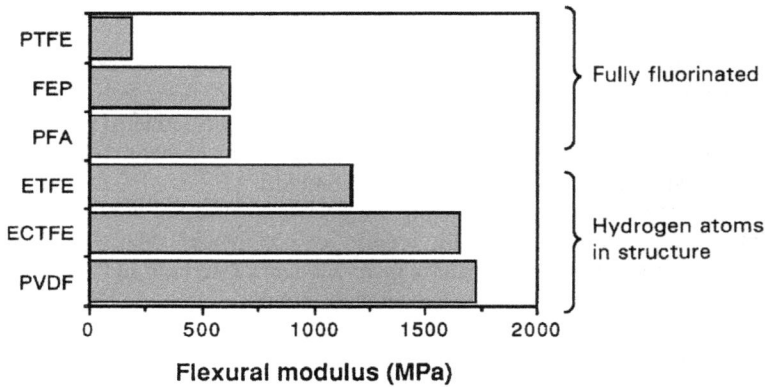

FIGURE 5.4 Flexure modulus values of commercial fluoropolymers. (Data from Imbalzano, J. F., *Chemical Engineering Progress*, April 1991, p. 69).

makes the preparation of thin optical coatings exceedingly difficult [6, p. 7]. On the other hand, TEFLON AF (amorphous fluoropolymer) contains in its molecule a bulky dioxole ring, which hinders crystallization (see Chapter 14). As a result, the polymer has an exceptionally high clarity and excellent optical properties. Its refractive index is the lowest of any plastic [7]. Dyneon's terpolymer of tetrafluoroethylene, hexafluoropropylene, and vinylidene (THV Fluoroplastic) is transparent to a broad band of light (UV to infrared), with an extremely low haze. Its refractive index is very low and depends on the grade. For the values of refractive indices for selected fluoroplastics see Tables 5.2 and 5.13.

5.1.2 Fluoroelastomers

Fluoroelastomers, which in this case are *fluorocarbon elastomers*, for the most part are based on a combination of vinylidene fluoride (VDF) and other monomers that disrupt the high crystallinity typical for the PVDF homopolymer. The properties of the resulting elastomeric materials are determined by the short VDF sequences and low or negligible crystallinity. As pointed out earlier, the properties of fluoroelastomers will be discussed in more detail in Part III (Chapters 9 to 11).

Elastomers based on VDF and TFE-VDF-HFP consist of fine particles 16 to 30 nm in diameter in contrast to PTFE, which has a rod-like microstructure in which the elementary fibrils are approximately 6 nm

TABLE 5.2

Typical Values of Refractive Index of Different Fluoroplastics

Polymer (100 μm thick film)	Refractive Index
PFA	1.340–1.346
FEP	1.342
ETFE	1.395
THV Fluoroplastic grades	1.350–1.363
PVDF	1.410–1.420

Sources: Dyneon™ Fluoroplastics, Product Comparison Guide, 99-0504-1501-1, Dyneon LLC, 2003; Dyneon Fluoroplastics, Product Comparison Guide, 5845 HB 98-0504-1611-8, Dyneon LLC, 2007

Note: ETFE, copolymer of ethylene and tetrafluoroethylene; FEP, fluorinated ethylene-propylene copolymer; PVDF, poly (vinylidene fluoride); THV, terpolymer of tetrafluoroethylene, hexafluoropropylene, and vinylidene fluoride.

wide and the molecular chains are all extended [8]. For example, the properties of a VDF/HFP elastomer such as their resilience and flexibility can be related to spherical domains with a diameter approximately 25 nm and that are interconnected. The diameter of these particles was found to be proportional to the molecular weight of the elastomer [8]. Additional fluorocarbon elastomers are based on other combinations of monomers, including ethylene and propylene (see Section 9.2.1) and fluorinated thermoplastic elastomers (FTPEs) which are processed as thermoplastics and are elastic within a specific temperature range (mostly at ambient temperature). These are the subject of Chapter 10. Yet another type of fluoroelastomer is the fluoro-inorganic elastomer, based on polymeric systems containing inorganic atoms, as discussed in Chapter 11.

5.2 Properties of Individual Commercial Fluoroplastics

As shown in the previous section, many of the fundamental properties of polymers depend on their structure, mainly on the nature of the monomeric units composing them. This section concentrates on the specific properties of individual fluoropolymers – more specifically fluoroplastics – and how they relate to their utility in practical applications.

5.2.1 Polytetrafluoroethylene

5.2.1.1 Molecular Weight

The molecular weight of standard PTFE is rather high, in the range 1 to 5×10^6 [9]. Such a high value is the main reason for the extremely high melt viscosity, which is about 1 million times higher than that of most polymers (see Section 5.1.1) and consequently too high for melt-processing methods used in the fabrication of common polymers. However, this high melt viscosity is also a reason why PTFE has an exceptionally high continuous service temperature of 260°C (500°F). Molecular weight also affects the crystallization rate (it decreases with increasing molecular weight) [10] and specific gravity. So-called standard specific gravity (SSG) is calculated from the number minus the average molecular weight (M_n) using the mathematical expression [11]:

$$SSG = 2.612 - 0.058\log_{10} M_n$$

5.2.1.2 Molecular Conformation

The molecules of PTFE are very long and unbranched. Consequently, the virgin polymer (i.e., the powder produced by polymerization) is highly crystalline with values of degree of crystallinity of 92 to 98% [12].

The CF_2 groups are along the polymer chain, and the whole chain twists into a helix as in a length of a rope; the fluorine atoms are simply too large to allow a planar zigzag conformation, so the carbon chain twists to accommodate them [13], as pointed out in Section 5.1.1.

5.2.1.3 Crystallinity and Melting Behavior

The initial high degree of crystallinity, reported to be well over 90%, can never be completely recovered after melting (i.e., sintering), presumably because of entanglements and other impediments caused by the extremely high molecular weight [14]. However, it has been established that rapidly cooled PTFE, although lower in degree of crystallinity, has the same molecular conformation and basic crystalline structure as slowly cooled PTFE [15].

As pointed out earlier, the fluorine atoms are too large to allow a planar zigzag structure, which confers rigidity on the polymer [16]. The PTFE molecule has a regular folded structure, which produces a laminar crystal [12].

The true densities of crystalline and amorphous PTFE differ considerably, and 100% crystalline PTFE densities of 2.347 at 0°C (32°F) and 2.302 at 25°C (77°F) were calculated from X-ray crystallographic data [17]. The density decrease of about 2% between these temperatures includes the decrease of approximately 1% due to the transition at 19°C (66°F), which results from a slight uncoiling of the helical conformation of molecules on heating through the transition. By contrast, the density of amorphous PTFE is not affected by the transition at 19°C, and values around 2.00 have been reported from extrapolations of specific volume measurements to zero crystallinity [15].

The density of PTFE undergoes complicated changes during processing and can be monitored by the values of the true specific volume. Discontinuity in such data show the transitions at 19°C and 30°C (66°F and 86°F) and also the very pronounced transition at the crystalline melting point of 327°C (621°F), the latter of which is due to the destruction of crystallinity [18]. The melting of the polymer is accompanied by a volume increase of approximately 30% [19]. The coefficient of linear expansion of PTFE has been determined at temperatures ranging from −190°C (−310°F) to +300°C (+572°F) [20].

The effects of structural changes on properties, such as specific heat, specific volume, or dynamic mechanical and electrical properties, are observed at various temperatures. A number of transitions were observed by various investigators; their interpretation and the modes of identification are listed in Table 5.3.

Besides the transition at the melting point, the transition at 19°C is of great consequence because it occurs around the ambient temperature and significantly affects product behavior. Above 19°C, the triclinic pattern changes to a hexagonal unit cell. Around 19°C, a slight untwisting of the molecule from a 180° twist per 13 CF_2 groups to a 180° twist per 15 CF_2 groups occurs. At the first-order transition at 30°C, the hexagonal unit disappears, and the rod-like hexagonal packing of the chains in the lateral

TABLE 5.3

Transitions in Polytetrafluoroethylene

Type of Transition	Temperature (°C)	Region Affected	Technique Used
First order	19	Crystalline, angular displacement causing disorder	Thermal methods, X-ray, NMR
	30	Crystalline, crystal disordering	Thermal methods, X-ray, NMR
	90 (80 to 110)	Crystalline	Stress relaxation, Young's modulus, dynamic methods
Second order	−90	Amorphous, onset of rotational motion around C–C bond	Thermal methods, dynamic methods
	−30 (−40 to −15)	Amorphous	Stress relaxation, thermal expansion, dynamic methods
	130 (120 to 140)	Amorphous	Stress relaxation, Young's modulus, dynamic methods

Source: Data from [10].
NMR = Nuclear magnetic resonance.

direction is retained [21]. Below 19°C there is almost a perfect three-dimensional order; between 19 and 30°C the chain segments are disordered; and above 30°C the preferred crystallographic direction is lost and the molecular segments oscillate above their long axes with a random angular orientation of the lattice [22, 23]. PTFE transitions occur at specific combinations of temperatures and mechanical or electrical vibrations. As dielectric relaxations, they can cause wide fluctuations in the values of the dissipation factor (see Section 5.2.1.7).

5.2.1.4 Mechanical Properties

The mechanical properties of PTFE at room temperature are similar to those of medium-density polyethylene, that is, relatively soft with high elongation and remaining useful over a wide range of temperatures, from cryogenic (just above absolute zero) to 260°C (500°F), which is its recommended upper use temperature [24]. Stress–strain curves are strongly affected by the temperature; however, even at 260°C (500°F) the tensile strength is about 6.5 MPa (942 psi) [25].

Under a sustained load, PTFE will creep (exhibit cold flow), which imposes limitations on PTFE in such applications as gasket material between bolted flange faces [25]. This tendency can be greatly reduced by the addition of mineral fillers, such as chopped glass fibers or bronze or graphite particles. These fillers also improve its wear resistance but do not have any significant effect on its tensile strength [26]. Fillers can improve the impact strength of the polymer significantly but they reduce its elongation [27].

In general, the mechanical properties of PTFE depend on the processing variables; for example, preforming pressure, sintering temperature and time, cooling rate, and the degree of crystallinity. Some properties, such as flexibility at low temperatures, coefficient of friction, and stability at high temperature, are relatively independent of the conditions during fabrication. Flex life, stiffness, impact strength, resilience, and permeability depend greatly on the molding and sintering conditions [28]. A summary of the mechanical properties of PTFE is given in Table 5.4.

5.2.1.5 Surface Properties

The surface of PTFE material is smooth and slippery. It is considered to be a very low energy surface with $\gamma_c = 18.5$ dyne/cm (mN/m) [29] and can be, therefore, completely wetted by liquids with surface tensions below 18 mN/m; for example, solutions of perfluorocarbon acids in water [30]. The PTFE surface can be treated by alkali metals to improve this wettability and consequently the adhesion to other substrates [31], but this increases its coefficient of friction [32].

5.2.1.6 Absorption and Permeation

Because of the high chemical inertness of PTFE to the majority of industrial chemicals and solvents and its low wettability, it absorbs only small amounts of liquids at ambient temperatures and atmospheric pressure [33].

TABLE 5.4

Typical Mechanical Properties of Polytetrafluoroethylene

Property	ASTM Test Method	Value
Tensile strength, MPa	D638	20 to 35
Elongation at break, %	D638	300 to 550
Tensile modulus, MPa	D638	550
Flexural strength, MPa	D790	No break
Flexural modulus at 23°C, MPa	D790	340 to 620
Impact strength, Izod, notched, J/m	D256	188
Compressive strength, MPa	D695	34.5

Source: Adopted from Ebnesajjad, S., *Fluoroplastics, Vol. 1*, William Andrew/Plastics Design Library, p. 19 (2000).

Gases and vapors diffuse through PTFE much more slowly than through most other polymers. The higher the degree of crystallinity, the lower the rate of permeation. Voids greater than molecular size increase the permeability. Thus, it can be controlled by molding PTFE articles to low porosity and high density. The optimum density for that is 2.16 to 2.195 [33]. Permeability increases with temperature due to an increase in the activity of the solvent molecules and the increase in vapor pressure of the liquids. The swelling of PTFE resin and films in any liquid is very low.

5.2.1.7 Electrical Properties

The dielectric *constant* of polytetrafluoroethylene is 2.1 and remains constant within a temperature range from –40 to 250°C (–40 to 482°F) within a frequency range from 5 Hz to 10 GHz. It changes somewhat, however, with density, and factors that affect density. The dielectric constant was found not to change over two to three years of measurements [33, p. 589].

The *dissipation factor* is affected by the frequency, temperature, crystallinity, and void content of the fabricated structure. At certain temperatures the crystalline and amorphous regions become resonant. Because of the molecular vibrations, the applied electrical energy is lost by internal friction within the polymer, and this leads to an increase in the dissipation factor, which peaks for these resins and corresponds to well-defined transitions [33, p. 589].

The *volume resistivity* of PTFE remains unchanged even after a prolonged soaking in water, because it does not absorb water. The *surface arc-resistance* of PTFE resins is high and is not affected by heat aging. They do not track or form a carbonized path when subjected to a surface arc in air [33, p. 590]. The electrical properties of PTFE are summarized in Table 5.5.

5.2.2 Modified Polytetrafluoroethylene

PTFE has many remarkable properties (see Section 5.2.1), but it has several shortcomings that limit its utility as an engineering material. It exhibits a significant cold flow (low creep resistance), is difficult to weld, and contains a large number of microvoids due to a rather poor coalescence of particles during the sintering process. The weaknesses result from the combination of a high molecular weight (extremely high melt viscosity) and a high degree of crystallinity. Major research efforts have resulted in the development of a modified PTFE, which contains a small amount (0.01 to 0.1 mole %) of a comonomer (see Table 5.7). The most suitable comonomer was found to be PPVE [34]. The comonomer reduces the degree of crystallinity and the size of the lamellae [35]. The polymerization process is similar to that for standard PTFE except additives to control the molecular weight are used [35, p. 243]. Based on the above findings a commercial *modified PTFE* was developed.

The resulting polymer has a melt viscosity lower by one order of magnitude, and because of that the particles coalesce better during sintering. Moreover, it has a markedly improved weldability [35, p. 251]. All the important physical properties of commercially modified PTFE are significantly improved without any noticeable reduction of other properties [36]. Modified PTFE is still processed by the same techniques as the standard polymer, and no adjustments to the processing techniques are necessary [37]. Because of the

TABLE 5.5

Typical Electrical Properties of Polytetrafluoroethylene

Property	ASTM Test Method	Value
Dielectric strength, short time, 0.08 in., V/mil	D149	>600
Surface arc resistance (time), s	D495	>300
Volume resistivity, Ω-cm	D257	>10^{18}
Surface resistivity Ω/sq	D257	>10^{16}
Dielectric constant at 60 to 2×10^9 Hz	D150	2.1
Dissipation factor at 60 to 2×10^9 Hz	D150	0.0003

Source: *Mechanical Design Bulletin* (DuPont), E. I. du Pont de Nemours & Co. (2005).

TABLE 5.6

Comparison of Modified PTFE and Conventional PTFE

Property	ASTM Test Method	Modified PTFE	Conventional PTFE
Tensile strength, MPa	D4894	33	34
Elongation at break, %	D4894	440	375
Specific gravity	D4894	2.17	2.16
Deformation under load at 23°C, %	D695		
3.4 MPa		0.2	0.7
6.9 MPa		0.4	1.0
13.8 MPa		3.2	8.2
Deformation under load, %	D695		
6.9 MPa at 25°C		5.3	6.7
3.4 MPa at 100°C		5.4	8.5
1.4 MPa at 200°C		3.6	6.4
Void content of typical parts, %	FTIR	0.5	1.5
Dielectric strength, kV/mm[a]	D149	208	140
Weld strength[b]	D4894	66 to 87	Very low
Permeation of perchloroethylene	Comparative rates		
Vapor		2	5
Liquid		4	13
Permeation of hexane	Comparative rates		
Vapor		0.2	3.4
Liquid		0	23.4

Source: *DuPont Teflon™ NXT Data Sheets* (2010).
FTIR = Fourier transform infrared spectroscopy.
[a] 76.2 μm film;
[b] specimens welded after sintering.

lower melt viscosity, modified PTFE performs better in coined molding, blow molding, and thermoforming [38]. Currently, granular molding resins are readily available from several manufacturers; an example is provided in [39, 40]. A comparison of modified and conventional PTFE resins is shown in Table 5.6.

Modified PTFE can be used in practically all applications where the conventional polymer is used. In addition to that, new applications are possible because of its improved flow and overall performance. In the chemical process industry, it is used for equipment linings, seals, gaskets, and other parts, where its improved resistance to creep is an asset. In semiconductor manufacturing, modified PTFE is used in fluid handling components and in wafer processing components. Typical applications in the electrical and electronic industries are connectors and capacitor films. Other applications are in unlubricated bearings, laboratory equipment, seal rings for hydraulic systems, and antistick components [40].

5.2.3 Copolymers of Tetrafluoroethylene and Hexafluoropropylene (FEP)

The copolymerization of TFE with HFP introduces a branched structure and results in the reduction of the melting point from the original 325°C (617°F) to about 260°C (500°F). Another consequence of that is a significant reduction of crystallinity, which may vary between 70% for virgin polymer and 30 to 50% for molded parts [41], depending on processing conditions, mainly the cooling rate after melting. The melting point is the only first-order transition observed in FEP. Melting increases the volume by 8% [42].

5.2.3.1 Mechanical Properties

The mechanical properties of FEP are in general similar to those of PTFE with the exception of the continuous service temperature, 204°C (400°F) compared with that of PTFE (260°C, or 500°F). Unlike

PTFE, FEP does not exhibit a marked volume change at room temperature because it is lacking the first-order transition at 19°C. FEP resins are useful above −267°C (−449°F) and are highly flexible above −79°C (−110°F) [43]. The static *friction* decreases with increasing load, and the static coefficient of friction is lower than the dynamic coefficient [44]. The coefficients of friction are independent of fabrication conditions.

Perfluorinated ethylene propylene tends to creep, and this has to be considered when designing parts for service under continuous stress. Creep can be reduced significantly by the use of suitable fillers, such as glass fibers or graphite. Graphite, bronze, and glass fibers also improve wear resistance and stiffness of the resin. The choice of fillers improving the properties of FEP and their amounts are limited, however, because of the processing difficulties of such mixtures [45].

FEP resins have a *very low energy surface* and are, therefore, very difficult to wet. Surface preparation for improved wetting and bonding of FEP can be done by a solution of sodium in liquid ammonia or naphthalenylsodium in tetrahydrofuran [45] by the exposure to corona discharge [46] or to amines at elevated temperatures in an oxidizing atmosphere [47]. FEP resins exhibit very good *vibration damping* at sonic and ultrasonic frequencies. However, to use this property for the welding of parts, the thickness of the resin must be sufficient to absorb the energy produced [45].

5.2.3.2 Electrical Properties

Perfluorinated ethylene propylene has outstanding electrical properties, practically identical to those of PTFE within its recommended service temperature. Its volume resistivity remains unchanged even after prolonged soaking in water.

The dielectric constant of FEP is constant at lower frequencies, but at frequencies of 100 MHz and higher it drops slightly with increasing frequency. Its dissipation factor has several peaks as a function of temperature and frequency. The magnitude of the dissipation peak is greater for FEP than for PTFE because the FEP structure is less symmetrical. The dielectric strength is high and unaffected by heat aging at 200°C (392°F) [48]. Typical values of the electrical properties of FEP resin are listed in Table 5.7.

5.2.3.3 Chemical Properties

FEP resists most chemicals and solvents, even at elevated temperatures and pressures. Acid and bases are not absorbed at 200°C (392°F) and exposures of one year. Organic solvents are absorbed only a little,

TABLE 5.7

Typical Values of Electrical Properties of Fluorinated Ethylene Propylene

Property	Value	ASTM Method
Dielectric strength, kV/mm		D149
0.254 mm	79	
3.18 mm	20–21	
Arc resistance (time), s	165	D495
Volume resistivity, ohm-cm	10^{17}	D257
Dielectric constant at 21°C		D1531
1 kHz-500 MHz	2.01–2.05	
13 GHz	2.02–2.04	
Dissipation factor, 21°C		D1531
1 kHz	0.00006	
100 kHz	0.0003	
1 MHz	0.0006	
1 GHz	0.0011	
13 GHz	0.0007	
Surface resistivity, Ω/sq.	$>10^{18}$	D257

Source: Teflon FEP Fluoropolymer Resin, Product and Properties Handbook, 220338 (9/94) DuPont (1994).

typically 1% or less, even at elevated temperatures and long exposure times. The absorption does not affect the resin and its properties and is completely reversible. The only chemicals reacting with FEP resins are fluorine, molten alkali metal, and molten sodium hydroxide [49].

Gases and vapors permeate FEP at a rate that is lower than for most plastics. It occurs only by molecular diffusion, because the polymer was melt processed. Because of the low permeability and chemical inertness, FEP is widely used in the chemical industry. Its permeation characteristics are similar to those of PTFE, with some advantage because of the absence of microporosity often present in PTFE. For permeation through FEP films, an inverse relationship between permeability and film thickness applies [50].

5.2.3.4 Optical Properties

FEP films transmit more UV, visible, and infrared radiation than ordinary window glass. They are considerably more transparent to the infrared and UV spectra than glass. The refractive index of FEP films is in the range 1.341 to 1.347 [51].

5.2.3.5 Other Properties

Products made from FEP resins resist the effects of weather, extreme heat, and UV radiation. This subject is covered in more detail in Chapters 16 to 18.

5.2.4 Copolymers of Tetrafluoroethylene and Perfluoroalkyl Ethers (PFA and MFA)

Because of the high bond strength among carbon, fluorine, and oxygen atoms, PFA and MFA exhibit nearly the same unique properties as PTFE at temperatures ranging from extremely low to extremely high. Since they can be relatively easily processed by conventional methods for thermoplastics into film and sheets without microporosity, they have a distinct advantage over PTFE in certain applications, such as corrosion protection and antistick coatings [52]. These polymers are semicrystalline, and the degree of crystallinity depends on the fabrication conditions, particularly on the cooling rate. The general properties of PFA and MFA are listed and compared with FEP in Table 5.8

5.2.4.1 Physical and Mechanical Properties

Commercial grades of PFA melt typically in a temperature range from 300 to 315°C (572 to 599°F) depending on the content of PPVE. The degree of crystallinity is typically 60% [53].

There is only one first-order transition, at −5°C (23°F), and there are two second-order transitions, one at 85°C (185°F) and one at −90°C (−130°F) [53].

TABLE 5.8

General Properties of Perfluorinated Melt-Processible Polymer

General Properties	ASTM Method	Unit	PFA	MFA	FEP
Specific gravity	D792	–	2.12–2.17	2.12–2.17	2.12–2.17
Melting temperature	D2116	°C	300–310	280–290	260–270
Coefficient of linear thermal expansion	E831	1/K 10^{-5}	12–20	12–20	12–20
Specific heat	—	kJ/kg K	1.0	1.1	1.2
Thermal conductivity	D696	W/K.m	0.19	0.19	0.19
Flammability	(UL 94)		V-O	V-O	V-O
Oxygen index	D2863	%	>95	>95	>95
Hardness Shore D	D2240	—	55–60	55–60	55–60
Friction coefficient on steel	—	—	0.2	0.2	0.3

Source: Modified data from [50].

In general, the mechanical properties of PFA are very similar to those of PTFE within a range from −200 to +250°C (−328 to +482°F). The mechanical properties of PFA and MFA at room temperature are practically identical; differences become obvious only at elevated temperatures, because of the lower melting point of MFA.

In contrast to PTFE with measurable void content, the melt-processed PFA is intrinsically void free. As a result, lower permeation coefficients should result because permeation occurs by molecular diffusion. This is indeed the case, but the effect levels off at higher temperatures [53, p. 233].

The most remarkable difference between PTFE and PFA is the considerably lower resistance to *deformation under load* (cold flow) of the latter. In fact, addition of even minute amounts of PFA to PTFE improves its resistance to cold flow [53] (see also Section 5.2.2).

5.2.4.2 Electrical Properties

PFA and MFA exhibit considerably better electrical properties than most traditional plastics. In comparison with the partially fluorinated polymers, they are only slightly affected by temperature up to their maximum service temperature [52, p. 384].

The dielectric constant remains at 2.04 over a wide range of temperatures and frequencies (from 100 Hz to 1 GHz). The dissipation factor at low frequencies (from 10 Hz to 10 kHz) decreases with increasing frequency and decreasing temperature. In the range from 10 kHz to 1 MHz, temperature and frequency have little effect, whereas above 1 MHz the dissipation factor increases with the frequency [52, p. 385].

5.2.4.3 Optical Properties

Generally, fluorocarbon films exhibit high transmittance in the UV, visible, and infrared regions of the spectrum. This property depends on the degree of crystallinity and the crystal morphology in the polymer. For example, 0.025 mm (0.001 in.) thick PFA film transmits more than 90% of visible light (wavelength 400 to 700 nm). A 0.2 mm (0.008 in.) thick MFA film was found to have a high transmittance in the UV region (wavelength 200 to 400 nm). The refractive indexes of these films are close to 1.3 [52, p. 386].

5.2.4.4 Chemical Properties

PFA and MFA have an outstanding chemical resistance even at elevated temperatures. They are resistant to strong mineral acids, inorganic bases, and inorganic oxidizing agents and to most of the organic compounds and their mixtures common in the chemical industry. However, they react with fluorine and molten alkali [52, p. 385].

Elemental sodium, as well as other alkali metals, react with perfluorocarbon polymers by removing fluorine from them. This reaction has a practical application for improving surface wettability and the adhesive bonding of perfluorocarbon polymers to other substrates [52, p. 387].

The absorption of water and solvents by perfluoropolymers is in general very low [52, p. 387]. Permeability is closely related to absorption and depends on temperature, pressure, and the degree of crystallinity. Since these resins are melt processed, they are usually free of voids and, therefore, exhibit much lower permeability than PTFE. Permeation through PFA occurs via molecular diffusion [53, p. 233].

5.2.5 Copolymers of Ethylene and Tetrafluoroethylene (ETFE)

Copolymers of ethylene and tetrafluoroethylene essentially comprise alternating ethylene and TFE units. They have an excellent balance of physical, chemical, mechanical, and electrical properties and are easily fabricated by melt-processing techniques but have found little commercial utility because they exhibit a poor resistance to cracking at elevated temperatures [54]. Incorporation of certain termonomers, so-called modifiers, in amounts from 1 to 10 mol% markedly improves the cracking resistance, while maintaining the desirable properties of the copolymer [54, 55]. ETFE resins are manufactured by several companies under different trade names.

5.2.5.1 *Structure and Related Properties*

The carbon chain is in a planar zigzag orientation and forms an orthorhombic lattice with the interpenetration of adjacent chains [54]. As a result of this structure, ETFE has exceptionally low creep, high tensile strength, and a high modulus compared with other thermoplastic fluoropolymers. Interchain forces hold this matrix until the alpha transition occurs at about 110°C (230°F), where the physical properties of ETFE begin to decline and more closely resemble perfluoropolymer properties at the same temperature. Other transitions occur at −120°C (−184°F) (gamma) and at about −25°C (−13°F) (beta) [54, p. 303].

The monomer ratio in the copolymer has an effect on the polymer structure and properties, mainly on the degree of crystallinity and on the melting point. As normally produced, ETFE has about 88% of alternating sequences and a melting point of 270°C (518°F) [54, p. 304].

5.2.5.2 *Mechanical, Chemical, and Other Properties*

ETFE exhibits exceptional toughness and abrasion resistance over a wide temperature range and a good combination of high tensile strength, high impact strength, flex, and creep resistance, and combines the mechanical properties of hydrocarbon engineering polymers with the chemical and thermal resistance of perfluoropolymers. Friction and wear properties are good and can be improved by incorporating fillers such as fiberglass or bronze powders. Fillers also improve creep resistance and increase the softening temperature [54].

The *continuous upper service temperature* of commercial ETFE is 150°C (302°F) [54, p. 305]. Physical strength can be maintained at even higher temperatures when the polymer is cross-linked by peroxide or ionizing radiation [56]. Highly cross-linked resins can be subjected to temperatures up to 240°C (464°F) for short periods of time [54, p. 305].

ETFE exhibits excellent *dielectric properties*. Its dielectric constant is low and essentially independent of frequency. The dissipation factor is low but increases with frequency and can be also increased by cross-linking. Dielectric strength and resistivity are high and are unaffected by water. Irradiation and cross-linking increase dielectric loss [54, p. 305].

Modified ETFE has excellent resistance to most common solvents and chemicals [57]. It is not hydrolyzed by boiling water, and weight gain is less than 0.03% in water at room temperature. Strong oxidizing acids, such as nitric acid, and some organic bases cause depolymerization at high concentrations and high temperatures [54, p. 305]. ETFE is also an excellent barrier to hydrocarbons and oxygenated components of automotive fuels [54, p. 305].

ETFE resins have good thermal stability; however, for high-temperature applications thermal stabilizers are often added [58]. A wide variety of compounds, mostly metal salts (e.g., copper oxides and halides, aluminum oxide, and calcium salts), will act as sacrificial sites for oxidation. Addition of certain salts can alter the decomposition from oligomer formation to dehydrofluorination. Iron and other transition metal salts accelerate the dehydrofluorination process. Hydrofluoric acid itself destabilizes ETFE at elevated temperatures, and the degradation becomes self-accelerating. For that reason, extrusion temperatures higher than 380°C (716°F) should be avoided [54, p. 306].

Ionizing radiation at lower levels affects ETFE polymers very little; therefore, they are used for wire coatings and molded parts in the nuclear energy industry [54, p. 306]. More details are provided in Chapter 8.

ETFE resins do not support combustion in air and have a typical limiting oxygen index (LOI) of about 30 to 31. The LOI depends on the monomer ratio in the polymer, and it increases gradually as the fluorocarbon content is increased to the alternating composition and then increases more rapidly to the LOI values for PTFE [54, p. 306].

ETFE resins are very often compounded with varied ingredients, such as glass fibers and bronze powder, to attain certain mechanical properties. For example, glass fibers are added at 25 to 35 wt. % levels to increase the modulus and to improve wear and friction characteristics. By adding 25% glass fibers, for example, the dynamic coefficient of friction is reduced from about 0.5 to about 0.3 [54, p. 307].

5.2.6 Polyvinylidene Fluoride (PVDF)

PVDF homopolymer is a semicrystalline polymer. Its degree of crystallinity can vary from 35% to more than 70%, depending on the method of preparation and thermomechanical history [59]. The degree of crystallinity greatly affects the toughness and mechanical strength as well as the impact resistance of the polymer. Other major factors influencing the properties of PVDF are molecular weight, molecular weight distribution, and extent of irregularities along the polymer chain and the crystalline form. Similar to other linear polyolefins, crystalline forms of polyvinylidene fluoride involve lamellar and spherulitic forms. The differences in the size and distribution of the domains as well as the kinetics of crystal growth are related to the method of polymerization [59].

PVDF exhibits a complex *crystalline polymorphism*, which cannot be found in other known synthetic polymers. There are a total of four distinct crystalline forms: alpha, beta, gamma, and delta. These are present in different proportions in the material, depending on a variety of factors that affect the development of the crystalline structure, such as pressure, intensity of the electric field, controlled melt crystallization, precipitation from different solvents, or seeding crystallization (e.g., surfactants). The alpha and beta forms are most common in practical situations. Generally, the alpha form is generated in normal melt processing; the beta form develops under mechanical deformation of melt-fabricated specimens. The gamma form arises under special circumstances, and the delta form is obtained by distortion of one of the phases under high electrical fields [59]. The density of PVDF in the alpha crystal form is 1.98 g/cm^3; the density of amorphous PVDF is 1.68 g/cm^3. Thus, the typical density of commercial products in a range from 1.75 to 1.78 g/cm^3 reflects a degree of crystallinity around 40%.

The structure of the polyvinylidene fluoride chain, namely, alternating CH_2 and CF_2 groups, has an effect on its properties which combine some of the best performance characteristics of both polyethylene ($-CH_2-CH_2-$)$_n$ and polytetrafluoroethylene ($-CF_2-CF_2-$)$_n$. Certain commercial grades of PVDF are copolymers of VDF with small amounts (typically less than 6%) of other fluorinated monomers, such as HFP, CTFE, and TFE. These exhibit somewhat different properties than the homopolymer.

5.2.6.1 Mechanical Properties

PVDF exhibits excellent mechanical properties (see Table 5.9), and when compared with perfluorinated polymers it has much higher resistance to elastic deformation under load (creep), much longer life in repeated flexing, and improved fatigue resistance [60, 61]. Its mechanical strength can be greatly increased by orientation [59]. Some additives, such as glass spheres and carbon fibers [62], increase the strength of the base polymer.

5.2.6.2 Electrical Properties

Typical values of electrical properties of the homopolymer without additives and treatments are listed in Table 5.10. The values can be substantially changed by the type of cooling and post-treatments, which determine the morphological state of the polymer. Dielectric constants as high as 17 have been measured on oriented samples that have been subjected to high electrical fields (poled) under various conditions to orient the polar crystalline form [63].

The unique dielectric properties and polymorphism of PVDF are the source of its high piezoelectric and pyroelectric activity [64]. The relationship between ferroelectric behavior, which includes piezoelectric and pyroelectric phenomena, and other electrical properties of the polymorphs of polyvinylidene fluoride, is discussed in [65].

The structure yielding a high dielectric constant and a complex polymorphism also exhibits a high dielectric loss factor. This excludes PVDF from applications as an insulator for conductors of high frequency currents since the insulation could heat up and possibly even melt. On the other hand, because of that PVDF can be readily melted by radiofrequency or dielectric heating, and this can be utilized for certain fabrication processes or joining [66]. High-energy radiation cross links polyvinylidene fluoride, and the result is the enhancement of mechanical properties (see Chapter 18). This feature makes it unique among vinylidene polymers, which typically are degraded by high-energy radiation [63].

TABLE 5.9

Typical Properties of Polyvinylidene Fluoride

Properties	Value or Description
Clarity	Transparent to translucent
Melting point, crystalline, °C	155–192
Specific gravity	1.75–1.80
Refractive index n^{25}_D	1.42
Mold shrinkage, average, %	2–3
Color possibilities	Unlimited
Machining qualities	Excellent
Flammability	Self-extinguishing, non-dripping
Tensile strength, MPa[a]	
At 25°C	42.0–58.5
At 100°C	34.5
Elongation, %	
At 25°C	50–300
At 100°C	200–500
Yield point, MPa[a]	
At 25°C	38–52
At 100°C	17
Creep, at 13.79 MPa[a] and 25°C for 10,000 h, %	2–4
Compressive strength, at 25°C, MPa[a]	55–90
Modulus of elasticity, at 25°C, GPa[b]	
In tension	1.0–2.3
In flexure	1.1–2.5
In compression	1.0–2.3
Izod impact, at 25°C, J/m[c]	
Notched	75–235
Unnotched	700–2300
Durometer hardness, Shore D scale	77–80
Heat-distortion temperature, °C	
At 0.455 MPa[a]	140–168
At 1.82 MPa[a]	80–128
Abrasion resistance, Taber CS-17, 0.5 kg load, mg/1000 cycles	17.6
Coefficient of sliding friction to steel	0.14–0.17
Thermal coefficient of linear expansion, per °C	$0.7–1.5 \times 10^{-4}$
Thermal conductivity, at 25–160°C, W/(m·K)	0.17–0.19
Specific heat, (J/(kg K)[d]	1255–1425
Thermal degradation temperature, °C	390
Low temperature embrittlement, °C	−60
Water absorption, %	0.04
Moisture vapor permeability, for 1 mm thickness, g/(24 h)(m^2)	2.5×10^{-2}
Radiation resistance (^{60}CO), MGy[e]	10–12

Source: Modified data from [12].

[a] To convert MPa to psi, multiply by 145;

[b] to convert GPa to psi, multiply by 145,000;

[c] to convert J/m to ftlbf/in., divide by 53.38;

[d] to convert J to cal, divide by 4.184;

[e] retains tensile strength of about 85% of its original value.

TABLE 5.10

Typical Electrical Properties of Polyvinylidene Homopolymer

Property	60 Hz	10^3 Hz	10^6 Hz	10^9 Hz
Dielectric constant at 25°C	9–10	8–9	8–9	3–4
Dissipation factor	0.03–0.05	0.005–0.02	0.03–0.05	0.09–0.11
Volume resistivity, Ω-m	2×10^{12}			
Dielectric strength, short time, kV/mm				
3.0 mm thickness	260			
0.2 mm thickness	1300			

Source: Modified data from [12].

5.2.6.3 Chemical Properties

Polyvinylidene fluoride exhibits an excellent resistance to most inorganic acids, weak bases, halogens, oxidizing agents (even at elevated temperatures), and to aliphatic, aromatic, and chlorinated solvents. Strong bases, amines, esters, and ketones cause it to swell, soften, and dissolve, depending on conditions [67]. Certain esters and ketones can act as latent solvents for PVDF in dispersions. Such systems solvate the polymer as the temperature is raised during the fusion of the coating, resulting in a cohesive film [68].

PVDF is among the few semicrystalline polymers that exhibit thermodynamic compatibility with other polymers [69], in particular with acrylic or methacrylic resins [70]. The morphology, properties, and performance of these blends depend on the structure and composition of the additive polymer, as well as on the particular PVDF resin. These aspects have been studied and are reported in some detail in [71]. For example, poly(ethyl acrylate) is miscible with polyvinylidene fluoride, but poly(isopropyl acrylate) and homologues are not. Strong dipolar interactions are important to achieve miscibility with PVDF, as suggested by the observation that polyvinyl fluoride is incompatible with polyvinylidene fluoride [72].

5.2.7 Polychlorotrifluoroethylene (PCTFE)

The inclusion of the relatively large chlorine into the polymeric chain reduces the tendency to crystallize. Commercially available grades include a homopolymer, which is mainly used for special applications, and copolymers with small amounts (less than 5%) of vinylidene fluoride [73]. The products are supplied as powder, pellets, pellets containing 15% glass fiber, and dispersions. A low-molecular-weight polymer is available as oil or grease. The oil is used to plasticize PCTFE [73, p. 104].

5.2.7.1 Thermal Properties

PCTFE is highly suitable for applications at extremely low temperatures; however, at elevated temperatures it is inferior to other fluoropolymers with the exception of PVDF. It has a relatively low melting point of 211°C (412°F), and it exhibits thermally induced crystallization at temperatures below its melting point, which results in brittleness [73, p. 106].

5.2.7.2 Mechanical, Chemical, and Other Properties

As long as thermally induced crystallization (see previous section) is avoided, PCTFE exhibits excellent mechanical properties. It also has an excellent resistance to creep [72]. The addition of glass fibers (typically 15%) improves high-temperature properties and increases hardness but also increases brittleness [73].

PCTFE has excellent chemical resistance, especially the resistance to most very harsh environments, particularly to strong oxidizing agents (e.g., fuming oxidizing acids, liquid oxygen, and ozone) and to sunlight. PCTFE alone has good resistance to ionizing radiation that is further improved by copolymerization

TABLE 5.11

Typical Properties of PCTFE

Property	Unit	Value
Specific gravity	—	2.1–2.16
Tensile strength	MPa (psi)	31–45 (4495–6525)
Elongation at break	%	50–150
Modulus of elasticity	GPa (kpsi)	1–1.6 (145–178)
Hardness, Shore D	—	75
Melting point	°C (°F)	210–212 (410–414)
Thermal conductivity	W/(mK)	0.35
Volume resistivity	Ω•cm	$>10^{18}$
Dielectric strength	kV/mm	20–24
Dielectric constant at 1 kHz	—	2.5
Dissipation factor at 1 kHz	—	0.01
Critical surface tension	mN/m	30.8
Refractive index	—	1.435
Service temperature range	°C	−240 to +150

Sources: 1. Polyfluor, PCTFE Data Sheet, polyfluor.nl (2022) 2. Drobny, J.G., Landolt-Börnstein, Specialty Thermoplastics, Springer, p. 41 (2015).

with small amounts of VDF (see previous section) [73, p. 103]. More on the resistance to ionizing radiation of PCTFE is included in Section 18.2.2.

Homopolymers and copolymers with VDF exhibit outstanding barrier properties [73, p. 106]. PCTFE does not absorb visible light, and it is possible to produce optically clear sheets and parts up to 3.2 mm (1/8 in.) thick by quenching from melt [73, p. 103 and 106]. Typical properties of PCTFE are shown in Table 5.11.

The disadvantage of PCTFE is that it is attacked by many organic materials and has a *low thermal stability in the molten state*. The latter requires great care during processing to maintain a high enough molecular weight necessary for good mechanical properties of the fabricated parts [73, p. 103].

5.2.8 Copolymer of Ethylene and Chlorotrifluoroethylene (ECTFE)

ECTFE is a semicrystalline (50 to 60%) polymer with a CTFE-to-ethylene ratio of 1:1 and contains ethylene blocks and CTFE blocks of less than 10 mole % each. It has alpha relaxation at 140°C, beta relaxation at 90°C, and gamma relaxation at −65°C. The conformation of ECTFE is the zigzag in which ethylene and CTFE alternate. The unit cell of an ECTFE crystal is hexagonal. The dielectric constant of ECTFE is 2.5 to 2.6, independent of temperature and frequency; its dissipation factor is 0.02. ECTFE is resistant to most chemicals except hot polar and chlorinated solvents. It does not stress crack or dissolve in any solvents. It has better barrier properties against SO_2, Ci_2, and water than FEP and PVDF [74].

The modified copolymers produced commercially exhibit improved high-temperature stress cracking. Typically, they are less crystalline and have lower melting points [75, 76]. Modifying monomers are hexafluoroisobutylene (HFIB), perfluorohexylethylene, and perfluoropropylvinyl ether (PPVE), and are produced for coating grades [77]. Coating grades are made in either powder form for electrostatic powder coating or very fine pellets for rotomolding and lining.

5.2.8.1 Properties of ECTFE

ECTFE resins are tough, moderately stiff, and creep resistant with service temperatures from −80 to +150°C (−112 to +302°F). The melt temperature depends on the monomer ratio in the polymer and is in the range of 235 to 245°C (455 to 473°F) [74, 75]. Its chemical resistance is good and similar to

TABLE 5.12

Typical Properties of ECTFE

Property	Unit	Value
Specific gravity	—	1.68
Tensile strength	MPa (psi)	54 (7500)
Elongation at break	%	250
Modulus of elasticity	MPa (kpsi)	1655 (240)
Hardness, Shore D	—	75
Service temperature range	°C (°F)	−80 to 150 (−112 to 302)
Melting point	°C (°F)	242 (468)
Deflection temperature*	°C	65 (149)
Thermal conductivity	W/(mK)	0.20
Volume resistivity	Ω•cm	>10^{14}
Dielectric strength	kV/mm	30–36
Dielectric constant at 1 kHz	—	2.5
Dissipation factor at 1 kHz	—	0.0016
Critical surface tension	mN/m	32
Refractive index	—	1.44

Sources: Halar® ECTFE, Properties; Halar® ECTFE, Design and Processing Guide, Solvay, 2022 (www.solvay.com) (2022).
* At 1.82 MPa (264 psi).

PCTFE. ECTFE, as most fluoropolymers, has outstanding weathering resistance. It also resists high-energy gamma and beta radiation up to 100 Mrad (1000 kGy) [77]. Typical properties of ECTFE are shown in Table 5.12.

5.2.9 Terpolymer of Tetrafluoroethylene, Hexafluoropropylene, and Vinylidene Fluoride (THV Fluoroplastic)

The driving force for the development of THV Fluoroplastic was the requirement for a fluoropolymer that could be used as a coating for polyester fabrics and provide protection similar to that of PTFE or ETFE in outdoor exposure. An additional requirement was that it could be used with PVC-coated polyester fabric without significantly compromising overall flexibility [77].

Chemically, THV Fluoroplastic is a terpolymer of TFE, HFP, and VDF, produced by emulsion polymerization. The resulting dispersion is either processed into powders and pellets or concentrated with an emulsifier and supplied in that form to the market [77, p. 258]. Currently, the manufacturer is 3M Dyneon, offering at the time of writing six commercial grades (five dry grades and one aqueous dispersion) that differ in their monomer ratios and consequently in their melting points, chemical resistance, optical properties, and flexibility. One granular grade exhibits electrostatic dissipative properties [78].

5.2.9.1 Properties

THV Fluoroplastic has a unique combination of properties that include relatively low processing temperatures, bondability (to itself and other substrates), high flexibility, excellent clarity, a low refractive index, and efficient electron-beam cross-linking [77, p. 258] . It also exhibits properties associated with fluoroplastics, namely, very good chemical resistance, weatherability, low friction, and low flammability. Typical properties of the dry commercial grades are summarized in Table 5.13.

The melting temperatures of THV Fluoroplastic commercial products range from 120°C (248°F) to 225°C (437°F). The lowest melting grade has the lowest chemical resistance and is easily soluble in ketones and ethyl acetate. It is also the most flexible of all grades, and the easiest to be cross linked by an electron beam. On the other hand, the highest melting grade has the highest chemical resistance and the highest resistance to permeation [77, p. 258].

TABLE 5.13

Typical Values of Properties of Dry Commercial Grades of Dyneon™ THV Fluoroplastics (Nominal Values, Not for Specification Purposes)

Properties	Unit	THV 2030G	THV 220	THV 500*	THV 610	THV 815
Physical Properties						
Physical form		G	A, G	A, G	A, G	G
Specific gravity	—	1.98	1.95	1.98	2.04	2.06
Melting point	°C (°F)	130 (266)	120 (248)	165 (329)	185 (365)	225 (437)
Mechanical Properties						
Hardness, Shore D	—	—	44	54	56	58
Tensile strength	MPa (psi)	23 (3,335)	20 (2,900)	28 (4,060)	28 (4,060)	29 (4,210)
Elongation at break	%	535	600	500	500	420
Flexural modulus	MPa (kpsi)	32 (4.6)	80 (12)	210 (30)	490 (71)	525 (76.2)
Electrical Properties						
Dielectric constant (1 MHz)	—	—	5.72	4.82	4.86	—
Dissipation factor (1 MHz)	—	—	0.14	0.10	0.09	+
Dielectric strength	kV/mm	—	62	48	56	—
Thermal Properties						
Limiting oxygen index	%	—	>65	>75	>75	—
Glass transition temperature	°C	—	5	26	34	36
Heat deflection temperature	°C	—	30	34	37	—

Source: 3M™ Dyneon™ Fluoropolymers, Product Comparison Guide (2022).
* Also available in electrostatic dissipative grade;
** ASTM D648, 0.45 MPa; G = pellet; A = agglomerate.

THV Fluoroplastic can be readily bonded to itself and to many plastics and elastomers and, unlike other fluoroplastics, does not require surface treatment, such as chemical etching or corona treatment. However, in some cases tie layers are required to achieve a good bonding to other materials [77, p. 260].

THV Fluoroplastic grades are transparent to a broad band of light (UV to infrared) with an extremely low haze; transparency depends on the grade. Their refractive index values are very low and also depend on the grade [77, p. 261] as shown in Table 5.14.

5.2.10 Terpolymer of Ethylene, Tetrafluoroethylene, and Hexafluoropropylene (EFEP)

EFEP is a terpolymer of ethylene, tetrafluoroethylene (TFE), and hexafluoropropylene (HFP). It was designed to have many of the mechanical properties of ETFE, but with a lower processing temperature,

TABLE 5.14

Typical Values of Properties of Dry Commercial Grades of Dyneon™ THV Fluoroplastics (Nominal values, Not for Specification Purposes)

Property	Test Conditions	Unit	THV Fluoroplastic Grade				
			2030G	220	500	610	815
Refractive index	Mericon Prism Coupler	n_D	1.350	1.363	1.355	1.353	1.350
UV-visible light transmission	300 nm	%T	90	87	85	82	—
	600 nm	%T	91**	93	93	93	—

Source: 3M™ Dyneon Fluoroplastics, Product Comparison Guide, (2020).
250 μm film.

TABLE 5.15

General Properties of Commercial NEOFLON™ EFEP

Property	ASTM Test Method	Grade	
		RP-4020	**RP-5000**
Specific gravity	D792	1.72–1.76	1.72–1.76
Melting point, °C	—	155–170	190–200
Decomposition temperature, °C	1 wt. % (air)	355	380
MFR* g/10 min, at 265°C	5 kg load	25-50	20-30
Tensile strength, MPa	D638	355	380
Elongation at break, %	D638	420–530	390–440
Flexural modulus, MPa	D790	1300	1000
Haze value	3 mm thick sheet	12	47
Contact angle, degrees	With water	96	96
Dielectric constant at 2.454 GHz	D150	2.07	2.14
Dissipation factor at v2,45 GHz	D150	0.014	0.014

Source: Adhesive fluoropolymer NEOFLON™ EFEP, Daikin, Document May.2003 EG–37 (0005) AI.
* Melt flow rate.

allowing it to be coextruded with conventional thermoplastic polymers, such as polyamide, ethylene-vinyl alcohol copolymer (EVOH), ETFE, and modified polyethylene without the use of adhesives or tie layers. EFEP also adheres to glass and metals and has a higher degree of transparency than most fluoropolymers. It resists strong acids and alkali and many organic liquids, including isooctane, toluene, acetone, esters, amines, dimethylformamide (DMF), and N-methylpyrrolidone (NMP) [79]. The general properties of commercial NEOFLON™ EFEP are shown in Table 5.14.

5.2.11 Polyvinyl Fluoride (PVF)

5.2.11.1 Properties of the PVF Polymer

Although PVF resembles polyvinyl chloride (PVC) in its low water absorption, resistance to hydrolysis, insolubility to common solvents at room temperature, and a tendency to split off hydrogen halides at elevated temperature, it has a much greater tendency to crystallize [80]. PVF is a semicrystalline polymer with a planar, zigzag conformation [81]. The degree of crystallinity can vary significantly from 20 to 60% and is a function of defect structures [82]. Commercial PVF is atactic, containing approximately 12% head-to-head linkages and displays a peak melting point at about 190°C (374°F) [83].

The polymer displays several transitions below the melting temperature. The lower T_g occurs at –15 to –20°C and the upper T_g is in the range of –40 to –50°C. Two other transitions, at –80 and 150°C have been reported [84]. PVF has low solubility in all regular solvents below about 100°C [85].

High molecular-weight PVF is reported to degrade in an inert atmosphere with concurrent hydrogen fluoride (HF) loss and backbone cleavage both occurring at about 450°C (842°F). In air, HF loss occurs at 350°C and backbone cleavage around 450°C. HF and a mixture of aromatic and aliphatic hydrocarbons are generated from the thermal degradation of polyvinyl fluoride [80, 81]. The self-ignition temperature of a PVF film is 390°C. The limiting oxygen index (LOI) for PVF is 22.6%. HF and a mixture of aromatic and aliphatic hydrocarbons are generated from the thermal degradation of PVF [82].

PVF is transparent to radiation in the UV, visible, and near-infrared regions, transmitting 90% of the radiation from 350 to 2,500 nm. PVF becomes embrittled upon exposure to an electron beam radiation of 1,000 Mrad (10,000 kGy) but resists breakdown at lower doses. While PTFE is degraded at 0.2 Mrad (2kGy), PVF retains its strength at 32 Mrad (320 kGy) [83].

In general, PVF in the form of homopolymers and copolymers exhibits: excellent resistance to weathering; outstanding mechanical properties; inertness toward a wide variety of chemicals, solvents, and staining agents; excellent hydrolytic stability; nonstick properties; and high dielectric strength and dielectric constant [85]. PVF is commercially used mainly in the form of films from the DuPont Chemical Company (www.dupont.com).

5.2.11.2 Polyvinyl Fluoride Films and Their Properties

PVF homopolymers and copolymers containing a large amount of vinyl fluoride cannot be processed from the melt due to the thermal decomposition of PVF because the resin degrades prior to reaching its melting point. Moreover PVF does not dissolve in most solvents at room temperature and pressure due to its high crystallinity and to large amounts of intermolecular hydrogen bonding. The absence of the adequate solubility of PVF in solvents rules out casting a film from a solution. Consequently PVF is not processible by melt or by solution techniques and a special method was developed to fabricate PVF into coatings and films. The polymer is dispersed in a polar solvent with a high boiling point to coalesce and form a film below the melting point of the polymer used [80, p. 117]. The manufacturing methods for PVF films may also include orientation of the film to obtain a product with certain required properties.

PVF films are available in a large variety under the trade name DuPont™ TEDLAR® PVF. They are produced from the PVF resin by the specific processes mentioned above and described below and are offered in both unoriented and oriented forms [80, p. 141]. The manufacturing methods are described in some detail in [80, p. 117].

Unoriented PVF films are made by casting a dispersion of PVF in a latent solvent onto a polyethylene terephthalate (PET) carrier web during which negligible stretching occurs [80, p. 141]. The lack of orientation makes the film more formable and compliant than the oriented film. *Oriented* PVF films are produced by a method using an extrusion of paste also based on the dispersion of PVF resin in latent solvents followed by orientation. The orientation takes place by biaxial stretching that enhances the mechanical properties of the film in both machine and cross-machine directions. Details are in [80, p. 117]. Unoriented films, in addition to being formable, have lower tensile strength and higher elongation at break than the oriented ones [80, p. 142]. They are supplied as TEDLAR® SP films with the PET web attached [86]. A comparison of the basic properties of 25 μm (0.001 in.) thick oriented and unoriented films is shown in Table 5.16 [80, p. 142].

5.2.11.2.1 Chemical Properties

Films of PVF retain their form and strength even when boiled in strong acids and bases. At ordinary temperatures, the films are not affected by many classes of common solvents, including hydrocarbon and chlorinated solvents and are impermeable to grease and oils [87]. They are partially soluble in a few highly polar solvents at temperatures above 149°C (300°F) [86]. Moreover they exhibit stain resistance to a few common and potent agents such as iodine and inks. PVF films have excellent hydrolytic stability as demonstrated by the retention of flex life, impact strength, and break elongation after 1,500 hours of exposure to steam at 100°C [79, p. 171].

5.2.11.2.2 Optical Properties

Transparent grades of PVF films are basically transparent to solar radiation in the near ultraviolet, visible, and near infrared ranges of the light spectrum [87]. Ultraviolet absorbing types of PVF films protect

TABLE 5.16

Comparison of Unoriented and Oriented PVF Transparent Films at Room Temperature

Property	Unoriented Film	Oriented Film
Ultimate tensile strength, MPa	41	90
Tensile modulus, MPa	—	2075
Elongation at break, %	200	95
Tear strength, initial, g/25 μm	212	423

Source: Data from [79, p. 114].

TABLE 5.17

Refractive Index Values of Selected Fluoropolymer Films at D-Line at 589.3 nm Wavelength

Property	Film				
	PVF Grade UT[a]	**PVF Grade TR[b]**	**ETFE**	**FEP**	**PFA**
Fluorine content, %	41	41	59	76	76
Refractive index	1.474	1.478	1.398	1.350	1.343

Sources: Data from [80, p. 171] and [88].
a UT grade, transparent film, opaque to UV;
b TR grade: transparent film, transparent to UV.

TABLE 5.18

Typical Electrical Properties of Oriented Polyvinyl Fluoride Films

Property	Transparent 0.002 in. Glossy Film[1]	White 0.002 in. Satin Film[2]	Test Method
Corona endurance, 60 Hz at1 kV/mil (hrs.)	2.5	6.0	ASTM D2275
Dielectric constant, 1 kHz at 22°C	8.5	11.0	ASTM D150
Dielectric strength, 60Hz kV/mm	130	140	ASTM D150
Dissipation factor, 1 kHz at 22°C (%)	1.6	1.4	ASTM D150
Dissipation factor, 1 kHz at 70°C (%)	2.7	1.7	ASTM D150
Dissipation factor, 10 kHz at 22°C (%)	4.2	3.4	ASTM D150
Volume resistivity at 22°C (Ω cm)	4×10^{13}	7×10^{14}	ASTM D257
Volume resistivity at 100°C (Ω cm)	2×10^{10}	1.5×10^{11}	ASTM D257
Surface resistivity at 23°C (Ω /sq.)	6.1×10^{15}	1.6×10^{15}	ASTM D257
Surface resistivity at 100°C (Ω /sq.)	7.2×10^{11}	1.6×10^{12}	ASTM D257
Hot wire ignition	0 PLC		UL746A
High amp arc ignition	0 PLC		UL746A
Comparative tracking index	0 PLC		UL746

Sources: *DuPont™ TEDLAR® General Properties*, Publication H-49725-5 Ltr04/14 (2014); *Typical Properties of TEDLAR® Films* (Tedlar.com, 2020) and [80, p. 179].
Note: Film[1] has a somewhat lower orientation than film[2].

substrates against ultraviolet light attack [79, p. 171]. The refractive indexes of PVF and other fluoropolymer films are shown in Table 5.17 [79, p. 171]. The UT grade is transparent PVF film which is opaque to UV light, and TR grade is the transparent grade of PVF that is transparent to UV. An increase in the fluorine content of the fluoropolymer decreases its refractive index; thus, FEP and PFA have the lowest index values. The fluoropolymers in this study are semicrystalline materials, but only films of PVF UT grade and ethylene-tetrafluoroethylene copolymer exhibit significant haze. Fluorinated ethylene-propylene copolymer, perfluoroalkoxy polymer, and polyvinyl fluoride TR grade films are very clear and show much less haze than the UV-opaque Tedlar grade [88].

5.2.11.2.3 Weathering Performance

PVF films exhibit an outstanding resistance to solar degradation. Unsupported transparent PVF films retained at least 50% of their tensile strength after ten years in Florida facing south at 45°. Pigmented films properly laminated to a variety of substrates impart a long service life. Most colors exhibit no more than five NBS-unit (Modified Adams Color Coordinates) color change after 20 years of vertical outdoor exposure. Additional protection of various substrates against UV attack can be achieved with UV-absorbing PVF films [87].

5.2.11.2.4 Electrical Properties

PVF films exhibit a high dielectric constant and a high dielectric strength [79, p. 173]. Typical electrical properties for standard oriented PVF films are shown in Table 5.18. PVF is one of the few materials that have the unusual and interesting properties called piezoelectricity and pyroelectricity. *Piezoelectricity* is the ability of a material to develop an electric charge in response to applied mechanical stress (the so-called direct piezoelectric effect) and vice versa. The direct piezoelectric effect was discovered when electric charges were created by mechanical stress (pressure) on the surface of tourmaline crystals. A concomitant property of piezo-crystals and piezo-materials is *pyroelectricity*, which is defined as the ability of certain materials to generate a temporary voltage when they are heated or cooled [79, p. 173].

5.2.11.2.5 Thermal Stability

The polymer is processed into films routinely at temperatures near or above 204°C (400°F) and for short times as high as 232 to 249°C (450 to 480°F) using ordinary industrial ventilation. At temperatures above 204°C (400°F) or upon prolonged heating, film discoloration and evolution of small amounts of hydrogen fluoride vapor will occur. The presence of Lewis acids (e.g., BF_3 complexes) in contact with PVF is known to catalyze the decomposition of the polymer at lower-than-normal temperatures. A thorough study of the degradation of polyvinyl fluoride films is reported in [84]. PVF performs well in temperatures ranging from −72 to 107°C (− 98 to 225°F), with intermittent short-term peaking up to 204°C (400°F) [89].

Figure 5.5 shows the effect of thermal aging on mechanical properties of PVF films when aged at 149°C (300°F). These properties include tensile strength, elongation at break, impact strength, and flex life to fatigue failure [86, 88].

PVF films are available in a large variety under the trade name DuPont™ TEDLAR® PVF. The amount of orientation is a key parameter for the film. Consequently, films are classified into specific types. *Type 1* films have controlled shrinkage and are used when high shrinkage is needed for processing film, for example with a high shrinkage substrate. *Type 2* clear films exhibit high tensile strength and high flexibility; *Type 3*, the standard film, is available in clear and pigmented forms. A clear film *Type 4* has high elongation and high tear resistance. *Type 5* TEDLAR film has been developed for applications where deep draw and texturing are required. Its ultimate elongation is almost twice that of standard Type 3 film.

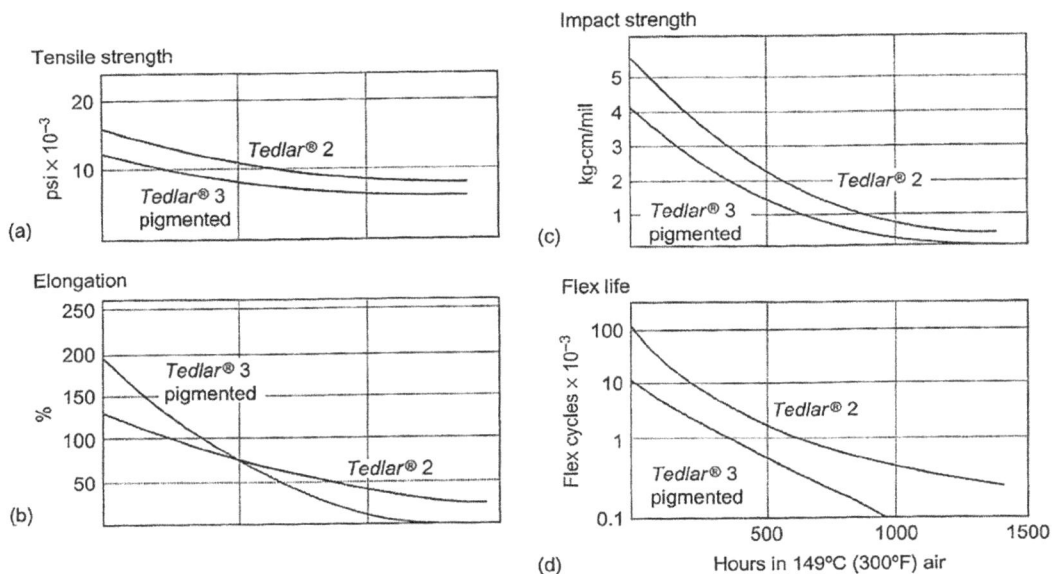

FIGURE 5.5 Effect of thermal aging on mechanical properties of polyvinyl fluoride films: (a) tensile strength; (b) elongation at break; (c) impact strength; (d) flex life to fatigue failure (Tedlar® 2 is more oriented than Tedlar® 3). (Courtesy of DuPont).

In classification, film Types 1 to 5 are oriented; Type 6 films have different surface treatments; Type 7 films have different gloss levels; and Type 8 is unoriented. Lower numbers indicate higher orientation [86]. The thickness of commercially available PVF films ranges from 0.5 to 2.0 mil (12.5 to 50 µm).

 TEDLAR films are supplied with different surface characteristics. "A" (one side adherable) and "B" (both sides adherable) surfaces are used with adhesives for bonding to a wide variety of substrates. These surfaces can be bonded with a variety of adhesives, including acrylics, polyesters, epoxies, elastomeric adhesives, and pressure-sensitive mastics. The "S" surface has excellent antistick properties and is used as a mold-release film for parts made from epoxies, phenolics, elastomers, and other polymeric materials [90]. A detailed explanation of the codes for commercial grades of PVF films is provided in Appendix 5.

REFERENCES

1. Pozzoli, M., Vita, G., Arcella, V. *Modern Fluoropolymers* (Scheirs, J., Ed.), John Wiley & Sons, Ltd, Chichester, UK, p. 374 (1997).
2. Koo, G. P.. *Fluoropolymers* (Wall. L. A., Ed.), Wiley-Interscience, New York, p. 508 (1972)
3. Sheratt, S. *Kirk-Othmer Encyclopedia of Chemical Technology*, Vol. 9, John Wiley & Sons, New York, p. 826 (1966).
4. Kerbow, D. L. *Modern Fluoropolymers* (Scheirs, J., Ed.), John Wiley & Sons, Ltd, Chichester, UK, p. 307 (1997).
5. Saunders, K. J., *Organic Polymer Chemistry*, Chapman & Hall, London, 2nd ed., Chapter 7, p. 149 (1988).
6. Scheirs, J. *Modern Fluoropolymers* (Scheirs, J., Ed.), John Wiley & Sons, Ltd, Chichester, UK, p. 3 (1997).
7. Resnick, P. R., Buck, W. H., Teflon® AF Amorphous Fluoropolymers, in *Modern Fluoropolymers* (Scheirs, J., Ed.) John Wiley & Sons, Ltd, Chichester, UK, Chapter 22 (1997).
8. Yamaguchi, S., Tatemoto, M., Fine structure of fluorine elastomer emulsions. *Sen'i Gakkaishi*, 50:414, (1994).
9. Sheratt, S. *Kirk-Othmer Encyclopedia of Chemical Technology*, Vol. 9, John Wiley & Sons, New York, p. 817 (1966).
10. Gangal, S. V. *Kirk-Othmer Encyclopedia of Chemical Technology*, Vol. 11, 3rd ed., John Wiley & Sons, New York, p. 7 (1980).
11. Sperati, A. C., Starkweather, H. W., *Fluorine containing polymers II. Polytetrafluoroethylene, Fortschr. Hochpolym. Forsch.*, 2:465 (1961).
12. Gangal, S. V. *Kirk-Othmer Encyclopedia of Chemical Technology*, Vol. 11, 3rd ed., John Wiley & Sons, New York, p. 8, (1980).
13. Bunn, C. W., Howels, E. R., *Structure of molecules and crystals of fluorocarbons, Nature*, 174:549, (1954).
14. Sheratt, S.. *Kirk-Othmer Encyclopedia of Chemical Technology*, Vol. 9, John Wiley & Sons, New York, p. 819 (1966).
15. Clark, E. S., Starkweather, H. W., *The crystal structure of quenched polytetrafluoroethylene, J. Appl. Polym. Sci.*, 6:S41, (1962).
16. Bunn, C. W., *J. Polym. Sci.*, 16:332, (1955).
17. Clark, E. S., Muus, L. T., *Z. Krystallogr.*, 117:119, (1962).
18. Sheratt, S.. *Kirk-Othmer Encyclopedia of Chemical Technology*, Vol. 9, John Wiley & Sons, New York, p. 820 (1966).
19. Gangal, S. V.. *Kirk-Othmer Encyclopedia of Chemical Technology*, Vol. 11, 3rd ed., John Wiley & Sons, New York, p. 8 (1980).
20. Kirby, R. K. Z., *J. Res. Natl. Bur. Std.*, 57:91, (1956).
21. Doban, R. C., Knight, A. C., Peterson, A. H., Sperati, C. A., paper presented at the 130th Meeting of the American Chemical Society, Atlantic City, NJ, September 1956.
22. Clark, E. S., Muus, L. T., paper presented at the 133rd Meeting of the American Chemical Society, New York, September 1957.
23. Clark, E. S., paper presented at the Symposium on Helics in Macromolecular Systems, Polytechnic Institute of Brooklyn, Brooklyn, NY, May 16, 1959.

24. Sperati, C. A. *Handbook of Plastic Materials and Technology* (Rubin, I. I., Ed.), John Wiley & Sons, New York, p. 119, (1990).
25. Blanchet, T. A.. *Handbook of Thermoplastics* (Olabisi, O., Ed.), Marcel Dekker, New York, p. 988, (1997).
26. Blanchet, T. A. *Handbook of Thermoplastics* (Olabisi, O., Ed.), Marcel Dekker, New York, p. 987, (1997).
27. TEFLON® Fluoropolymer Resin Properties Handbook, DuPont Polymers, Wilmington, DE, Publication H-37051 (1992).
28. Gangal, S. V. *Encyclopedia of Polymer Science and Technology*, Vol. *16* (Mark, H. F. and Kroschwitz, J. I., Eds.), John Wiley & Sons, New York, p. 584, (1989).
29. Pittman, A., *Fluoropolymers* (Wall, L. A., Ed.), Wiley-Interscience, New York, p. 426 (1972).
30. Bernett, M. K., Zisman, W. A., *Relation of wettability by aqueius solutions on the surface of low-energy solids, J. Phys. Chem. 63*, p. 1911 (1951).
31. Doban, R. C., U.S. Patent 2,871,144 (January 27, 1959) to E.I. du Pont de Nemours & Co.
32. Allan, A. J. G., Roberts, R., *Wettability of perfluorocarbon films, J. Polym. Sci. 39*, p. 1 (1959).
33. Gangal, S. V. *Encyclopedia of Polymer Science and Technology*, Vol. *16* (Mark, H. F. and Kroschwitz, J. I., Eds.), John Wiley & Sons, New York, p. 587 (1989).
34. Doughty, T. R., Jr., Sperati, C. A., U.S. Patent 3,855,191 (December 17, 1974) to E. I. Du Pont de Nemours & Co.
35. Hintzer, K., Löhr, G. *Modern Fluoropolymers* (Scheirs, J., Ed.), John Wiley & Sons, Ltd, Chichester, UK, p. 240 (1997).
36. Sulzbach, R., Tschacher, M., *Angew. Makromol. Chem. 109/110*, p. 113 (1982).
37. Hintzer, K., Löhr, G. *Modern Fluoropolymers* (Scheirs, J., Ed.), John Wiley & Sons, Ltd, Chichester, UK, p. 254 (1997).
38. Michel, W, *Kunststoffe/German Plastics*, *79*(10):984 (1989).
39. *TEFLON®NXT 70 Fluoropolymer Resin, Modified PTFE Granular Molding Resin*, Publication 2474696 C (4/98) DuPont Fluoroproducts, Wilmington, DE (1998).
40. *TEFLON® NXT Benefits*, www.teflon.com/nxt, DuPont Fluoroproducts, Wilmington, DE (2008).
41. Carlson, D. P., Schmiegel, W. *Ullman's Encyclopedia of Industrial Chemistry* (Gerharz, W., Ed.), Vol. *A11*, VCH Publishers, Weinheim, Germany, p. 403, (1988).
42. Gangal, S. V. *Encyclopedia of Polymer Science and Technology*, Vol. *16* (Mark, H. F. and Kroschwitz, J. I., Eds.), John Wiley & Sons, New York, p. 604 (1989).
43. Benderly, A. A. *Treatment of teflon to promote bondability*, J. Appl. Polym. Sci. 6:221, (1962).
44. Gangal, S. V. *Encyclopedia of Polymer Science and Technology*, Vol. *16* (Mark, H. F. and Kroschwitz, J. I., Eds.), John Wiley & Sons, New York, p. 605, (1989).
45. Gangal, S. V. *Encyclopedia of Chemical Technology* (Grayson, M., Ed.), John Wiley & Sons, New York, p. 29, (1978).
46. Ryan, D. L., British Patent 897,466 (February 28, 1962).
47. Chesire, J. R., U.S. Patent 3,063,882 (November 13, 1962) to E.I. du Pont de Nemours & Co.
48. Gangal, S. V. *Encyclopedia of Chemical Technology*, Vol. *11* (Grayson, M., Ed.), John Wiley & Sons, New York, p. 30, (1978).
49. Gangal, S. V. *Encyclopedia of Polymer Science and Technology*, Vol. *16* (Mark, H. F. and Kroschwitz, J. I., Eds.), John Wiley & Sons, New York, p. 608, (1989).
50. Gangal, S. V. *Encyclopedia of Polymer Science and Technology*, Vol. *16* (Mark, H. F. and Kroschwitz, J. I., Eds.), John Wiley & Sons, New York, p. 609, (1989).
51. TEFLON® FEP-Fluorocarbon Film, Bulletin 5A, Optical, E. I du Pont de Nemours & Co., Wilmington, DE.
52. Pozzoli, M., Vita, G., Arcella, V. *Modern Fluoropolymers* (Scheirs, J., Ed.), John Wiley & Sons, Ltd, Chichester, UK, p. 380, (1997).
53. Hintzer, K., Löhr, G. *Modern Fluoropolymers* (Scheirs, J., Ed.), John Wiley & Sons, Ltd, Chichester, UK, p. 230, (1997).
54. Kerbow, D. L. *Modern Fluoropolymers* (Scheirs, J., Ed.), John Wiley & Sons, Ltd, Chichester, UK, p. 302 (1997).
55. Carlson, D. P., U.S. Patent 3,624,250 (November 30, 1971) to E.I. du Pont de Nemours & Co.
56. Gotcher, A. J., Gameraad, P. B., U.S. Patent 4,155,823 (May 22, 1979).
57. TEFZEL® *Chemical Use Temperature Guide*, Bulletin E-18663-1, E. I. du Pont de Nemours & Co. Wilmington, DE (1990).

58. Anderson, J. C., U.S. Patent 4,390,655 (June 28, 1983).

59. Dohany, J. E., Humphrey, J. S. *Encyclopedia of Polymer Science and Technology*, Vol. *17* (Mark, H. F. and Kroschwitz, J. I., Eds.), John Wiley & Sons, New York, p. 536, (1989).

60. Hertzberg, R. W., Manson, J. A., Wu, W. C., *ASTM Bull*, STP 536, p. 391 (1973).

61. Hertzberg, R. W., Manson, J. A., *Fatigue of Engineering Plastics*, Academic Press, Orlando, FL, pp. 82, 85, 90, 109, 131, 150, 151 (1980).

62. Seiler, D. A. *Modern Fluoropolymers* (Scheirs, J., Ed.), John Wiley & Sons, Ltd, Chichester, UK, p. 490, (1997).

63. Dohany, J. E., Humphrey, J. S. in *Encyclopedia of Polymer Science and Technology*, Vol. *17* (Mark, H. F. and Kroschwitz, J. I., Eds.), John Wiley & Sons, New York, p. 539, (1989).

64. Bloomfield, P. E., Ferren, R. A., Radice, P. F., Stefanou, H., Sprout, O. S., *U.S. Naval Res. Rev. 31*, 1 (1978); Robinson, A. L., *Science* 200, 1371 (1978).

65. Lovinger, A. J., *Science 220*:1115. (1983).

66. Nakagava, K., Amano, M., *Polym. Comm.*, 27:310.

67. Dohany, J. E., Humphrey, J. S. *Encyclopedia of Polymer Science and Technology*, Vol. *17* (Mark, H. F. and Kroschwitz, J. I., Eds.), John Wiley & Sons, New York, p. 540, (1989).

68. Dohany, J. E., Stefanou, H., *Am. Chem. Soc. Div. Org. Coat. Plast. Chem. Pap.*, *35*(2):83, (1975).

69. Paul, D. R., Barlow, J. W., *J. Macromol Sci. Rev. Macromol. Chem.*, *C18*:109, (1980).

70. Mijovic, J., Luo, H. L., Han, C. D., *Polym. Eng. Sci.*, 22(4):234, (1982).

71. Wahrmund, D. C., Bernstein, R. E., Barlow, J. W., Barlow, D. R., *Polym. Eng. Sci.*, *18*:877 (1978).

72. Guerra, G., Karasz, F. E., Mac Knight, W. J., *Surface hydrolyzation of poly(vinylidene fluoride)*, *Macromolecules*, *19*:1935 (1986).

73. Sperati, C. A. *Handbook of Plastic Materials and Technology* (Rubin, I. I., Ed.), John Wiley & Sons, New York, p. 102, (1990).

74. Ebnesajjad, S, *Introduction to Fluoropolymers*, 2nd ed., Elsevier, Oxford, UK, p. 55, (2021).

75. Reimschuessel, H. K., Marti, J., Murthy, N. S., *J. Polym. Sci., Part A, Polym. Chem.*, 26:43, (1988).

76. Stanitis, G. *Modern Fluoropolymers* (Scheirs, J., Ed.), John Wiley & Sons, Ltd, Chichester, UK, p. 529, (1997).

77. Hull, D. E., Johnson, B. V., Rodricks, I. P., Staley, J. B. *Modern Fluoropolymers* (Scheirs, J., Ed.), John Wiley & Sons, Ltd, Chichester, UK, p. 257, (1997).

78. 3M™ Dyneon™ Fluoroplastics Comparison Guide, Document 2018 Dyneon FTP PCG 10p en_us.pdf (2018).

79. *Adhesive Fluoropolymer NEOFLON™ EFEP*, Document May. 2003 EG-37 (0005) AI, Daikin (2003), daikin-america.com

80. Ebnesajjad, S. *Polyvinyl Fluoride, Technology and Application of PVF*, Elsevier, Oxford, UK, p. 152, (2013).

81. Nata, G. *Macromol. Chem.*, *35*:94, (1960).

82. Goerlitz, M. et al., *Angew. Makromol. Chem.*, *29/30*:137, (1973).

83. Ebnesajjad, S., Snow, L. G. *Kirk-Othmer Encyclopedia of Chemical Technology*, 4th ed., John Wiley & Sons, New York, p. 683, (1994).

84. Farneth, W. F., Aronson, M. T., Uschold, R. E., *Macromolecules*, *26*, 4765 (1993).

85. Ebnesajjad, S., Introduction to Fluoropolymers, in: *Applied Plastics Engineering Handbook*, Chapter 4, (Kutz, M., Ed.), Elsevier, Oxford, UK, 2016.

86. *Mechanical Properties of Tedlar® PVF Films*, Technical Bulletin, DuPont, Tedlar.com, (2020).

87. *TEDLAR®*, *Technical Information*, Publication H-49725 (6/93), E. I. du Pont de Nemours & Company, Wilmington, DE (1993).

88. French, R. H. et al. Optical properties of materials for concentrator photovoltaic. *Solar Energy Materials and Solar Cells*, *95*, 2077, (2011).

89. *General Properties of Tedlar® PVF Films*, Technical Bulletin, DuPont, Tedlar.com, (2020).

90. *TEDLAR®*, *Technical Information*, Publication H-49719 (6/93), E. I. du Pont de Nemours & Company, Wilmington, DE, (1993).

6

Processing of Polytetrafluoroethylene Resins

Jiri G. Drobny

Polytetrafluoroethylene (PTFE) is manufactured and offered to the market in essentially three forms, namely as granular resins, as fine powders, and as aqueous dispersions. Although they are all chemically high-molecular-weight PTFE with an extremely high melt viscosity, each of them requires a different processing technique. This chapter deals with the fabrication methods used for granular resins and fine powders. The applications of PTFE and modified PTFE are covered in some detail in Chapter 8. The technology specific to aqueous dispersions is discussed in Chapters 12 and 13.

6.1 Processing of Granular Resins

Granular PTFE resins are most frequently processed by compression molding using a technique similar to that common in powder metallurgy and by ram extrusion. Each of these processes requires a specific type of granular resin. All major manufacturers of PTFE resins offer several grades of granular resins for this purpose. Chemours offers them as Teflon™ granular molding powders. Resin grades currently available from this manufacturer, their descriptions, and typical applications are listed in Table 6.1.

AGC Chemicals Americas, Inc. markets the Fluon® Granular PTFE grades listed in Table 6.2.

3M™ Dyneon™ markets granular resins as standard grades, modified grades of PTFE TFM™, extensive line filled materials, and Dyneon™ Custom PTFE compounds. Filled compounds are made with glass fiber, carbon fiber, mineral fillers, graphite, bronze, and polyimide with amounts ranging from 10 to 60 weight % depending on the type of filler and application [1]. Examples of the typical properties of unmodified and modified unfilled granular grades are shown in Table 6.3.

Daikin markets Polyflon™ PTFE granular molding powders, which are fine cut resins suited for demanding chemical, mechanical, electrical, and nonstick surface applications [2]. There are essentially three types of granular molding powders:

1. PTFE, a standard homopolymer.
2. Modified PTFE.
3. Free-flowing granular molding powders.

Examples of each of these types, their typical properties, and processing techniques used for them are shown in Table 6.4.

Solvay offers two types of Algoflon® PTFE Granulars [3]:

- Algoflon® F, which consists of granular "fine cut" PTFE powders with a typical average particle size in the range 15–35 mμ. These are processed by normal compression molding techniques and compounding fillers such as carbon, graphite, and glass fibers.
- Algoflon® S, which consists of free-flowing powders with increased bulk density and flowability.

These grades can be processed by the usual molding techniques and also by isostatic and ram molding as well as ram extrusion.

DOI: 10.1201/9781003204275-8

TABLE 6.1

Available Teflon™ Granular Molding Powders

Grade	Description	Type/Grade*	Applications
Fine Cut Resins			
7AX	A resin designed to be molded into shapes	II	High-performance mechanical/electrical applications; skived tapes, films, sheets; machined gaskets, packings, mechanical seals.
7CX	A resin used in molding applications requiring excellent flex life	II	Expansion joints; bellows; piston rings
Pelletized Resins			
8AX	A free-flow resin ideal for isostatic molding; designed for low preform pressure; improved surface smoothness	IV/1	Shallow molds; ball valve seals; pipe linings; seals; valves; valve plugs
807NX	A free-flow multipurpose resin (compression or automatic/isostatic molding and ram extrusion) whose high-fill density yields high productivity	IV/2	Seals; valve seals; bearing pads; gaskets; electrical/electronic parts
Modified Resins			
NXT 70	A fine-cut resin designed for compression molding	III/1	Seal rings; valve seals; bearing pads; linings; encapsulations; base resin for filled compounds; block and sheet compression molding
NXT 75	A fine-cut resin designed for compression molding with improved weldability	III/1	Seal rings; valve seats; bearing pads
NXT 85	A free-flow molding resin also convertible to parts via isostatic, billet, sheet molding methods, and ram extrusion	III/2	Valve seats; fittings; pipe valve and vessel linings; seal rings; bearing pads; electrical insulation; fluid handling systems

Source: Teflon™ PTFE Granular Molding Powders, Chemours Teflon.com/en/products (2022).
* As per ASTM D4894.

TABLE 6.2

Grades of Fluon® Granular PTFE

	Grade		
Property	**G155**	**G163**	**G201**
Type	Powder	Powder	Presintered for extrusion
Bulk density, g/L	400	325	650
Mean particle size, mμ	45	37	600
Preform pressure, MPa	16	16	Extrusion only
Shrinkage, %	5	5	12
Ultimate tensile strength, MPa	37	40	22
Elongation at break, %	400	380	330
Applications	Blending with fillers; general molding	Blending with fillers; billets for skived tape	Ram extrusion of small sections

Source: Fluon® High Performance Fluoropolymers, Document FP Grade Range E 06-32021, AGC Chemicals agcchem.com.

TABLE 6.3

Comparison of Unmodified and Modified Granular Grades of Dyneon™ PTFE Granular PTFE

Property	Grade	
	Unmodified (TF 1750)	Modified (TFM 1700)
Powder flow properties	Non-free flowing	Non-free flowing
Bulk density, g/L	370	420
Average particle size, mμ	25	25
Deformation under load, %*	4	2
Shrinkage, %	4.8	5
Tensile strength, psi (MPa)	4,000 (27.6)	4,800 (33.1)
Elongation at break, %	350	450
Applications	Blending with fillers; general molding, billets, rods, tubes, and sheets	Blending with fillers; suitable for manufacture of complex shapes; molding, billets, rods, tubes, and sheets

Source: Dyneon™, Polytetrafluoroethylene Product Comparison Guide, Document 98-0505-1448-5, 3M Dyneon, 2003, www.dyneon.com.
* 2,175 psi (15 MPa), 24h.

TABLE 6.4

Examples of Daikin Polyflon™ Granular Molding Powders, Their Typical Properties, and Processes Used for Them

Property	Grade		
	M-17 Homopolymer PTFE	M-111 Modified PTFE	M-531 Free Flowing Powder
Bulk density, g/L (ASTM D4894)*	420	360	740
Shrinkage, % (ASTM D4894)	3.1	4.4	3.0
ASTM type/grade (ASTM D4894)	II	III/1	IV/1
Tensile strength, MPa (psi) (ASTM D4894)	43 (6,237)	40 (5,802)	42 (6,090)
Elongation at break, % (ASTM D4894)	400	500	380
Compression creep characteristics			
Total deformation, % (25°C, 13.7 MPa)	17.12	10.6	16.1
Compression set, % (ASTM D621)	8.6	3.0	8.0
	Filled compounds; molding	Filled compounds; most molding techniques; ram extrusion	Excellent flow; isostatic molding; automatic molding; ram extrusion

Source: Daikin Polyflon™ PTFE Products, Document PB-PTFE-001-R0 11/23/2016 Daikin www.daikin-america.com.
* ASTM D4894; see Appendix 2.

The currently available Solvay products, their properties, and processing are listed in Table 6.5.

AGC Chemicals currently offers three grades of granular powders that are used for molding, ram extrusion, and as a feedstock for filled compounds (see Table 6.6).

Standard specifications for the properties and test methods of PTFE granular molding and ram extrusion materials are covered by ASTM D4894. Details pertaining to this standard are shown in Appendix 2.

When filled compounds are prepared from certain PTFE molding resins, it is necessary to mix the base PTFE resin and the filler to enable mixing with high shear systems. The temperature must also be controlled (it should not rise above 40°C, or 104°F) during the mixing process to avoid agglomeration of PTFE and the generation of white spots. In general, filled compounds are processed using normal compression molding, similar to virgin fine cut PTFE.

TABLE 6.5

Currently Available Algoflon® PTFE Granulars

Properties	Grade			
	F 5/S	F 5	F 7	S 121
Description	Non-free flowing powder	Non-free flowing powder	Non-free flowing powder	Non-free flowing powder
Average particle size[a], mμ	15	15	15	
Bulk density[b], g/L	410	380	410	
Shrinkage, radial[b], %	3.1	3.0	3.0	
Tensile strength[b], MPa	44	40	44	
Elongation at break[b], %	400	350	400	
Processing	Base for filled compounds, particularly suitable for high filler amounts; also used for normal compression molding	Base for filled compounds, particularly suitable for glass fiber; also used for normal compression molding	Normal compression molding, in particular billets for thin skived films down to 25 mμ	

Source: Data from *Algoflon® PTFE Granulars Technical Data Sheets*, Solvay Specialty Polymers, Rev. 7/14/2021.
a Internal test method;
b ASTM D4894.

TABLE 6.6

Fluon® Granular PTFE Powders, Their Properties, and Processing

Property	Grade		
	G155	G163	G201
Type	Fine powder	Fine powder	Presintered for extrusion
Bulk density, g/L	400	325	650
Mean particle size, mμ	45	37	600
Preform pressure, MPa (psi)	16	16	Extrusion only
Shrinkage, %	5	5	12
Tensile strength, MPa (psi)	37	40	22
Elongation at break, %	400	380	320
Process used	Blending with fillers; general molding	Blending with fillers; billets for skived films	Ram extrusion of small sections

Source: Fluon® High Performance Fluoropolymers, Document FP Grade Range E-06-2021, p.4, AGC Chemicals America, Inc. www.agcchem.com (2022).

PTFE fine cut molding powders and filled compounds must be stored and handled carefully to achieve defect-free molding. Before use the temperature of the powder must be conditioned above 19°C (66°F), typically between 21 and 27°C (70 and 81°F). Below this temperature, the powder is difficult to mold without cracks. Too high temperatures should also be avoided to prevent powder lumping and poor flow.

6.1.1 Compression Molding

The basic molding process for PTFE consists of the following three important steps: (1) preforming; (2) sintering; and (3) cooling. In the preforming step the PTFE molding powder is compressed in a mold at the ambient temperature into compacted form, with sufficient mechanical integrity for handling and sintering. The preform is then removed from the mold and sintered (heated above the crystalline melting point of the resin). During sintering, the resin particles coalesce into a strong homogeneous structure.

During the subsequent step – cooling – the product hardens while becoming highly crystalline. The degree of crystallinity depends mainly on the rate of cooling. The weight of the parts fabricated by compression molding may vary from less than 1 g to several hundred kg.

6.1.1.1 Preforming

The loose bed of the molding powder is compacted in a mold placed in a hydraulic press. The molds used for compression molding are similar to those for thermosets or powdered metals. The most common shape of a preform is a cylinder, commonly called a "billet". The assembly for smaller-size billets consists of the main mold and upper and lower end plates or pistons (Figure 6.1). Molds for larger billets are more complex. Molds and mandrels are normally made from tool steel and are nickel or chrome plated for corrosion protection. End plates can be made from tool steel, brass, or plastics (e.g., polyacetal or nylon) [4]. Presses used for compression molding must have good controls for a smooth pressure application, must be capable of applying specific pressures up to 100 N/mm² (14,500 psi), must have sufficient daylight, and must allow easy access to molds. Typically, virgin resins require specific pressures up to 60 N/mm² (8,700 psi) and filled compounds up to 100 N/mm² (14,500 psi). A press for large preforms is shown in Figure 6.2.

PTFE resins exhibit a first-order transition at 19°C (66°F) due to a change of crystalline structure from a triclinic to a hexagonal unit cell (see Section 5.2.1.3). A volume change of approximately 1% is associated with this transition (Figure 6.3). Another consequence is that the resin has a better powder flow below 19°C but responds more poorly to preform pressure. Billets prepared below this transition are weaker and tend to crack during sintering. For this reason, the resin should be conditioned at 21 to 25°C (70 to 77°F) overnight before preforming to prevent that. The preforming operation should be done at room temperature, preferably higher than 21°C (77°F). Preforming at higher temperatures is sometimes useful to overcome press capacity limitations. As the temperature is raised, the resin particles exhibit higher plastic flow and consequently can be more easily compacted and become more responsive to preform pressure.

Mold filling is another key factor in the quality of the final product. It has to be uniform, and this is achieved by breaking up lumps of resin with a scoop or screening. The full amount of the powder has to be charged into the mold before the pressure is applied; otherwise, contamination, layering, or cracking at the interfaces may occur on sintering.

During the compression of the PTFE powder both plastic and elastic deformations occur. At low pressures the particles slip, slide, and tumble in place to align themselves into the best possible array for packing. With increasing pressure, contact points between adjacent particles are established and further enlarged by plastic deformation. Plastic deformation also eliminates internal particle voids.

FIGURE 6.1　Mold assembly for small - to medium-size billets. (Courtesy of DuPont)

FIGURE 6.2   Compression molding press for PTFE billets (Courtesy DPA)

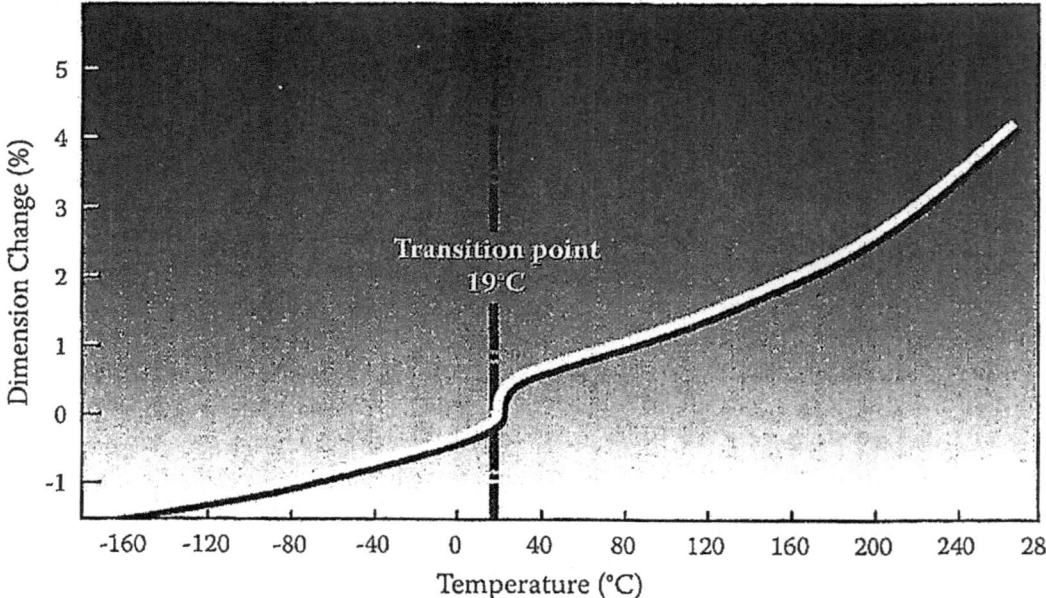

FIGURE 6.3   Transition point and linear thermal expansion of PTFE. (Courtesy of DuPont)

When the maximum pressure required for compression is reached, it is held for a certain time, which is referred to as "dwell time". This time is required for the transmission of pressure throughout the preform and for the removal of entrapped air. Too short a dwell time can cause density gradients in the preform, which may lead to *hour-glassing* and property variation in the sintered billet. Incomplete removal of trapped air can cause microfissures or worsened properties. Dwell time is dependent on the rate of pressure application and on the shape and mass of the billet. Generally, 2 to 5 min/10 mm or 0.5 in. finished height for small billets, and 1 to 1.5 min/10 mm or 0.5 in. finished height for large billets, give satisfactory results [4, p. 9]. Pressure release after the dwell period should be very slow until the initial expansion and relaxation have taken place. Typically, this is done with a bleeder valve or a capillary. Sudden pressure decay can result in microcracks or visible cracks as the still entrapped air expands.

After preforming, there are still residual stresses and entrapped air in the preformed part, which invariably causes cracking mainly during the initial stage of sintering. The removal of entrapped air, or "degassing", requires some time called "resting time", which depends mainly on the wall thickness.

6.1.1.2 *Sintering*

The purpose of the sintering operation is to convert the preform into a product with increased strength and a reduced fraction of voids. Massive billets are generally sintered in an air-circulating oven heated to 365 to 380°C (689 to 716°F). Both the sintering temperature and time have a critical effect on the degree of coalescence, which in turn affects the final properties of the product.

Sintering has two stages and consists of a variety of processes. During the first stage, the preform expands up to about 25% of its volume as its temperature is increased to and above the melting point of the virgin resin (about 342°C, or 648°F). The next stage is the coalescence of the particles in which voids are eliminated. After that, the contacting surfaces of adjacent particles fuse and eventually melt. The latter process gives the part its strength. After reaching the melting point, the resin changes from a highly crystalline material to an almost transparent amorphous gel. When the first stage is completed, the billet becomes translucent, though it requires additional time to become fully sintered. This time depends not only on the sintering temperature but also on the conditions when the part was preformed and on the type of resin. High pressures and small particle size facilitate fusion. As with preform pressure, sintering time eventually reaches a point beyond which there is no significant improvement in the physical properties.

During sintering some degradation of the polymer takes place. Prolonging the process beyond the required time or using a very high sintering temperature will invariably result in excessive degradation and considerable worsening of properties. To achieve a uniform heat distribution in the oven, turbulent airflow is required. Variability of heat distribution can cause billet distortion or even cracking. Massive billets should be loaded into the oven at a maximum of 100°C (182°F) to avoid thermal shock. A holding time of one to two hours before heat-up is usually required for the temperature to reach equilibrium.

The heating rate is very critical for the quality of the final product. Because of the very low thermal conductivity of PTFE resin, the billets have to be heated slowly to the sintering temperature, or cracking may occur even before the resin is fully melted. The highest heating rate a given preform will tolerate depends on a complex interaction of many factors. Major parameters include the thermal gradient (the difference between the ambient temperature in the oven and the temperature in the midpoint of the preform wall) and the rate of internal stress relaxation. The thermal gradient, in turn, is related to the heating rate and the wall thickness. Internal stresses in the preform originate in the preforming process and are dependent mainly on the preform pressure, closure rate, and preform temperature. The normal heating rate for large billets is 28°C (50°F) per hour up to 300°C (572°F), at which point it is reduced to 6 to 10°C (11 to 18°F) per hour.

A proper rate for a given set of billet geometry and preforming conditions is usually determined experimentally. Another way to minimize the thermal gradient is to introduce a series of hold periods. In this method a higher heating rate is used in the early phase of the heat-up cycle. Hold periods used between temperatures of 290 and 350°C (554 and 662°F) ensure a minimum temperature gradient through the melting transition and minimize any tendency to cracking due to the about 10% volume change associated with melting.

At the point when the billet is in the gel state, the particles coalesce and the voids are eliminated. However, the rate of sintering near the melting point is very slow. To achieve more commercially acceptable sintering rates, the temperatures used for that purpose are in the range of 365 to 380°C (689 to 716°F). For massive billets, temperatures above 385°C (725°F) for virgin PTFE and 370°C (698°F) for filled compounds should be avoided since thermal degradation above these temperatures becomes significant. The time at peak temperature depends on the wall thickness, type of resin used, and the method of sintering. Generally, prolonged sintering times have beneficial effects on properties, particularly on dielectric strength, provided that no significant degradation takes place. Typical times for a complete sintering are fairly constant – about two hours after the resin has reached its optimum sintering temperature. Once the oven has reached the sintering temperature, it takes about 1 to 1.5 hours to transmit the heat through each centimeter of thickness. As a rule of thumb, the time at sintering temperature should be 1.0 hour/cm or 0.4 in. of diameter for solid billets and 1.4 hour/cm or 0.4 in. of wall thickness for billets with a small hole in the middle. For small parts, sintering times of 0.8 hour/cm of wall thickness are adequate. Figure 6.4 illustrates production sintering ovens, and sintered billets are shown in Figure 6.5.

After sintering, the molten billet is cooled to room temperature in a controlled fashion. As the freezing range of 320 to 325°C (608 to 617°F) is reached, crystallization starts to happen. The degree of crystallization in the cooled-down part depends on the cooling rate. Since a majority of properties depends on the degree of crystallinity, the cooling rate has to be closely controlled to achieve the desired results. The effect of the cooling rate on crystallinity is shown in Table 6.7 [4, p. 17]. Melt strength and wall thickness are the key factors in determining the cooling rate. Typically, cooling rates between 8°C (14°F) per hour and 15°C (27°F) per hour are satisfactory for larger billets. Because of low thermal conductivity, slow cooling rates are necessary to avoid cracking due to excessive thermal gradients. This is particularly important during the transition in the freezing zone since stresses caused by a rapid volume change can tear the melt apart. Therefore, slower rates are maintained until the inside of the wall is below the freezing point and the center of the billet is crystallized. Then, faster cooling rates, about 50°C (90°F) per hour, can be used since the sintered part can tolerate higher thermal gradients [4, p. 17].

FIGURE 6.4   Typical production sintering oven. (Courtesy DPA)

FIGURE 6.5 Medium size sintered billets. (Courtesy DPA)

Massive moldings often require *annealing* during the cooling period to minimize thermal gradients and to relieve any residual stresses. The temperature range for annealing is typically from 290 to 325°C (554 to 617°F). The temperature at which the annealing is carried out is very critical. If annealing is done in the temperature range 310 to 325°C (590 to 617°F), the molding exhibits a high degree of crystallinity, which may not always be desirable. The product is highly opaque and has a low tensile strength, high stiffness, and a high specific gravity. If a lower degree of crystallinity is desired, annealing should be done at 290°C (554°F) [4, p. 17].

Pressure sintering and cooling is more often used for compounds than for virgin resins. In this process the preform is either sintered and cooled in a confined mold or placed into a special self-supporting frame. The frame is placed into an oven with the pressure cylinder outside, actuating the piston to compress the preform in the mold. Different sinter/cooling/pressure cycles can minimize certain physical properties for each different compound. The disadvantage of this method is that the product has a decreased dimensional stability due to internal stresses. Stress relieving can offset this disadvantage, but it compromises the desired higher properties. Since the melting point of PTFE increases with pressure (Figure 6.6), the sintering temperature has to be adjusted accordingly [4, p. 17].

TABLE 6.7

Effect of Cooling Rate on Polytetrafluoroethylene Crystallinity

Cooling Rate (°C/min)	Crystallinity (%)
Quenched in ice water	45
5	54
1	56
0.5	58
0.1	62

Source: Data from *Compression Moulding, Technical Information*, Publication H-59487, 05/95, DuPont, Wilmington, DE (1995).

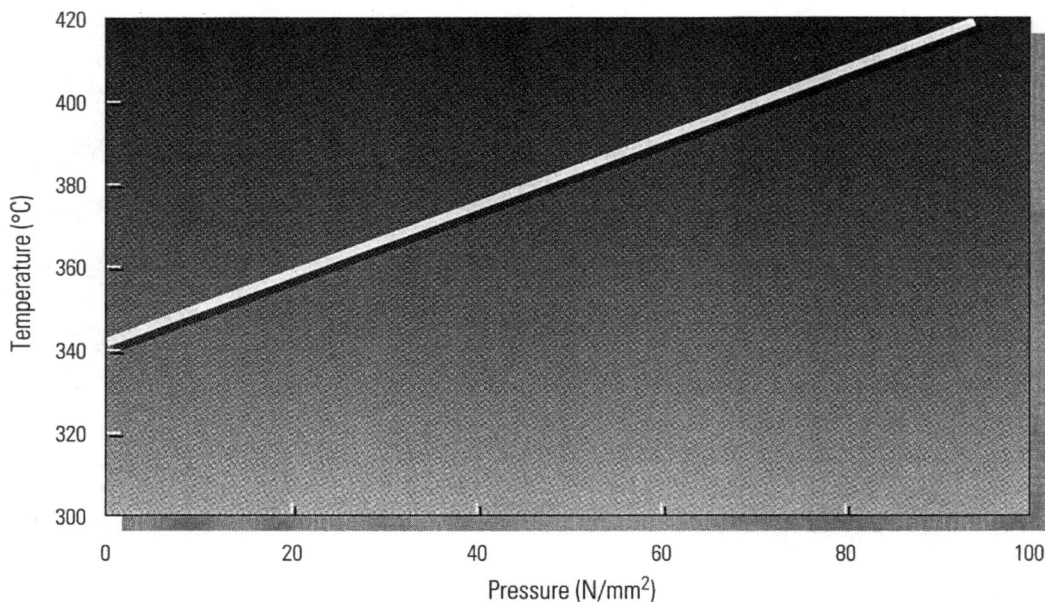

FIGURE 6.6   Dependence of PTFE melting point on pressure. (Courtesy of DuPont)

Billets are almost always subjected to some kind of finishing. The most frequent finishing is by machining. The machinery used for PTFE is the same as for other plastics. The achievable dimensional tolerances depend mainly on the quality of the cutting edge of the tools used, which controls heat generation. At any rate, cooling is necessary to remove excess heat. To achieve very close tolerances, parts have to be stress relieved prior to machining above the expected service temperature. The common practice is to use a holding time of 1 hour/25 mm (1 in.) of thickness, followed by slow cooling.

A compression molding method called "coining" is used for parts that are too complicated to be produced by machining. In coining, a sintered molding is heated to the melting point, and then it is quickly pressed into a mold cavity and held under pressure until it solidifies [5].

6.1.2 Other Molding Methods

Other molding methods for granular PTFE resins are automatic molding and isostatic molding. *Automatic molding* is used to produce small parts (e.g., gaskets, bearings, seals, valve seats) in automatic presses, with preform pressures and speeds higher than in compression molding. Molded parts made by this technique do not require additional finishing. High-flow resins are used in automatic molding [5]. *Isostatic molding* allows uniform compression from all directions. It uses a flexible mold, which is filled with a free-flowing granular powder and then is evacuated and tightly sealed. The mold is then placed into an autoclave containing a liquid that can be compressed to the pressure required for preforming. Isostatic molding is used to make complicated shapes that otherwise would require expensive machining, such as large tubes, valves, pumps, and thin-walled small tubes [5]. If close tolerances are required, the molded part must be finished to the required dimensions.

Films and sheets are produced by *skiving*, which is "peeling" of the billet in a similar fashion to a wood veneer. A grooved mandrel is pressed into a billet, and the assembly is mounted onto a lathe. A sharp cutting tool is used to skive a continuous tape of a constant thickness. The arrangement is shown in Figure 6.7. The range of thickness of films and sheets produced by skiving is typically from about 25 µm to 3 mm (0.001 to 0.125 in.). A modern, high-performance skiving machine capable of machining billets up to 1,500 mm (60 in.) wide is shown in Figure 6.8.

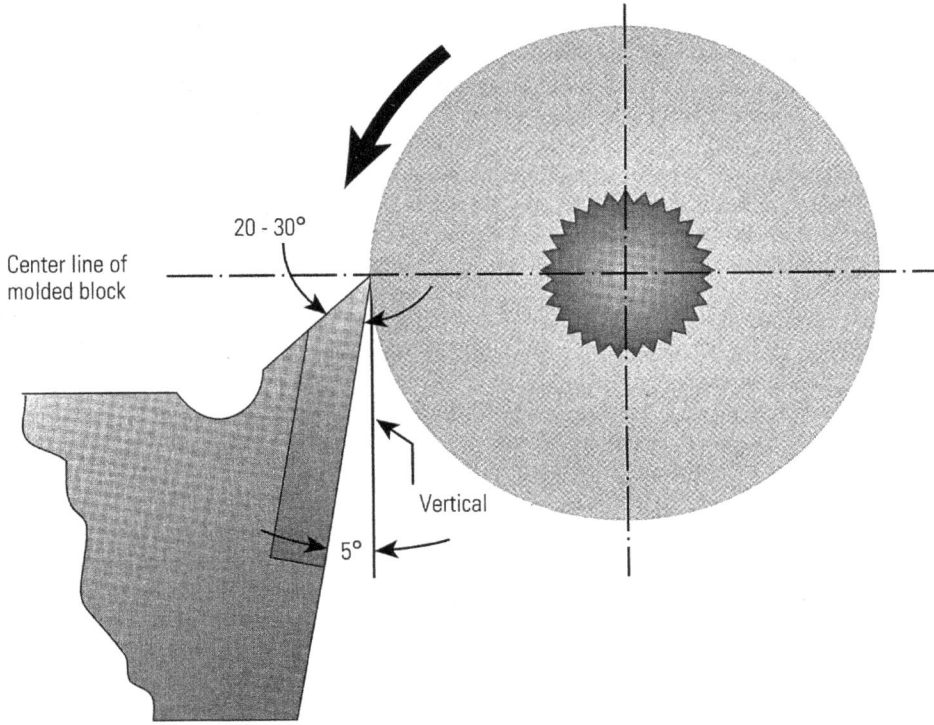

FIGURE 6.7 Typical arrangement of skiving knife. (Courtesy of DuPont)

FIGURE 6.8 Modern skiving machine. (Courtesy of Dalau Inc.)

6.1.3 Ram Extrusion

Ram extrusion is a process to produce PTFE extrudates of continuous lengths. Granulated resins used in ram extrusion must have good flow characteristics so they can be fed readily to the extruder die tube. Presintered and agglomerated PTFE powders with a bulk density ranging typically from 675 to 725 g/L are used for this process [6].

The resin is fed into one end of a straight die tube of uniform diameter, where it is compacted by a ram and forced through the tube, which incorporates a heated sintering zone. The ram is then withdrawn, the die tube refilled by the resin, and the cycle repeated. Thus, the compacted powder is forced stepwise through the die to its heated section where it is sintered and then through a cooler section where it is cooled and eventually emerges in a continuous length. A rise in temperature and excessive working, such as severe shearing or agitation, have an adverse effect on the flow of the powder. The temperature adjacent to the top of the die tube and in the feed system should be in the range of 21 to 30°C (70 to 85°F) [6, p. 4]. A vertical ram extruder is shown in Figure 6.9.

The powder is preformed to a void-free condition during the preforming stage and, as such, is moved through the die tube. The process maintains pressure on the molten PTFE in the sintering zone to coalesce the resin particles. The rate of compaction has to be slow enough to allow the air mixed with the resin to escape.

During sintering, the powder is heated by conduction. Thus, the time needed to heat it to the sintering temperature depends on the size and shape of the extrudate as well as on its heat transfer properties. Temperature settings in the sintering zone are from 380 to 400°C (716 to 752°F), although for large-diameter rods, where the center takes much more time than the surface to reach sintering temperature, the setting can be as low as 370°C (700°F) to avoid degradation of the surface [6, p. 4].

The rate of cooling determines the degree of crystallinity of the extrudate (see Table 6.7) and consequently its dimensions and properties. Very rapid cooling, especially of large-diameter rods, will produce a high internal stress in them. Such parts have to be annealed before they can be machined to close tolerances.

Granular PTFE resins are most frequently extruded as rods or tubes, but it is possible to produce extrudates of noncircular cross-sections. Typical conditions for the extrusion of a rod are shown in Table 6.8 [7].

6.2 Processing of Fine Powders

6.2.1 Introduction

PTFE fine powders (also called "coagulated dispersion powders") are a highly crystalline form of PTFE materials susceptible to shear damage, particularly above the 19°C transition point. At 30°C a change to a different phase (phase I) occurs that softens the particles. This change allows the fine powders to be processed at temperatures above 30°C. Because PTFE fine powders do not melt and flow, they are commonly fabricated by a technology adopted from ceramic processing called "paste extrusion". In paste extrusion, the paste is prepared by mixing the powder with a 15 to 25% hydrocarbon lubricant, such as kerosene, white oil, or naphtha, with the resultant blend appearing much like the powder on its own [8].

A major requirement for paste extrusion is that, up to the point of sintering and coalescence, the extrudate must possess sufficient strength to withstand the extensive handling that takes place during the process. The tendency of PTFE fine powders to fibrillate while extruded provides the required strength and the unique characteristics of fine powder articles [9]. Fibrillation is present in a final product such as thread sealant tape which has been stretched monoaxially, or in filtration membranes, which have been stretched biaxially.

A crucial factor in the selection of a powder for paste extrusion is its reduction ratio range, which is often specified by the manufacturer. The reduction ratio (RR) is defined as the ratio of the cross-sectional surface areas of the preform to the extrudate. It is an important variable impacting the pressure during the operation [9, p. 126]. For a given extruder barrel, the smaller the cross-section of the final product is, the higher the reduction ratio will be. If for example a thin wire is extruded from a large diameter extruder

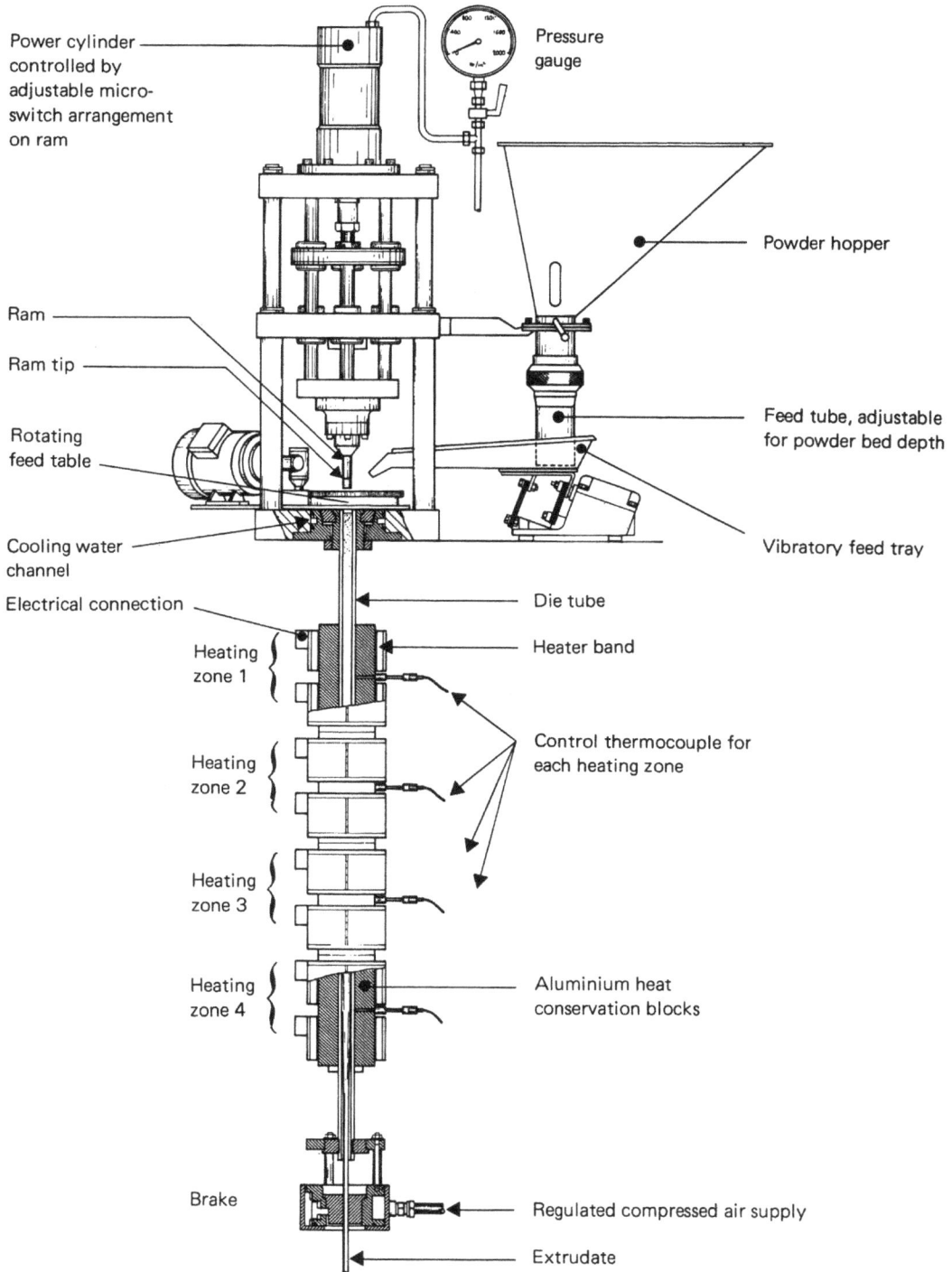

Power cylinder controlled by adjustable micro-switch arrangement on ram

Pressure gauge

Powder hopper

Ram

Ram tip

Rotating feed table

Feed tube, adjustable for powder bed depth

Cooling water channel

Vibratory feed tray

Electrical connection

Die tube

Heating zone 1

Heater band

Heating zone 2

Control thermocouple for each heating zone

Heating zone 3

Heating zone 4

Aluminium heat conservation blocks

Brake

Regulated compressed air supply

Extrudate

FIGURE 6.9 Vertical ram extruder, main components. (Courtesy of AGC Chemicals America)

TABLE 6.8

Example of Extrusion Process Conditions for a Rod (Diameter 10 mm)

Die Tube	
Diameter	10.6 mm
Unheated length at top of die tube	90 mm
Heated length	900 mm
Unheated length at bottom of die tube	400 mm
Total length	1,550 mm
Heated length/diameter	85:1
Water cooling	Over top 60 mm
Heating Arrangements	Four separately controlled heated zones, each with two 1.5 kW heater bands
Temperature profile:	
Zone 1 (top)	380°C
Zone 2	400°C
Zone 3	400°C
Zone 4 (bottom)	350°C
Extrusion Conditions	
Powder type	Presintered
Approximate powder charge weight	1.5 g
Total cycle time	5 seconds
Powder compaction rate	32 mm/s
Ram dwell time at bottom of stroke	1.5 seconds
Brake	None
Extrusion pressure	82 MPa
Extrusion rate	6.8 m/h (1 kg/h)
Extrudate Properties	
Diameter	9.4 mm
Shrinkage	11.7%
Relative density	2.16
Tensile strength	22 MPa

Source: Fluon® Extrusion of Granular Powders, Technical Note F2, AGCFP, September 2002.

then the RR is also quite large. So it is important to have products that develop a lower extrusion pressure under these conditions. On the other hand, if a large diameter pipelining tube is extruded then the RR will be very low, because there are practical limits as to the size and performance of extruders. In a very broad manner it can be said that the great majority of RRs lie between 10:1 and 4,000:1. The choice of a material often depends on the equipment available, the dimensions of the product being extruded, as well as the final properties of the product. The exact same end product can be produced at different RRs, depending on what size of extruder is being used. The resin selected to make a particular part must be converted into the extrudate at a reasonable pressure in the extruder. This means that it should generate sufficient pressure for fibrillation of the resin, yet the pressure must not exceed the normal range of commercial equipment [9]. Examples of different reduction ratios are [10]:

• Thick walled liners	RR = 10 to50.
• Thin walled liners	RR = 50 to 500.
• Micro-tubes	RR = 500 to 2,000.
• Cable insulations	RR = 300 to 3,000.

TABLE 6.9

Typical Properties and Applications of Fluon® Coagulated Dispersions

Grade	SSG[a]	Extrusion Pressure, MPa	Mean Particle Size, μm	Reduction Ratio	Bulk Density, g/L	Applications
Homopolymers						
CD122E	2.16	50	500	400:1	545	Electrical tape; fibers; liner of large diameter pipes
CD123E	2.16	43	530	400:1	545	Electrical tape; liner of medium diameter pipes
CD127E	2.16	35	530	400:1	545	Electrical tape; liner of small diameter pipes
CD143E	2.18	30	530	400:1	570	Low density thread seal tape; tubing; antidrip additives for thermoplastics
Trace Copolymer[b]						
CD086EL	2.15	28	475	400:1	470	High performance hose; gaskets
CD086EH	2.15	35	475	400:1	470	High performance hose; gaskets
CD090E	2.19	40	600	1,600:1	500	Wire coating; small diameter transparent tubing
CD097E	2.18	34	580	1,600:1	510	Wire coating; heat shrinkable tubing

[a] SSG = standard specific gravity;

[b] modified PTFE.

Major PTFE manufacturers offer several grades of fine powders. For example AGC Chemicals Americas, Inc. offers eight grades of coagulated dispersion powders, four from homopolymer and four from modified PTFE. The typical properties and applications for these products are shown in Table 6.9.

Chemours lists 16 grades in its brochure *Teflon™ PTFE Powders* [11] with detailed information about their properties and specific applications. Displaying all that information here could be confusing and the reader is advised to review the referenced document or contact the supplier. Nevertheless, we include a table displaying data from two different types of Teflon™ products Table 6.10). The grade *6CN X* is a general purpose fine powder used mainly for wire and cable insulation and tubes with thin walls, such as spaghetti tubing. The grade *62 X* is a premium product that compared with other PTFE fine powders has increased thermal stability, superior flex life, considerably improved stress crack resistance, low permeability, and high clarity. It is used for tubing installed in demanding applications such as reinforced hoses required for maximum reliability and performance in the chemical, pharmaceutical, and automotive industries and which are used with hydraulic fluid, hydrocarbon fuel, or reactive gas. Examples are overbraided hoses for fuel assemblies and brake systems. Moreover, it is suitable for after-processing technologies, such as flanging, welding, blow molding, and convoluting. Both grades, when properly processed, can qualify for use in contact with food in compliance with FDA 21 CFR 177 1550 and European Regulation (EU) No. 10/2011. More details are available from the supplier.

Daikin markets three distinct series of their Polyflon™ PTFE fine powder including a total of 12 grades. Typical properties and applications of these products are shown in Table 6.11.

3M Dyneon manufactures and sells a broad line of fine powder PTFE grades for paste extrusion processes. Various grades are tailored to the extrusion of wire and cable insulation tapes, tubes, liners, and profiles. The product line includes standard grades of unmodified PTFE as well as modified grades of TFM™ PTFE and special compounds. Typical properties as well as major applications of these products are listed in Table 6.12.

Solvay offers a wide range of products as Algoflon® DF Fine Coagulated Powders including homopolymers, modified PTFE polymers, as well as antistatic compounds. These powders may differ in molecular weight, molecular weight distribution, modifier (type, content, distribution inside the particle), and particle size: the overall RR range is 10:1 up to 3,000:1 [11].

TABLE 6.10

Typical Properties and Applications of Selected Grades of Teflon™ PTFE Fine Powders

Property/Test	Test Method	Unit	Grade 6CN X	Grade 62 X
Average particle size, d50	ASTM D4895/ISO12086	μm	400	480
Bulk density	ASTM D4895/ISO12086	g/L	470	495
Standard specific gravity	ASTM D4895/ISO12086		2.1852	2.152
Thermal instability index	ASTM D4895/ISO12086		<50	<7
Stretch void index			-	<50
Extrusion pressure at RR =1,600:1	ASTM D4895/ISO12086	MPa (psi)	50 (7,252)	23 (3,335)
Melt peak temperature:				
Initial	ASTM D4895/ISO12086	°C (°F)	341 (646)	341 (646)
Second	ASTM D4895/ISO12086	°C (°F)	326 (620)	322 (612)
Typical applications			Wire and cable insulation; tubing with thin walls, such as spaghetti tubing	Tubing installed in demanding applications, such as overbraided hoses for fuel assemblies and brake systems

Source: *Teflon™ PTFE 6CN X Fluoroplastic Resin*, C-10118 (2/16) and *Teflon™ PTFE 62 X Fluoroplastic Resin*,C-10127 (2/16), The Chemours Company, 2016.
Notes: Teflon ™ 6CN X meets the requirements for ASTM D 4895 Type 1, Grade 4, Class B; Teflon ™ 62 X meets the requirements for ASTM D 4895 Type 1, Grade 2, Class C.

TABLE 6.11

Typical Properties and Applications of Polyflon™ PTFE Fine Powders

Property	Grade F-100 Series	F-200 Series	F-300 Series
Type of polymer	Homopolymer	Modified	Modified
Average particle size, μm	500	500	500
Apparent (bulk) density, g/L	450	450	450
Melting point, °C (°F)	326–328 (619–622)	322–328 (612–622)	322–328 (612–622)
Tensile strength, MPa (psi)	>25 (3,626)	>25 (3,626)	>25 (3,626)
Elongation at break, %	>300	>300	>250
Reduction ratio*	1,000 maximum	4,000 maximum	1,500 maximum
Typical applications	Unsintered tape; sealing tape; low specific gravity tape; flat cable; tube wrap; coaxial cable; expanded	Small thin wall tubing; thin wall electric wire; heat shrinkable tubing	Tube wrap; thick wall electric wire; large thick wall tubing; heat shrinkable tubing; coaxial cable

Source: Daikin Polyflon™ PTFE Products: Product Information for Aqueous Dispersions, Fine Powders, Granular Molding Powders, Daikin America, Inc., 845-365-9500 www.daikin-america.com 2022.
* The reduction ratio (RR) refers to the cross-sectional area of the resin in the cylinder of the extruder (S1) and the cross-sectional area of the resin in the die land (S2). RR = S1/S2.

Their typical properties as well as major applications are listed in Table 6.13.

Standard specifications for properties and test methods of PTFE fine powders produced from a PTFE aqueous dispersion are covered by ASTM D4895. Details pertaining to this standard are in Appendix 3.

Fine powder resins are shipped in specially constructed drums that typically hold 23 kg (50 lb.) of resin. These shallow, cylindrical drums are designed to minimize the compaction and shearing of the resin

TABLE 6.12

Dyneon™ PTFE Fine Powders, Their Typical Properties, and Major Applications

Property	Grade					
	TF 2025Z	**TF 2029Z**	**TF 2053Z**	**TF 2072Z**	**TF 2073Z**	**TFM 2033Z**
Type of polymer	Homopolymer	Homopolymer	Homopolymer	Homopolymer	Homopolymer	Modified
Average particle size, μm	500	500	520	440	400	520
Bulk density, g/L	480	480	500	470	490	460
Tensile strength, MPa (psi)	34 (4,921)	28 (4,061)	32 (4,640)	35 (5,076)	35 (5,076)	37 (5,365)
Elongation at break, %	360	340	340	470	460	430
Specific gravity	2.17	2.15	2.16	2.17	2.17	2.16
Extrusion pressure at RR 400:1, MPa (psi)	40 (5,800)	50 (7250)	65 (9,425)	40 (5,800)	26 (3,770)	32 (4,640)
Minimum reduction rate	20:1	5:1	20:1	50:1	50:1	20:1
Industries	CPI/semicon; electrical; rubber and plastics; transportation	CPI/semicon; defense; electrical; rubber and plastics; transportation	CPI/semicon; electrical; rubber and plastics; transportation	CPI/semicon; electrical; rubber and plastics; transportation	CPI/semicon; electrical; rubber and plastics; transportation	CPI/semicon; defense; electrical;
Typical applications	Thread sealing tape; wire and cable insulation (wrap); yarns; mid-size tubing	Thread sealing tape; wire and cable insulation (wrap); pipe liners; yarns; expanded sealing joints; mid-size tubing	Ideal grade for extrusion of tubing and cables requiring only moderately high reduction ratio.	Efficient grade for wire and cable insulation; small diameter tubing; thin wall tubing; spaghetti tubing	Thin wall; smallest diameter tubing; wire and cable insulation with a maximum reduction ratio of up to 4,400:1	High performance pipes; fitting; wire and cable insulations

Source: Data from *3M Dyneon Data Sheets,* (2022); *MatWeb Data Sheets* (2022); *Omnexus, Universal Selector,* (2022).

Note: The products listed in this table meet the following ASTM D4895 classifications:

TF 2025Z: Type 1, Grade1, Class B; **TF 2029 Z:** Type 1, Grade 1, Class C; **TF 2053Z:** Type 1, Grade 1, Class C
TF 2072Z: Type 1, Grade 1, Class C; **TF 2073Z:** Type1, Grade 1, Class C; **TFM 2033Z:** Type 1, Grade 1, Class B.

TABLE 6.13

Typical Properties and Applications of Algoflon® DF Fine Coagulated Powders

Property	Test Method	Grade				
		DF 120F	DF 130F	DF 261F	DF 330F	DF681F
Physical						
Specific gravity	ASTM D792	2.18	2.16	2.15	2.14	2.17
Average particle size, μm	Internal	450	600	600	600	550
Bulk density, g/L	ASTM D4895	500	500	475	475	500
Shrinkage @RR 20:1, %	Internal	33	—	—	—	—
Stretchability, %	Internal	—	1,500	1,500	—	—
Mechanical						
Tensile strength, MPa	ASTM D4895	30.0	30.0	35.0	35.0	35.0
Elongation at break, %	ASTM D4895	400	300	350	350	400
Thermal						
Thermal instability index	ASTM D4895	<25	<15.0	<15.0	<15.0	<10.0
Other						
Reduction ratio range	Internal	10 to 300	30 to 300	10 to 300	10 to 300	100 to 2500
Rheometer pressure @ RR 100:1, MPa	ASTM D4895	8.00	9.50	50.0	10.0	48.0
Main features		Good green strength; low memory effects; low radial shrinkage	High green strength; high and homogeneous ability to be stretched	Lower sintering temperature; good dimension stability; low friction	Very high green strength; good welding; low permeability; good processing stability; low friction	Smooth surface; excellent clarity; good adhesion; optimal electrical properties; low permeability; low friction
Typical applications		Gaskets; liner; tape; industrial applications	Aerospace; automotive; electrical and electronics; fibers; tape; membranes; wire and cable	Aerospace; automotive; electrical and electronics; fibers; yarn; membranes; wire and cable	Aerospace; automotive; electrical and electronics; industrial; tubings; wire and cable	Aerospace; automotive; electrical and electronics; industrial; tubings; wire and cable

Source: Algoflon® DF 120F, 130F, 261F, 330F and 681F, Technical Data Sheets, Solvay Specialty Polymers (3/7/2022).

during shipment and storage. To further assure that the compaction is kept at an absolute minimum, the resin must be kept at a temperature below 19°C (66°F), its transition point, during shipping and warehouse storage. Prior to blending with lubricants, the resin should be stored below its transition temperature for 24 hours. A safe storage temperature for most resins is 15°C (60°F). Generally, the particles form agglomerates, spherical in shape with an average size of 500 μm. If lumps have formed during shipping, the resin should be poured through a four-mesh screen immediately prior to blending. To prevent shearing, the screen should be vibrated gently up and down. Sharp objects, such as scoops, should not be used to remove the resin because they could shear and ruin the soft resin particles. It is best to avoid screening unlubricated powder unless it is absolutely necessary.

6.2.2 Fabrication Methods for Products from Fine Powders

The current products from PTFE fine powders include:

- Films, tapes, and sealing cords.
- Tubing and hoses.
- Thick walled pipes and liners.
- Wire and cable insulation.

In general, the technology consists of the following key processing steps:

- Preparation of the extrusion mix.
- Preforming.
- Extrusion.
- Drying.

Additional steps that are specific to the different products may include calandering, monoaxial or biaxial stretching, sintering, and cutting when required.

6.2.2.1 Preparation of the Extrusion Mix

Lubricants enable PTFE fine powders to be processed on commercial equipment. Liquids with a viscosity between 0.5 and 5.0 cP (0.5 to 5.0 mPa•s) are preferred, although more viscous liquids are used occasionally. When selecting a lubricant, its ability to be incorporated easily into the blend and to vaporize completely and rapidly in a later processing step without leaving residues that would discolor the product or adversely affect its properties is important. The amounts of lubricant added are typically 16 to 19% of the total weight of the mix. The type and amount of the lubricant vary with the type of final product.

For colored products, pigments may be added during blending. These can be added dry and directly to the powder prior to the addition of the lubricant or as a wet blend in the lubricant. In the latter case, the pigment dispersion is added with the remaining lubricant. All blending operations must be performed in an area in which the temperature is maintained below the PTFE transition temperature of 19°C (66°F), and the relative humidity should be kept at approximately 50%. A high level of cleanliness and an explosion-proof environment are additional requirements to assure high quality and safety.

The blending of small batches, up to about 4.5 kg (10 lb), is most frequently done in wide-mouth jars that are placed on horizontal rollers tumbling at approximately 15 rpm for about 20 min. Larger, commercial batches, of sizes typically 10 to 136 kg (22 to 300 lb) are often prepared in twin-shell blenders (Figure 6.10) by tumbling for 15 minutes at 24 rpm. The blend is screened again, transferred to a storage vessel, and allowed to age for at least 12 hours [12]. There exists other, larger equipment, such as the Turbula® Shaker Mixer (Willy A. Bachofen, A. G. Maschinenfabrik), which consists of a cylindrical container held in mechanical arms with a complex motion pattern [13].

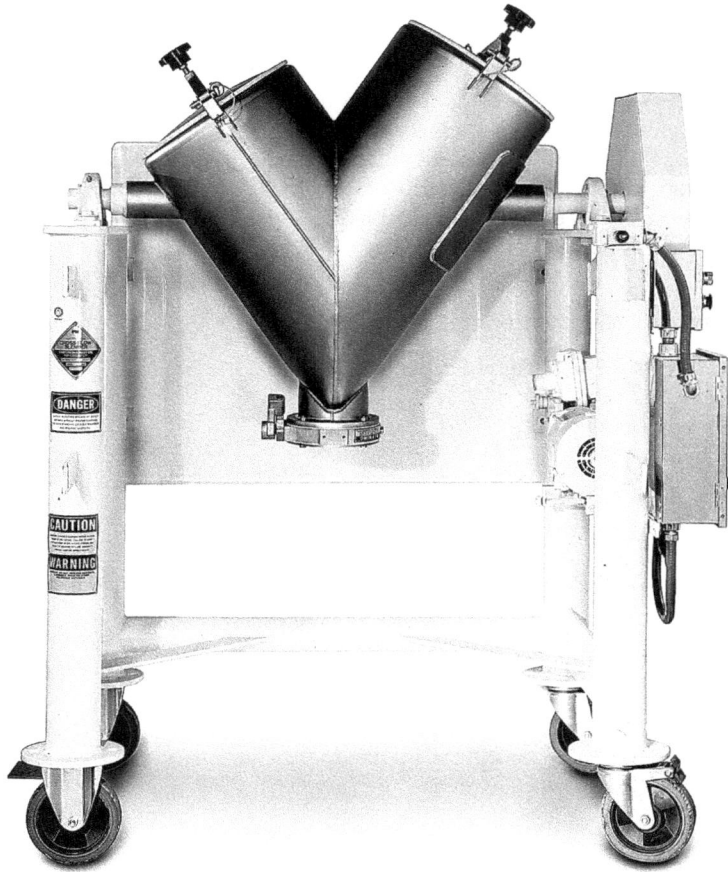

FIGURE 6.10   Twin shell blender. (Courtesy of Patterson-Kelley Co.)

6.2.2.2 Preforming

The properly aged lubricated powder is usually preformed at room temperature into a billet of the size required by the equipment in which it is to be later processed. Preforming removes air from the material and compacts it so that it has a sufficient integrity for handling during the manufacturing process. Preforming pressures are of the order of 0.7 MPa (100 psig) in the initial stage of the cycle and may increase up to 2 MPa (300 psig). Higher pressures do not increase compaction and may cause the lubricant to be squeezed out [12]. The compaction rate is initially up to 250 mm (10 in.) per minute and is reduced toward the end. The finished preform is rather fragile and must be stored in a polyvinyl chloride (PVC) or polymethylmethacrylate (PMMA) tube for protection against damage and contamination. A schematic of the preforming equipment is shown in Figure 6.11.

6.2.2.3 Extrusion

The extrusion step is performed at temperatures above 19°C (66°F), the first transition point of the resin, where it is highly deformable and can be extruded smoothly. The resin preform is placed in the extrusion cylinder, which is kept at a temperature of 38°C (100°F) for several minutes to be heated up to the higher temperature [10].

Unsintered rod and tape for packing and unsintered tape for thread sealing and cable wrapping together represent one of the largest applications of this technology. Extrusion of an unsintered rod is the simplest

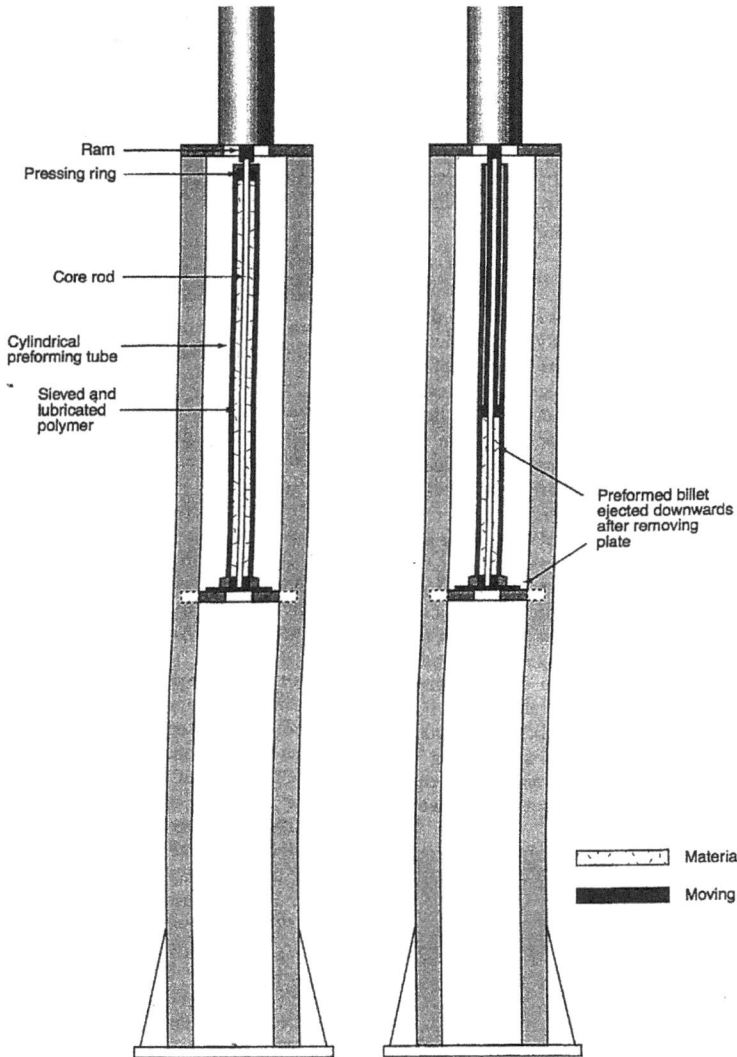

FIGURE 6.11 Preforming equipment for molding billets for paste extrusion. (Courtesy of Dyneon GmbH)

process for paste extrusion of PTFE fine powders. The extrusion is done by a simple hydraulic ram extruder with a total available thrust ranging from 10 to 20 tons. The ram speed is adjustable up to a maximum of 50 to 100 mm/min (2 to 4 in./min). The ram forces the lubricated powder through the orifice of the die. The head of the ram is usually fitted with a PTFE seal to prevent the polymer from flowing back along the ram. The surfaces of the extrusion cylinder and die are made from corrosion-resistant steel and are highly polished. The pressures during the rod extrusion are normally in the range of 10 to 15 MN/m² (1,500 to 2,000 psi). The diameter of the extrusion cylinder is designed to accommodate the required size of the extrudate and the type of polymer used. The usual extrusion cylinder diameters are between 40 and 150 mm (1.5 and 6 in.). The die consists of two parts: a conical and a parallel section. The conical part is more important, as the reduction in area between cone entry and exit determines the amount of work done on the polymer and, hence, to a high degree, the properties of the extrudate. The cone angle has also some effect on the extrudate and is most commonly 30°, although larger angles are used for large-diameter cylinders. The length/diameter ratio of the die parallel (die land), which also has some effect on the properties of the extrudate, may vary from five to ten times the exit diameter.

6.2.3 Fabrication of Films, Tapes, and Sealing Cords

Films, tapes, and sealing cords can be fabricated in unscratched, stretched, sintered, or unsintered forms [10]. The common applications are:

- Thread sealing tapes.
- Flat or plate seals.
- Electrical insulation tapes for wound insulators.
- Tape cable.

The key processing steps are:

1. Preparation of the extrusion mix.
2. Profile extrusion.
3. Film calandering.
4. Film drying.
5. Monoaxial and biaxial stretching.
6. Film sintering, and cutting if required.

In this particular technology, the content of lubricant is 21 to 25 parts by weight related to 100 parts by weight of PTFE. In contrast to cube extrusion, higher boiling lubricants (boiling range 180 to 250°C, or 356 to 482°F) are used in order to avoid lubricant loss during calandering. Extruders for profile fabrication are relatively simple in design. A cylinder with a nozzle or a ram of mechanical or hydraulic drive that runs at a constant velocity and independently of pressure is sufficient. The profiles are calandered by a two-roll calander with an appropriate feeding system. A temperature of 40°C (104°F) for the tool surface and a roll speed of about 30 revolutions per minute (depending on roll diameter) are recommended. The film or sheet is dried at a temperature of 160 to 200°C (320 to 302°F). Calandering and drying should be performed independently of each other as both steps achieve optimal results at different speeds.

The dried films or tapes are stretched for certain applications. In practical use, stretching of the free running PTFE sheets at temperatures between 280 and 300°C (536 and 572°F) is typical. The sheet is stretched by two different roll systems that run at different speeds in the running direction. The film can be stretched up to the ratios of 1:10 to 1:15. Biaxial stretching is used for the production of breathable waterproof membranes for clothing, microfiltration membranes, medical implants, microwave carriers, industrial sealants, and high tensile fabrics (more on this subject in Section 6.2.7). Film, sheets, or tapes can also be stretched after the sintering process in order to produce high tensile strength films, sheets, tapes, or yarns [10].

6.2.3.1 Manufacture of Unsintered Tape

The production of unsintered tape from an extruded rod normally consists of the following sequence of operations:

1. Calendering.
2. Removal of lubricant.
3. Slitting.
4. Reeling.

The rod used in this method has a relatively large diameter, typically 10 to 15 mm (0.4 to 0.6 in.), and is calandered in a single pass to a tape about 100 to 200 mm (4 to 8 in.) wide and 0.075 to 0.01 mm (0.003 to 0.004 in.) thick, which is subsequently slit to several tapes of desired width. No advantage has been found in using multiple-stage calendering [14].

The rod is fed into the nip of the calander by means of a guide tube that prevents it from wandering and a consequent variation in the width of the calendered tape. A tape with straight edges and a controlled width is produced when a fishtail guide made from metals (e.g., aluminum) or plastics (e.g., acetal) is used successfully [14].

The calander rolls are heated to temperatures up to 80°C (176°F) to produce a smooth, strong tape. They may be heated either electrically or by circulating hot water or oil. Production speeds of 0.05 to 0.5 m/s (10 to 100 ft/min) are quite common. The thickness of the calendered tape depends mainly on the calander nip setting. The tape width depends on several factors, such as lubricant type and content, reduction ratio and die geometry during the extrusion, and the speed and temperature of the calander rolls [14].

Normally, the lubricant is removed by passing the calendered tape through a heated tunnel oven. If the lubricant used in the mixture has a very high boiling point, such as mineral oil, it may be removed by passing the calendered tape through a degreasing bath containing a hot vapor of trichloroethylene or other suitable solvent. The slow output rate of this process and safety and health hazards associated with this method rarely offset the advantages of using heavy oils, such as virtually no loss of lubricant by evaporation between extrusion and calendering and the amount of work done on the polymer as a result of the much higher viscosity of the paste. Therefore, this technique is seldom used and only in cases where it is absolutely necessary.

After the lubricant is removed, the tape is slit to the desired width by leading it under a slight tension over stationary or rotating cutting blades. The slit tape is reeled onto small spools. An outline of the process is shown in Figure 6.12.

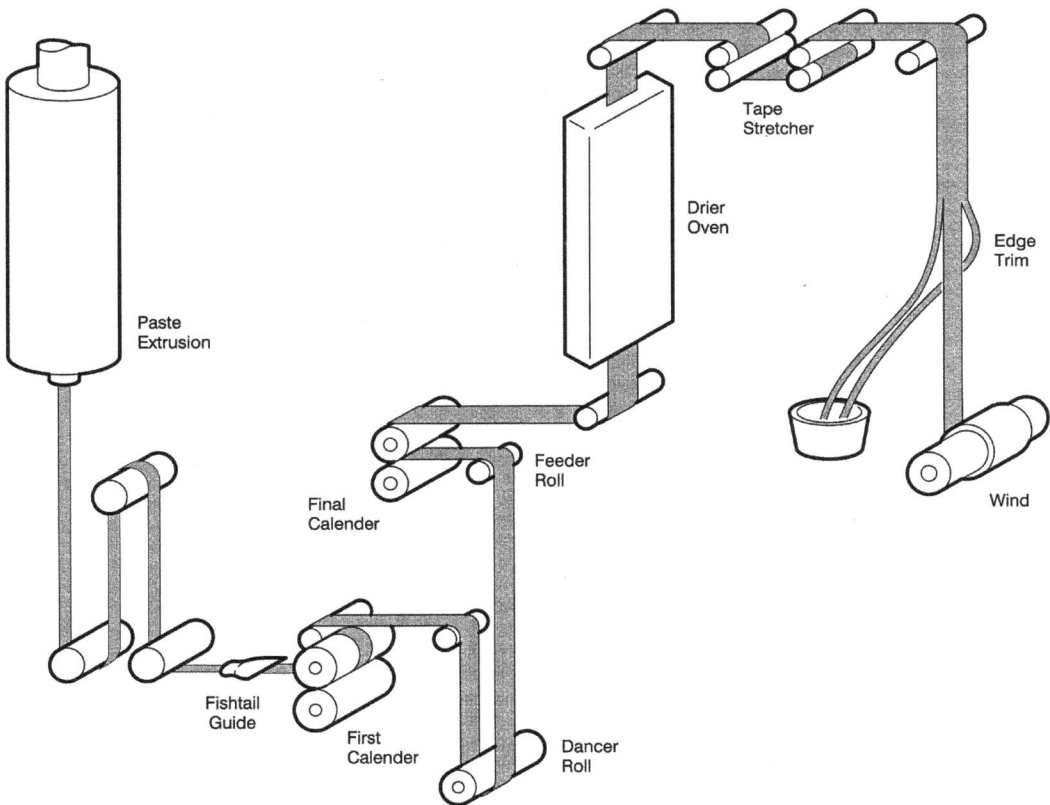

FIGURE 6.12 Schematic of the process for producing thread seal tape. (Courtesy of DuPont)

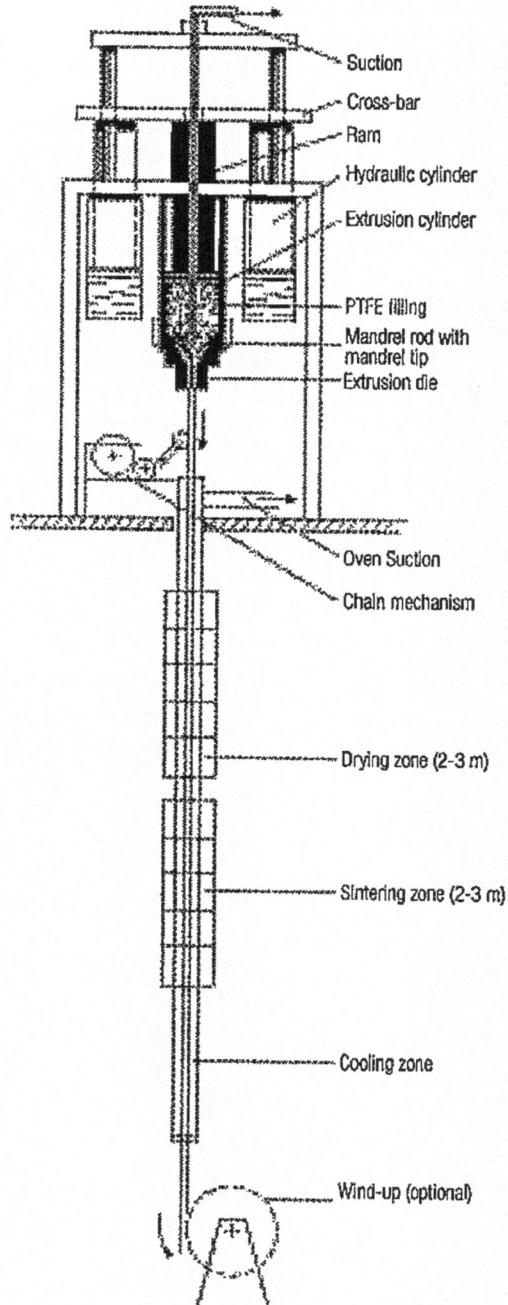

FIGURE 6.13 Hydraulic paste extruder for tube fabrication. (Courtesy of Dyneon GmbH)

6.2.4 Fabrication of Tubing and Hoses

Paste extrusion is used to fabricate extremely thin-walled micro- and spaghetti hose as well as thin-walled industrial tubes. Dimensions range from a 0.1 to approximately 25 mm (0.040 to 1.0 in.) interior diameter and wall thicknesses from approximately 0.1 to 2 mm (0.040 to 0.080 in.). The extruded tubes can be further processed to produce shrink tubes, corrugated tubes, and braided tubes [10].

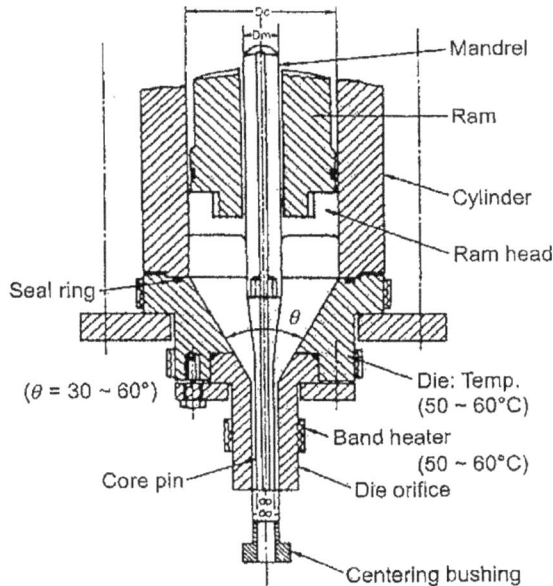

FIGURE 6.14 Extrusion die of an extruder for tube fabrication. (Courtesy of Dyneon GmbH)

6.2.4.1 Tube Extrusion

Tubes made from paste are fabricated by a special ram extruder with an inner mandrel that determines the inner diameter of the tube [10]. The extrudates are then dried. The extrusion process is discontinuous. Extrusion is stopped after each extruded preform to return the ram and insert a new preform. The extrudates are then dried and sintered. A schematic of a hydraulic paste extruder for tubing is shown in Figure 6.13 and that of a die for this extruder in Figure 6.14.

6.2.4.2 Drying and Sintering of Tubes

The extruded tube is dried and sintered in a continuous oven. In the drying zone, the lubricant is evaporated and removed at temperatures above the boiling point of the lubricant and below the sintering temperature of PTFE, typically at 150 to 200°C (302 to 392°F). After drying, the tube is sintered at a temperature between 360 and 380°C (680 and 716°F). Sintering conditions depend on the tube dimensions, extrusion velocity, and the temperature and length of the oven. The drying and sintering process changes the dimensions of the tube. Crosswise shrinkage of 0 to 15% and lengthwise shrinkage of 15 to 25% is expected. This dimensional change has to be taken into consideration when selecting the extrusion tool [10].

6.2.5 Fabrication of Thick-Walled Pipes and Liners

Liners are thick-walled pipes with wall thicknesses of about 2 to 15 mm (0.08 to 0.60 in.) used in the corrosion resistant lining of steel pipes in chemical plants [10].

6.2.5.1 Liner Extrusion

Liner extrusion is done on the same ram extruder as that described in Section 6.2.4.1. However, due to the heavy weight of the liner pipes, the extruder is set up horizontally. The extruded liner is drawn over an interior supporting pipe if required, and put into a supporting half-pipe if the extrudate has a low green strength.

6.2.5.2 Drying and Sintering the Liner

The tube sections that are typically up to 10 m (33 ft.) long are dried and sintered in a horizontal oven. For that process, the pipes are placed into metal supporting half-pipes with the interior supporting pipe so as to avoid deformation of the extrudate. The recommended conditions are shown below and are adjusted depending on wall thickness and the diameter of the pipe [10].

Drying: two to three hours at 150 to 200°C
Sintering: one to three hours at 300 to 380°C

Cooling can be performed quickly or slowly, depending on the desired crystallinity level and final properties. The level of crystallinity depends on the rate at which the product passes through the gelling point of 310 to 320°C (590 to 608°F). Fast cooling reduces the crystallinity and enhances flexibility, while slow cooling increases crystallinity and specific gravity and reduces permeability.

6.2.6 Fabrication of Wire and Cable Insulation

The use of wire and cable insulation based on PTFE fine powders is one of its major applications; the paste extrusion of wire and cable insulation using special extruders is a well-established technology. The current industrial applications include the automotive, aerospace, and other industries where a high temperature rating (more than 250°C or 480°F) and resistance to chemicals are required. The main use of PTFE wire insulations is for hook-up for electronic equipment in the aerospace and military industries. Other large volume use is in coaxial cable insulation made by paste extrusion or tape wrapping, and in airframe and computer applications [13, p. 242].

The preparation of the extrusion mix requires a great deal of attention. First, lubricants with a low boiling range are preferred because the dwell time in the drying oven is limited due to a high extrusion speed for this process. The next step is the screening of the mix containing a lubricant through a fine mesh to remove agglomerates that would introduce flaws to the insulation. Additional care must be taken that the compacting process in the prepress is slow enough to allow air completely to escape from the paste. The preform is then slowly relaxed to avoid cracks.

6.2.6.1 Wire Extruder System

The wire extrusion system consists of a wire unwind roll, a dancer roll, an extruder, drying and sintering ovens, a deflector wire puller, an electrical breakdown test device, and a wire take-up roll. The extruder can be set up either vertically or horizontally (the orientation refers to the direction of the ram). The extruder consists of a heated barrel, where the preform is loaded, and a screw-driven ram. The conductor is drawn by a power system through a hollow mandrel located in the center of the ram. The wire payoff system is usually motorized and equipped by a tensioning device [10, 13]. Because of the high extrusion speed in the production equipment, long drying and sintering ovens are required. The ovens are usually arranged parallel to each other in order to ensure better space utilization [10]. A schematic of a wire and cable insulation extrusion system is shown in Figure 6.15.

6.2.6.2 Wire Extrusion Process

The preform with a bore is inserted into the extrusion cylinder of the machine and then pressed by the ram through a die. The preform fibrillates in the die under ram pressure and the extruded paste material coats the wire that is guided through the extruder head at the same time. The cylinder and the extrusion die are heated to 40 to 60°C (104 to 140°F) to ensure that the surface of the extrudate is as smooth as possible. A sketch of the extrusion die for a wire and cable insulation extruder is given in Figure 6.16. The extrusion speed of this process is determined by the sensitivity of the material and the conditions in the downstream drying and sintering ovens [10].

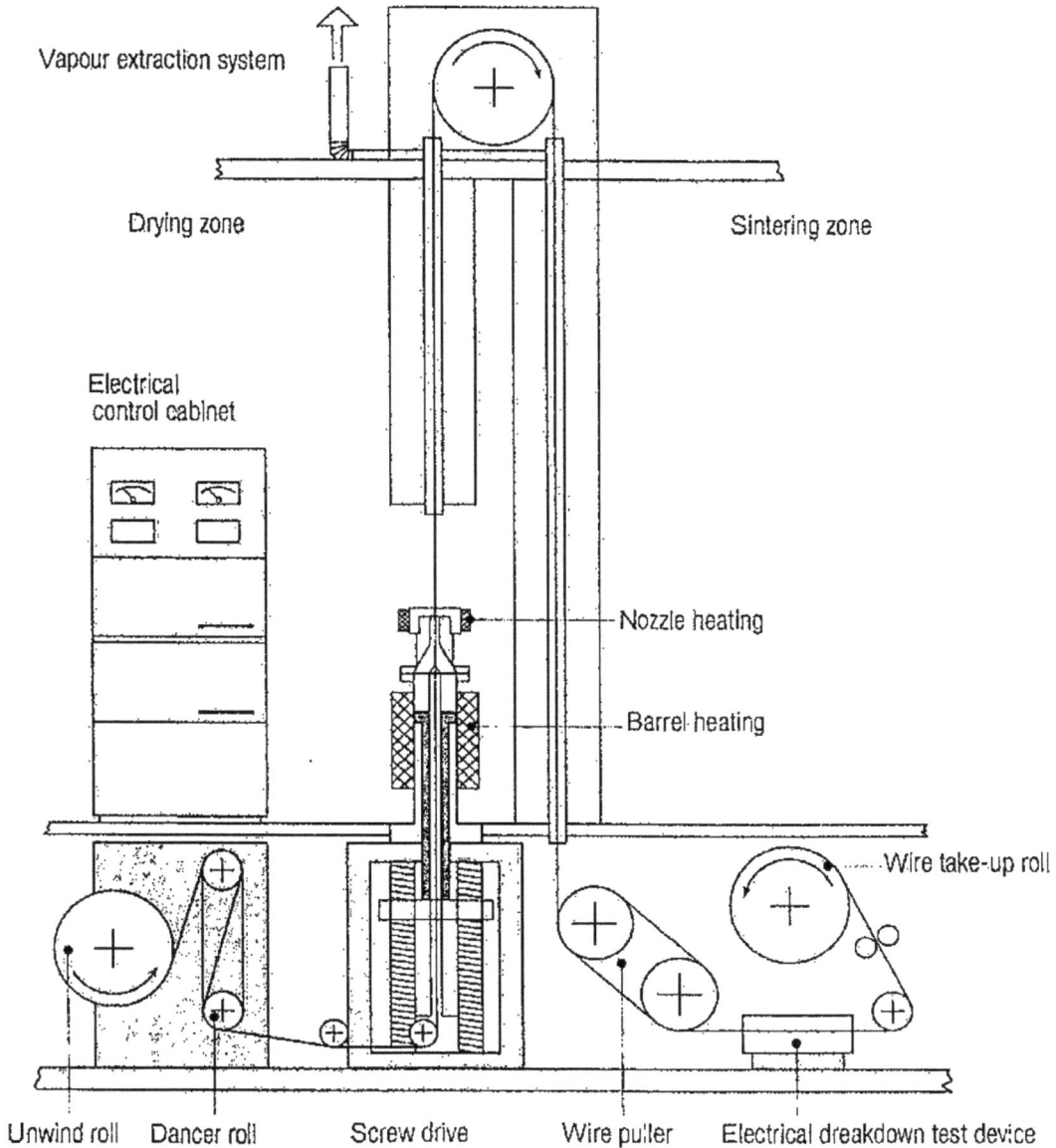

FIGURE 6.15 Extruder system for the fabrication of PTFE wire and cable insulation. (Courtesy of Dyneon GmbH)

6.2.6.3 Drying and Sintering of Wire Insulation

After extrusion, the insulation must be dried at a temperature ranging between 120 to 200°C (248 to 392°F). Any remaining lubricant in the extrudate might lead to brown discolorations, cracks, and electrical flaws during sintering. Sintering takes place at temperatures above 345°C (653°F), preferably at 360 to 380°C (680 to 716 °F). Multiple sintering ovens are used, sometimes as many as eight [13, p. 253]. Cooling coated wire is relatively easy because it is thin compared to parts from granular PTFE powders. The insulation exits the sintering oven in a molten state, solidifying upon contact with the ambient air. The wire is usually allowed to cool by natural convection in the ambient air. The crystallinity of the PTFE coating is about 50. After cooling, the wire enters the spark tester, which is a dielectric breakdown tester.

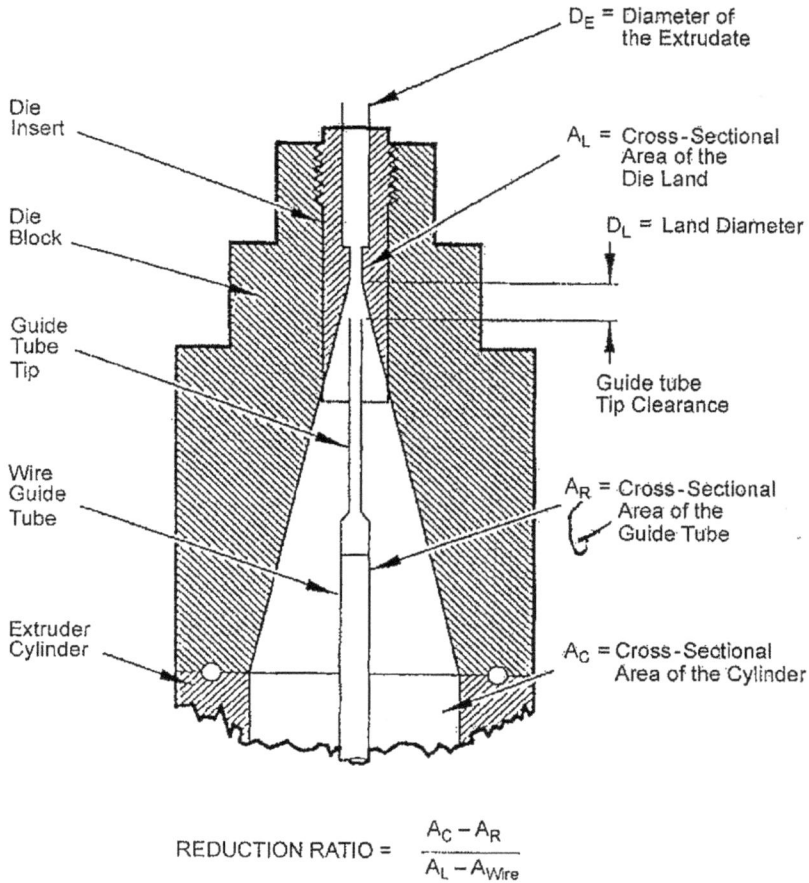

$$\text{REDUCTION RATIO} = \frac{A_C - A_R}{A_L - A_{Wire}}$$

FIGURE 6.16 Extrusion die for the wire and cable insulation extruder. (Courtesy of Dyneon GmbH)

The testing is performed continuously and the number of sparks is automatically recorded. The number of sparks is a measure of the quality of the wire insulation and represents the functional quality of the wire.

6.2.7 Expanded PTFE

Expanded PTFE (ePTFE) is another product produced from PTFE fine powders. It is based on the fibrillation of a high molecular weight PTFE matrix by controlled stretching [15]. This process induces physically a very large number of small pores into the structure of the PTFE material, resulting in new properties and significant changes in material consumption [13, p. 272]. One of the significant features of expanded PTFE is a drastic reduction in creep, which is the weakness of this polymer [13, p. 272]. A comparison of expanded and full density PTFE is shown in Table 6.14 [16].

This technology is based on the inventions of W.L. Gore Associates and enables the production of: lightweight membranes used for waterproof breathable fabrics, widely known as Gore-Tex®; ; filtration membranes; fuel cells; high-tensile strength fabrics; and other specialty products [13, p. 272], [15, 17, 18]. Examples of applications of components and parts from expanded PTFE are shown in Table 6.15.

6.2.7.1 The Expansion Process

The basic principles of the process are explained in this section. The technology of fabricating the entire large number of varied products is very complex and a detailed description of all aspects is beyond the scope of this book. The best known current source for further study of this subject is [17, Chapters 5 and 6].

TABLE 6.14

Comparison of Typical Properties of Expanded and Full Density PTFE

Property	Unit	Full Density	Expanded
Specific gravity	g/cm^3	0.1	0.1–1.0
Crystallinity	%	50–70	95
Porosity	%	<0.1	25–96
Matrix tensile strength	MPa	20–30	50–800
Flexure fatigue resistance	Cycles to failure	1x10^6	3x10^7
Maximum service temperature	°C	260	280
Thermal conductivity	kcal/m h °C	0.2	<0.1
Thermal expansion coefficient	Per °C	3×10^{-4}	1×10^{-4}
Resistance to flow	As creep	Poor	Excellent
Chemical resistance	—	Excellent	Excellent

Source: Data from [16].

TABLE 6.15

Examples of Components and Parts Made from ePTFE

ePTFE Form	Components Containing ePTFE
Membranes/films	Laminates; filtration laminates; stent grafts; fuel cell membranes; electrode assemblies; dielectric materials; battery/capacitor separators
Sheets	Electromagnetic interference gaskets; sealing gaskets; medical patches
Tapes	Electrical and electronic wire and cable
Fibers	Weaving/sewing threads; dental floss; packaging; filtration felts
Tubes	Peristaltic pump hoses; vascular grafts; environmental screening modules

Source: www.gore.com/about/technologies, May 2022.

The process begins with the paste extrusion of PTFE fine powder using typical lubricants and the subsequent removal of the lubricant, as discussed in Section 6.2.3. The lubricant is completely removed in a similar way as for unsintered tape. The lubricant-free extrudate, which can be in the shape of a rod, tube, or tape, is fed into the expansion process. The expansion can be done by uniaxial or biaxial stretching.

Uniaxial expansion (orientation) is a process in which a dried extruded film is stretched (oriented) in the machine direction (MD) using slow and fast rolls. The effects of stretching can be seen on images from a scanning electron microscope (SEM). These images show that the matrix is a complex interconnection of a large number of nodes by way of fibrils. The long axis of the nodes is oriented perpendicularly to the direction of the stretch. A schematic of a uniaxially oriented membrane is shown in Figure 6.17.

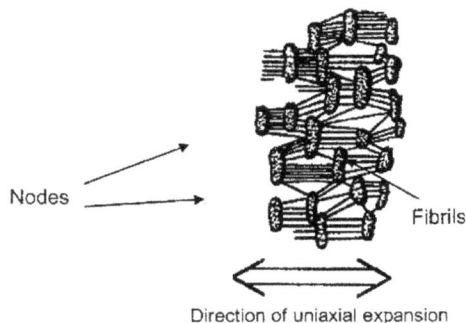

FIGURE 6.17 Schematic of a uniaxially oriented PTFE [15].

TABLE 6.16

Characteristics of the Machine Direction (MD) and Transverse
Direction (TD) Stretching Machinery

Specification	Value
MD	
Stretch ratio	≤20:1
Line speed, m/min	≤400
Width, m	≤4
Temperature capability, °C	≤400
TD	
Stretch ratio	≤30:1
Line speed, m/min	≤450
Initial width, m	0.1
Exit width, m	10

Source: Courtesy of Parkinson Technologies.

Biaxial expansion (orientation) is a process of stretching a membrane that has already been stretched in the MD (as described earlier) in the transverse or *cross direction* (CD). In other words, an extruded (and calendered) PTFE is first oriented in the MD uniaxially without being sintered. It is then oriented in the CD either sequentially (continuously) or wound up and oriented in the CD in a separate step, under heat. The commercial equipment used for the transverse orientation of thermoplastic films is a *tenter frame* [17, p. 131]. This consists of two horizontal chain tracks. A clip and chain assembly rides on these tracks, which can move at the required speed while gradually stretching the film. The tracks are enclosed in a heat oven. The amount of stretching is adjustable. The characteristics of a typical stretching machinery are listed in Table 6.16. The temperatures, speeds, and individual steps used for specific ePTFE products are outlined thoroughly in [17, Chapters 5 and 6]. A SEM image of the biaxially expanded membrane is shown in Figure 6.18.

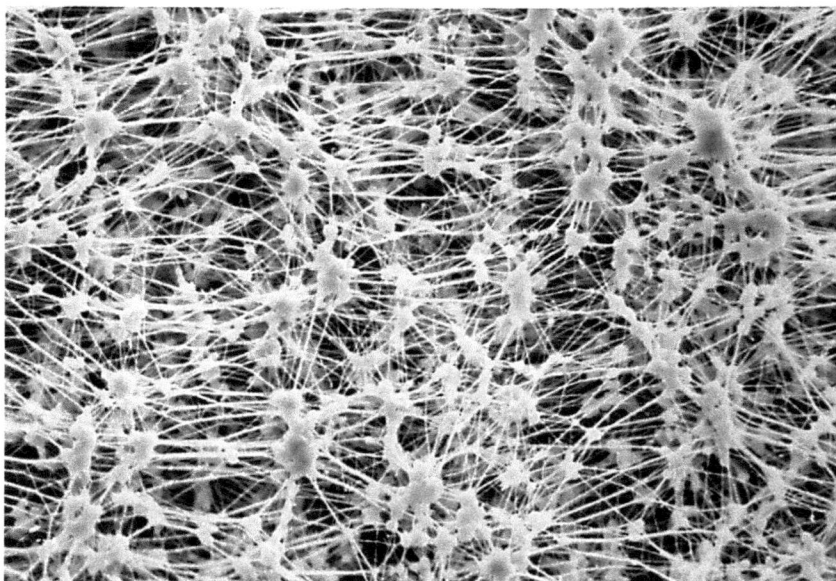

FIGURE 6.18 Scanning electron micrograph of a biaxially expanded PTFE membrane (5000x). (Courtesy of W. L. Gore & Associates)

REFERENCES

1. *Dyneon™ Tetrafluoroethylene, Product Comparison Guide*, Document 98-0505-1448-5, Dyneon (2003) www.dyneon.com
2. *Daikin Polyflon™ PTFE Products-Product Information Guide for Aqueous Dispersions, Fine Powders, Granular Molding Powders*, Document PB-PTFE-0001 R0 (2016).
3. *Algoflon® Granulars, Technical Data Sheets*, Solvay Specialty Polymers, Rev. (2021) www.solvay.com.
4. *Compression Moulding, Technical Information*, Publication H-59487, 05/95, E. I. du Pont de Nemours & Co, Wilmington, DE, p. 3 (1995).
5. Carlson, D. P., Schmiegel, N. *Ulmann's Encyclopedia of Industrial Chemistry* (Gerhartz, W., Ed.), Vol. A 11, VCH Publishers, Weinheim, Germany, p. 400, (1988).
6. *The Extrusion of PTFE Granular Powders*, Technical Service Note F2, 6th ed., Imperial Chemical Industries PLC, Blackpool, Lancs, U.K., p. 3, (1989).
7. *Fluon® Extrusion of PTFE Granular Powders, Technical Note F2*, AGCFP, September 2002 www.agcchem.com
8. Blanchet, T. A. *Handbook of Thermoplastics* (Olabisi, O., Ed.), Marcel Dekker, New York, p. 990, (1997).
9. Ebnesajjad, S. *Introduction to Fluoropolymers, Second Edition*, Elsevier, Oxford, UK, p. 125, (2021).
10. *Processing of Dyneon PTFE Fine Powder*, Dyneon GmbH, PTFEFP201503EN (2015), www.dyneon.eu
11. *Processing Guide for Fine Powder Resins*, Publication H-21211-2, E.I. du Pont de Nemours & Co., Inc., Wilmington, DE, p. 4. (1994).
12. *Algoflon® DF 120F, 130F, 261F, 330F and 681F, Technical Data Sheets*, Solvay Specialty Polymers (2022) www.solvay.com
13. Ebnesajjad, S. *Fluoroplastics, Vol. 1, Non-Melt Processible Fluoropolymers, Second Edition*, Elsevier, Oxford, UK, p. 245 (2015).
14. *The Processing of PTFE Coagulated Dispersion Powders*, Technical Service Note F3/4/5, 4th ed., ICI Fluoropolymers, Imperial Chemical Industries PLC, Blackpool, Lancs, U.K., p. 18, (1992).
15. Gore, R. W., US Patent 3,962,153 (June 8, 1976) to W. L. Gore & Associates.
16. Norman, E. G., Paper 6 at UMIST Conference Expanded PTFE Properties and Applications, RAPRA Technology, (1992).
17. Ebnesajjad, S. *Expanded PTFE Application Handbook: Technology, Manufacturing and Applications*, p. 99, Elsevier, Oxford, UK, (2017).
18. Hishinuma, Y., Chikahisa, T., Yoshikawa, H., in *Proceedings of Fuel Cell Seminar*, November 16–19, 1998, Palm Spring, CA (1998).

7

Fabrication of Melt-Processible Fluoropolymers

Jiri G. Drobny

The melt processability of this type of fluoropolymers allows for a wide variety of conventional melt-processing techniques to be used their fabrication. These techniques include:

- Injection molding.
- Transfer molding.
- Extrusion.
- Blow molding.
- Rotomolding/rotolining.
- Thermoforming.

Injection molding involves the slow injection of melted resin from an injection barrel into mold cavities. The molding equipment must be constructed of corrosion resistant metal in order to minimize corrosion and contamination for most of the fluoropolymer resins. More details will be provided for specific fluoropolymers.

Transfer molding is a technique using a mold with a valve or fitting. The preformed melt flows from a transfer pot, usually above the mold cavity, to the parts below the trough sprues. The transfer ram and pot can either be part of the mold or the press. Here also, the equipment must be made from corrosion resistant metal.

Extrusion is a processing method in which the resin is melted in the barrel of the extruder, pushed through a heated die of specific dimensions, and then quenched. Standard extrusion technology may be used to produce tubes, profiles, sheets, and films. More advanced versions of extrusion allow the fabrication of multilayer products. Special corrosion resistant metal alloys must be used for the equipment to prevent metal corrosion and contamination and equipment damage when processing most fluoropolymer resins.

Blow molding is the method used to produce hollow plastic products. While there are considerable differences in the processes available, all have in common the production of a parison (a tube-like plastic shape), the insertion of the parison into a closed mold, and the injection of air into the parison to blow it out against the sides of the mold, where it sets into the finished product. The parison is sealed at one end to allow the expansion. In most cases the parison is produced by either extrusion or injection molding. More advanced techniques allow the production of multilayer parisons. As mentioned earlier, parts of the equipment must be made from corrosion materials for most of fluoroplastics.

Rotomolding/rotolining is particularly useful for making free-standing, hollow parts and the seamless linings of metal vessels. It is a slow, nonshearing, processing technique in which the resin (typically in small bead or powder form) is heated above its melting point and then simultaneously rotated on two perpendicular axes inside the mold.

Thermoforming is a process of heating a thermoplastic sheet to its softening point. The sheet is stretched across a single-sided mold and then manipulated. Then, it cools into the desired shape. The

DOI: 10.1201/9781003204275-9

most common methods to get the sheet to conform to its final shape are vacuum forming, pressure forming, and mechanical forming.

7.1 Melt-Processible Perfluoroplastics

The need for highly fluorinated thermoplastic polymers that, unlike polytetrafluoroethylene (PTFE), could be fabricated by conventional melt-processing methods led to the development of a group of resins that are copolymers of tetrafluoroethylene (TFE) with other perfluorinated monomers. Commercially, the copolymer of TFE and hexafluoropropylene (HFP) is commonly known as fluorinated ethylene propylene (FEP). Copolymerization of TFE with perfluoropropylvinyl ether (PPVE) leads to perfluoroalkoxy (PFA) resins, and copolymerization of TFE with perfluoromethylvinyl ether (PMVE) produces methylfluoroalkoxy (MFA) resins (see Chapter 4).

7.1.1 Copolymers of Tetrafluoroethylene and Hexafluoropropylene: Fluorinated Ethylene Propylene

FEP resins are available at low melt viscosity, extrusion grade, intermediate viscosity, high melt viscosity, and as aqueous dispersions [1]. They can be processed by techniques commonly used for thermoplastics, such as extrusion, injection molding, rotational molding, dipping, slush molding, and powder and fluidized bed coating [2], and they can be expanded into foams [3]. Compression and transfer molding of FEP resins can be done, but with some difficulty. Extrusion of FEP is used for primary insulation or cable jackets and for tubing and films.

Processing temperatures used for FEP resins are usually up to 427°C (800°F), at which temperatures highly corrosive products are generated. Therefore, the parts of the processing equipment that are in contact with the melt must be made of special corrosion-resistant alloys such as high-nickel alloys (e.g., Inconel® 625, Haynes® 242, Hastelloy® C, and Reiloy®) to assure a trouble-free operation.

7.1.2 Copolymers of Tetrafluoroethylene and Perfluoroalkyl Ethers (PFA and MFA)

PFA can be processed by the standard techniques used for thermoplastics, such as extrusion, injection molding, and transfer molding at temperatures up to 425°C (797°F). High processing temperatures are required because PFA has a high melt viscosity with an activation energy lower than most thermoplastics, at 50 kJ/mole [4]. *Extrusion and injection molding* are done at temperatures typically above 390°C (734°F) and at relatively high shear rates. For these processing methods PFA grades with high melt flow indexes (MFIs) (i.e., with lower molecular weights) are used. Although PFA is thermally a very stable polymer, it is still subject to thermal degradation at processing temperatures, the extent of which depends on temperature, residence time, and the shear rate. Thermal degradation occurs mainly in the end groups; chain scission becomes evident at temperatures above 400°C (752°F) depending on the shear rate. Thermal degradation usually causes discoloration and bubbles [4]. PFA can be extruded into films, tubing, rods, and foams [3]. As in the case of FEP the components of the extrusion equipment must be made of special corrosion-resistant alloys such as high-nickel alloys (e.g., Inconel® 625, Haynes® 242, Hastelloy® C, and Reiloy®).

Transfer molding of PFA is done at temperatures in the range 350 to 380°C (662 to 716°F) and at lower shear rates. At these conditions chain scission does not occur. The gaseous products evolving from the thermal degradation of the end groups are practically completely dissolved in the melt since the molded parts are cooled under pressure. For transfer molding, PFA resins with lower MFIs (i.e., higher molecular weights) are preferred [4]. Because, at the high processing temperatures, large amounts of highly corrosive products are generated, the components of the equipment have to be made from corrosion-resistant alloys to assure a trouble-free operation, as is the case with extrusion. PFA can also be processed as an aqueous dispersion (see Chapter 12).

7.2 Processing of Other Melt-Processible Fluoroplastics

7.2.1 Copolymers of Ethylene and Tetrafluoroethylene (ETFE)

ETFE copolymers can be readily fabricated by a variety of melt-processing techniques [5]. They have a wide processing window, in the range 280 to 340°C (536 to 644°F), and can be *extruded* into films, tubing, and rods, or as thin coating on wire and cables. *Injection molding* of ETFE into thin sections is considerably easier than injection molding of melt-processible perfluoropolymers because the former has a critical shear rate at least two orders of magnitude greater than perfluoropolymers. When molding thick sections (thickness greater than 5 mm, or approximately 0.2 in.), it is important to consider that melt shrinkage occurs during freezing, which can be as great as 6% [6].

Coatings can be prepared by hot flocking, in which the heated part is dipped into a fluidized bed of ETFE powder and then removed to cool. ETFE coatings can also be applied by other powder coating methods (e.g., electrostatically) or by the spraying of water- or solvent-based suspensions, followed by drying and baking [6].

Welding of ETFE parts can be done easily by spin welding, ultrasonic welding, and conventional butt-welding using a flame and ETFE rod. The resins bond readily to untreated metals, but chemical etch corona and flame treatment can be used to increase adhesion further [7].

ETFE resins are very often *compounded* with varied ingredients (e.g., fiberglass, bronze powder) or modified during their processing. The most significant modification is *cross-linking* by peroxides or ionizing radiation. The cross-linking results in improved mechanical properties, higher upper-use temperatures, and a better cut-through resistance without significant sacrifice of electrical properties or chemical resistance [6]. The addition of fillers improves creep resistance, improves friction and wear properties, and increases the softening temperature [6, p. 304]. ETFE can also be processed as an aqueous dispersion; however, at the time of writing no ETFE dispersions are commercially available, as they were discontinued.

7.2.2 Polyvinylidene Fluoride (PVDF)

PVDF resins for melt processing are supplied as powders or pellets with a rather wide range of melt viscosities. Lower viscosity grades are used for *injection molding* of complex parts, while the low viscosity grades have a high enough melt strength for the *extrusion* of profiles, rods, tubing, pipes, film, wire insulation, and monofilaments. PVDF extrudes very well, and there is no need to use lubricants or heat stabilizers [8]. The equipment required for the melt processing of PVDF is the same as that for PVC or polyolefins, as during normal processing of PVDF no corrosive products are formed. Extrusion temperatures vary between 230 and 290°C (446 and 554°F), depending on the equipment and the profile being extruded. Water quenching is used for wire insulation, tubing, and pipes, whereas sheet and cast film from slit dies are cooled on polished steel rolls kept at temperatures between 65 and 140°C (149 and 284°F). A monofilament is extrusion-spun into a water bath and then oriented and heat-set at elevated temperatures [8, p. 541]. PVDF films can be *monoaxially and biaxially oriented*.

PVDF resins can be *molded by compression, transfer, and injection molding* in conventional molding equipment. The mold shrinkage can be as high as 3% due to the semicrystalline nature of the polymer. Molded parts often require annealing at temperatures between 135 and 150°C (275 and 302°F) to increase dimensional stability and to release internal stresses [8, p. 541].

Parts from PVDF can be machined, sawed, coined, metallized, and fusion bonded more easily than most other thermoplastics. Fusion bonding usually yields a weld line that is as strong as the part. The adhesive bonding of PVDF parts can be done, and epoxy resins produce good bonds [8, p. 541]. Because of a high dielectric constant and loss factor, PVDF can be readily melted by radiofrequency and dielectric heating. This is the basis for some fabrication and joining techniques [9].

PVDF can be *coextruded and laminated*, but the processes have their technical challenges in matching the coefficients of thermal expansion, melt viscosities, and layer adhesion. Special tie-layers, often from blends of polymers compatible with PVDF, are used to achieve bonding [10, 11].

7.2.3 Polychlorotrifluoroethylene (PCTFE)

PCTFE can be processed by most of the techniques used for thermoplastics. Processing temperatures can be as high as 350°C (662°F) for *injection molding* with melt temperatures leaving the nozzle in the range of 280 to 305°C (536 to 579°F). In *compression molding*, process temperatures up to 315°C (599°F) and pressures up to 69 MPa (10,000 psi) are required. Since relatively high molecular weight resins are required for adequate mechanical properties, the melt viscosities are somewhat higher than those usual in the processing of thermoplastics. The reason for this is a borderline thermal stability of the melt, which does not tolerate sufficiently high processing temperatures [13].

7.2.4 Copolymers of Ethylene and Chlorotrifluoroethylene (ECTFE)

The most common form of ECTFE is hot-cut pellets, which can be used in all melt-processing techniques, such as *extrusion, injection molding, blow molding, compression molding, and fiber spinning* [12]. ECTFE is corrosive in melt form; the surfaces of machinery that come in contact with the polymer must be lined with a highly corrosion-resistant alloy (e.g., Hastelloy C-276). Recently developed grades with improved thermal stability and acid scavenging have been processed on conventional equipment [13, p. 529].

Electrostatic powder coating using fine ECTFE powders and rotomolding and rotolining using very fine pellets are other processing methods. Formulated primers are used to improve adhesion and moisture permeability for powder-coated metal substrates. For *rotolining*, primers are usually not used [13, p. 530].

7.2.5 Terpolymers of Tetrafluoroethylene, Hexafluoropropylene, and Vinylidene Fluoride (THV Fluoroplastics)

THV Fluoroplastics can be processed by virtually any method used generally for thermoplastics, including *extrusion, coextrusion, tandem extrusion, blown film extrusion, blow molding, injection molding, and vacuum forming*, as well as *skived film and solvent casting* (only the grade THV Fluoroplastic 220 can be processed by solvent casting).

Generally, processing temperatures for THV Fluoroplastics are comparable to those used for most thermoplastics. In extrusion, melt temperatures at the die are in the 230 to 250°C (446 to 482°F) range. These relatively low processing temperatures open new options for combinations of different melts (coextrusion, cross-head extrusion, co-blow molding) with thermoplastics as well as with various elastomers [14]. Another advantage of low processing temperatures is that they are generally below the decomposition temperature of the polymer; thus, there is no need to protect equipment against corrosion. Yet, as with any fluoropolymer, it is necessary to prevent long residence times in equipment and to purge the equipment after the process is finished. Also, appropriate ventilation is necessary. THV Fluoroplastics have been found to be suitable for *coextrusion* with a variety of materials into multilayer structures [15].

THV Fluoroplastics can be readily processed by *blow molding* alone or with polyolefins. The olefin layer provides a structural integrity while the THV Fluoroplastic provides chemical resistance and considerably reduced permeation [14].

In *injection molding*, THV Fluoroplastics are processed at lower temperatures than other fluoropolymers, typically from 200 to 300°C (392 to 572°F) with mold temperatures being 60 to 100°C (140 to 212°F). Generally, standard injection molding equipment is used [14, p. 263].

A THV Fluoroplastic material can be readily bonded to itself and to other plastics and elastomers. It does not require surface treatment, such as chemical etch or corona treatment, to attain good adhesion to other polymers, although in some cases tie-layers are necessary. For bonding THV Fluoroplastics to elastomers, an adhesion promoter is often compounded into the elastomer substrate [14, p. 260].

7.2.6 Terpolymer of Ethylene, Tetrafluoroethylene, and Hexafluoropropylene: (EFEP)

EFEP, a terpolymer of ethylene, TFE, and HFP, has many of the mechanical properties of ETFE, but with a lower processing temperature, allowing it to be coextruded with conventional thermoplastic polymers,

TABLE 7.1

Processing Temperatures for EFEP

	Neoflon™ EFEP Grade	
Temperature	**RP 4020**	**5000**
Melting temperature, °C (°F)	160 (320)	195 (383)
Molding temperature, °C (°F)	200–240 (392–464)	220–260 (428–500)
Die temperature, °C (°F)	30–80 (86–176)	30–100 (86–212)

Source: [16].

such as polyamides, ethylene-vinyl alcohol copolymer (EVOH), ETFE, modified polyethylene, and other thermoplastics, without the use of adhesives or tie layers. EFEP also adheres to glass and metals and has a higher degree of transparency than most fluoropolymers [16, 17].

EFEP can be processed by *extrusion, compression molding, overmolding, transfer molding, injection molding, blow molding, and lamination* [16]. Typical processing temperatures are shown in Table 7.1.

REFERENCES

1. Gangal, S. V. *Encyclopedia of Polymer Science and Technology*, Vol. *16* (Mark, H. F. and Kroschwitz, J. I., Eds.), John Wiley & Sons, New York, p. 603, (1989).
2. Sperati, C. A. *Handbook of Plastics Materials and Technology* (Rubin, I. I., Ed.), John Wiley & Sons, New York, p. 96, (1990).
3. Gupta, C. V. *Polymeric Foams* (Frisch, K. C. and Klempner, D, Eds.), Chapter 15, Hanser Publishers, Munich, p. 349, (1991).
4. Hintzer, K., Lohr, G. in *Modern Fluoropolymers* (Scheirs, J., Ed.), John Wiley & Sons, New York, p. 234, (1997).
5. *Extrusion Guide for Melt Processible Fluoropolymers*, Bulletin E-85783, E. I. du Pont de Nemours & Co., Wilmington, DE.
6. Kerbow, D. L. *Modern Fluoropolymers* (Scheirs, J., Ed.), John Wiley & Sons, Ltd, Chichester, UK, p. 307, (1997).
7. TEFZEL® *Fluoropolymer Design Handbook*, E-31302-1, 7/90, E. I. Du Pont de Nemours & Co., Wilmington, DE, (1970).
8. Dohany, J. E., Humphrey, J. S. *Encyclopedia of Polymer Science and Engineering*, Vol. *17* (Mark, H. F. and Kroschwitz, J. I., Eds.), John Wiley & Sons, New York, p. 540, (1989).
9. Nakagawa, K., Amano, M., Microwave Heat-drawing of Poly(vinylidene fluoride), *Polym. Comm. 27*, p. 310, 1986.
10. Strassel, A., U.S. Patent 4,317,860 (March 2, 1982) to Produits Chimiques Ugine Kuhlmann.
11. Kitigawa, Y., Nishioka, A., Higuchi, Y., Tsutsumi, T., Yamaguchi, T., Kato, T., U.S. Patent 4,563,393 (1986) to Japan Synthetic Rubber Co., Ltd.
12. Sperati, C. A. *Handbook of Plastic Materials and Technology* (Rubin, I. I., Ed.), John Wiley & Sons, New York, p. 103, (1990).
13. Stanitis, G. *Modern Fluoropolymers* (Scheirs, J., Ed.), John Wiley & Sons, Ltd, Chichester, UK, p. 528, (1997).
14. Hull, D. E., Johnson, B. V., Rodricks, I. P., Staley, J. B. *Modern Fluoropolymers* (Scheirs, J., Ed.), John Wiley & Sons, Ltd, Chichester, UK, p. 262, (1997).
15. Lavallee, C., The 2nd International Fluoropolymers Symposium, SPI, Expanding Fluoropolymer Processing Options (1995).
16. *Adhesive Neoflon™ EFEP*, Brochure May. 2003 EG-37/0005/AI www.daikinchem.de
17. *Daikin America, Product Brochure*, daikin-america.com, (2022).

8

Applications of Commercial Thermoplastic Fluoropolymers

Jiri G. Drobny

Because of their unique properties, fluoroplastics are used in a great variety of applications (see Table 8.1). Details regarding the uses of these materials are covered in this chapter. The applications of polytetrafluoroethylene (PTFE) are summarized in Section 8.1 and those of the remaining fluoroplastics in Sections 8.2 to 8.10.

8.1 Applications of PTFE and Modified PTFE

A large proportion (about one half of the PTFE resin produced) is used in electrical and electronic applications with a major use for the insulation of hookup wire for military and aerospace electronic equipment. PTFE is also used as insulation for airframe and computer wires, as "spaghetti" tubing, and in electronic components. PTFE tape is used for wrapping coaxial cables. An example of an application of PTFE wrap tape is shown in Figure 8.1.

Other large quantities of PTFE are used in the chemical industry in fluid-conveying systems as gaskets, seals, molded packing, bellows, hoses, and lined pipes [1], as well as the lining of large tanks or process vessels. PTFE is also used in laboratory apparatus. Compression-molded parts are made in many sizes and shapes (Figure 8.2). Unsintered tape is used for the sealing threads of pipes for water and other liquids. Pressure-sensitive tapes with silicone or acrylic adhesives are made from skived or cast PTFE films and PTFE-coated fiberglass fabrics.

Because of its very low friction coefficient, PTFE is used for bearings, ball-and roller-bearing components, and sliding bearing pads in static and dynamic load supports [1]. Piston rings of filled PTFE in nonlubricated compressors permit operation at lower power consumption or at increased capacities [2].

Modified PTFE (e.g., Teflon NXT) – because of its improved processing, lower creep, improved permeation, reduced porosity, and better insulation than standard PTFE – finds use in pipe and vessel linings, gaskets and seals, fluid-handling components, wafer processing, and the electric and electronic industries. An example of a molded part from modified PTFE is shown in Figure 8.3.

Since PTFE is highly inert and nontoxic, it finds use in medical applications such as cardiovascular grafts, heart patches, and ligaments for knees [2].

Highly porous membranes and other expanded articles are prepared by a process based on the fibrillation of high-molecular-weight PTFE [3]. More details regarding expanded PTFE (ePTFE) are provided in Section 6.2.7. Since the porous ePTFE membranes have a high permeability to water vapor and none for liquid water, they are combined with fabrics and used for breathable waterproof garments and camping gear (made by W. L. Gore & Associates and sold under the brand name Gore-tex®). Other uses for these membranes are for special filters and analytical instruments and in fuel cells [4], membrane electrode assemblies, dielectric materials, and battery/capacitor separators.

Other expanded PTFE articles are suitable for a variety of applications:

- Expanded tubes: peristaltic pump tubes, vascular grafts, environmental screening modules.
- Expanded fibers: weaving/sewing threads, dental floss, packing, filtration felts.

DOI: 10.1201/9781003204275-10

TABLE 8.1

Typical Applications for Fluoroplastics

Industry	Applications
Aerospace	Wiring, special coatings, seals, flexible hose and tubing, foams in aircraft insulation, particularly for ducts and air conditioning, electroluminescent lamps
Architectural	Roofing materials, architectural fabric, protective and decorative coatings
Automotive	O-rings, gaskets, shaft seals, head gaskets, fuel hose linings, flexible hoses, valve stem seals, bearings and bushings, coatings, protective and decorative films, tubing, membranes in fuel cells
Chemical processing	Chemically resistant coatings and linings, pumps, pipe linings, impellers, tanks, heat exchangers, reaction vessels, autoclaves, valves and valve parts, flue duct expansion joints, solid pipes and fittings, bearings and bushings, sight glasses, flow meter tubes, flexible hoses, filtration membranes, and textiles
Domestic	Nonstick coatings for cookware, nonstick utensils, cooking sheets for cookies
Electrical/ electronics	Electrical insulation, wires and cables, flexible printed circuits, ultrapure components for semiconductor manufacture, condensers, batteries, barrier films for packaging of sensitive electronic parts, films for photovoltaic modules
Engineering	Bearings, bushings, gears, nonstick surfaces, low friction surfaces, pipes and pipe coatings, fittings, valves and valve parts, seals and sealants, foams, conveyor belts for the manufacture of ceramic tiles, outdoor signs
Food industry	Release sheets for fast food, conveyor belts for cooking and drying cookies and chips
Medical devices	Catheters, probes, cardiovascular grafts, heart patches, ligaments for knees, sutures, blood filters, tubings, dental floss, barrier packaging films for drugs
Sporting goods	High-performance breathable fabrics (e.g., Gore-tex), waterproofing, and ski waxes

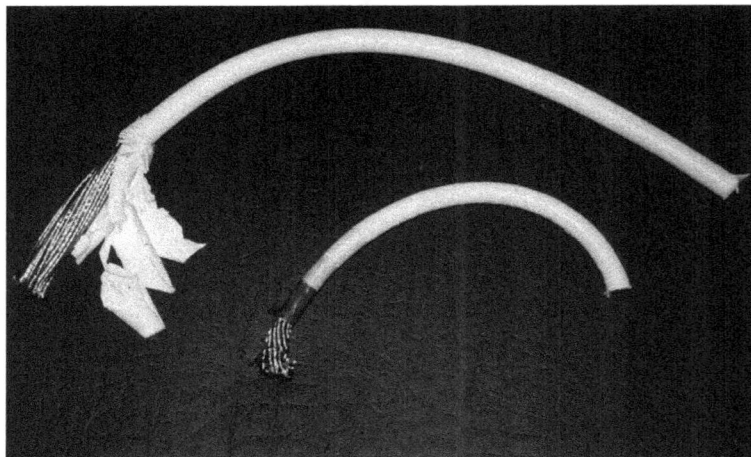

FIGURE 8.1   Cables with PTFE wrap. (Courtesy of DeWAL Industries)

- Densified ePTFE: barrier materials, gaskets.
- Densified sheets: sealing gaskets, medical patches, electromagnetic interference gaskets.
- Rods/tapes: gasket strips.

Because of its low surface energy and limited chemical reactivity, PTFE exhibits poor wettability and adhesive bonding. Surface modification of PTFE is an established and valuable technique to adjust the surface properties to improve wetting and adhesive bonding. Although many different methods for that have been developed, such as radiation grafting [5], plasma treatment [6], roughness [7], and impregnation with metal oxides [8], the well-established commercial method is the treatment with a sodium-naphthalene complex [9, 10].

FIGURE 8.2 PTFE molded parts. (Courtesy of Dalau Inc.)

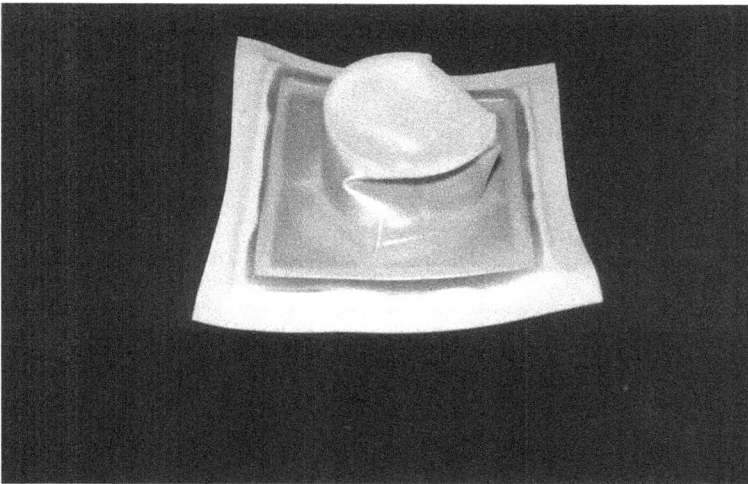

FIGURE 8.3 Molded part from modified PTFE. (Courtesy of DeWAL Industries)

Micropowders, PTFE homopolymers with a molecular weight significantly lower than normal PTFE, are commonly used as additives in a large number of applications, where they provide nonstick and sliding properties. They are added to plastics, inks, lubricants, and lacquers [11].

8.2 Applications of FEP

The largest proportion of the copolymer of tetrafluoroethylene and hexafluoropropylene (FEP) is used in electrical applications, such as hookup wire, interconnecting wire, thermocouple wire, computer wire, and molded parts for electrical and electronic components. Chemical applications include lined tanks,

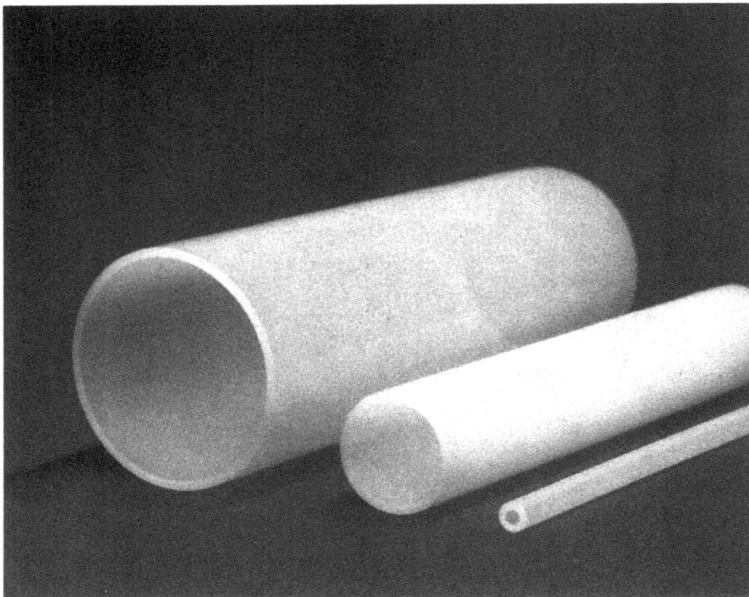

FIGURE 8.4   Extruded pipes from FEP. (Courtesy of DuPont)

lined pipes and fittings, heat exchangers, overbraided hoses, gaskets, component parts of valves, and laboratory ware [11, p. 611]. Mechanical uses include antistick applications such as conveyor belts and roll covers. FEP film is used in solar-collector windows because of its light weight, excellent weather resistance, high transparency, and easy installation [11, p. 611]. FEP film is also used for the heat-sealing of PTFE-coated fabrics (e.g., architectural fabric). An example of extruded pipes is shown in Figure 8.4.

8.3 Applications of PFA and MFA

Perfluoroalkoxy resins (PFA and MFA) are fabricated into high-temperature electrical insulation and into components and parts requiring a long flex life [11, p. 625]. Certain grades are used in the chemical industry for process equipment, liners, specialty tubing, and molded articles. Other uses are bellows and expansion joints, liners for valves, pipes, pumps, and fittings. Examples are shown in Figure 8.5. Extruded PFA films can be oriented and used as such for specialized applications [12]. PFA resins can be processed into injection-, blow-, and compression-molded components. High-purity grades are used in the semiconductor industry for demanding chemical applications [12]. Coated metal parts can be made by powder coating.

8.4 Applications of ETFE

Copolymer of ethylene and tetrafluoroethylene (ETFE) is used in electrical applications for heat-resistant insulation and the jackets of low-voltage power wiring for mass transport systems, for wiring in chemical plants, and for control and instrumentation wiring for utilities [11, p. 641]. Because ETFE exhibits an excellent cut-through and abrasion resistance, it is used in airframe wire and computer hookup wire. Electrical and electronic components, such as sockets, connectors, and switch components, are made by injection molding [13]. ETFE has excellent mechanical properties; therefore, it is used successfully in seal glands, pipe plugs, corrugated tubing, fasteners, and pump vanes [13]. Its radiation resistance is a reason for its use in nuclear industry wiring [14]. The lower density of ETFE provides an advantage over perfluoropolymers in aerospace wiring [14].

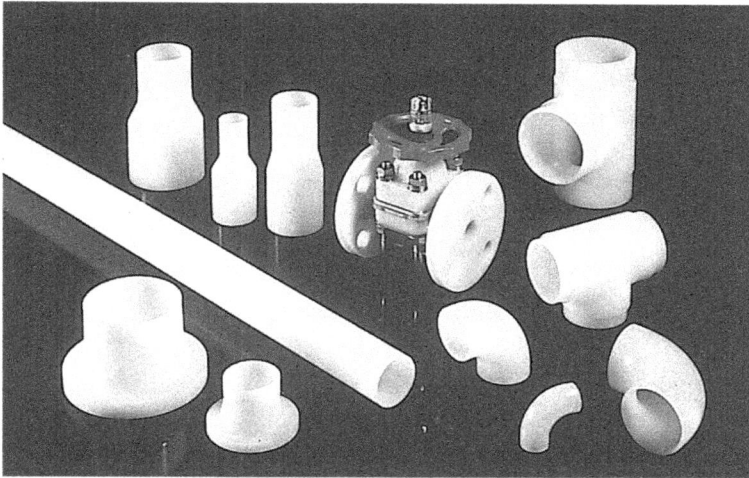

FIGURE 8.5 Variety of parts for chemical industry from MFA. (Courtesy of Solvay)

Because of its excellent chemical resistance, ETFE is used in the chemical industry for valve components, packings, pump impellers, laboratory ware, and battery and instrument components, as well as for oil well, down-hole cables [14]. Examples of such applications are shown in Figures 8.6 and 8.7.

Heat-resistant grades are used for insulation and jackets for heater cables and automotive wiring and for other heavy-wall applications where operating temperatures up to 200°C (392°F) are experienced for short periods of time or where repeated mechanical stress at 150°C (302°F) is encountered [11, p. 641]. Another use is wiring for high-rise building and skyscraper fire alarm systems. Thin ETFE films are used in greenhouse applications because of their good light transmission, toughness, and resistance to UV radiation [14, p. 308]. Biaxially oriented films have excellent physical properties and toughness equivalent to polyester films [15].

Injection-molded parts, such as electrical connectors and sockets, distillation column plates and packings, valve bodies, pipes, and fitting linings, are easily made because ETFE exhibits a low shear sensitivity and a wide processing window [14, p. 308].

ETFE can be extruded continuously into tubing, piping, and rod stock. An example of the application of extruded tubing is automotive tubing, which takes advantage of its chemical resistance, mechanical

FIGURE 8.6 Molded parts from ETFE. (Courtesy of DuPont)

FIGURE 8.7   ETFE lined part for the chemical process industry. (Courtesy of DuPont)

strength, and resistance to permeation by hydrocarbons. A high weld factor (more than 90%) is utilized in the butt welding of piping and the sheet lining of large vessels [14, p. 308].

ETFE resins in the powder and bead form are rotationally molded into varied structures, such as pump bodies, tanks, and fittings and linings, mostly for the chemical process industries. Inserts can be incorporated to provide attachment points or reinforcement [14]. Adhesion to steel, copper, and aluminum can be up to 3 kN/m (5.7 pli) peel force [16].

Carbon-filled ETFE resins (about 20% carbon) exhibit antistatic dissipation and are used in self-limiting heater cables and other antistatic or semiconductive applications [14, p. 308].

Certain grades of ETFE are used for extruded foams with void contents from 20 to 50%. The closed foam cells are 0.001 to 0.003 in. (0.02 to 0.08 mm) in diameter. Special grades of ETFE processed in a gas-injection foaming process may have void contents up to 70%. Foamed ETFE is used in electrical applications, mainly in cables, because it exhibits a lower apparent dielectric constant and dissipation factor and reduces cable weight.

8.5 Applications of PVDF

Polyvinylidene fluoride (PVDF) is widely used in the chemical industry in fluid-handling systems for solid and lined pipes, fittings, valves, pumps, tower packing, tank liners (Figure 8.8), and woven filter cloth. Because it is approved by the U.S. Food and Drug Administration for food contact, it can be used for fluid-handling equipment and filters in the food, pharmaceutical, and biochemical industries. It also meets high standards for purity, which is required in the manufacture of semiconductors, and therefore is used for fluid-handling systems in that industry [17]. Examples of high-purity water lines made from PVDF are shown in Figure 8.9. PVDF is also used for the manufacture of microporous and ultrafiltration membranes [18, 19].

In the electrical and electronic industries PVDF is used as a primary insulator on computer hookup wire. Irradiated (cross-linked) PVDF jackets are used for industrial control wiring [20] and self-limiting heat-tracing tapes used for controlling the temperature of process equipment as well as ordnance [21, 22]. Extruded and irradiated heat-shrinkable tubing is used to produce termination devices for aircraft and electronic equipment [23]. Because of its very high dielectric constant and dielectric loss factor, the use of PVDF insulation is limited to only low-frequency conductors. Under certain conditions PVDF films become piezoelectric and pyroelectric. The piezoelectric properties are utilized in soundproof telephone headsets, infrared sensing, respiration monitors, high-fidelity electric violins, hydrophones, keyboards, and printers [24].

FIGURE 8.8   Parts from PVDF. (Courtesy of Solvay)

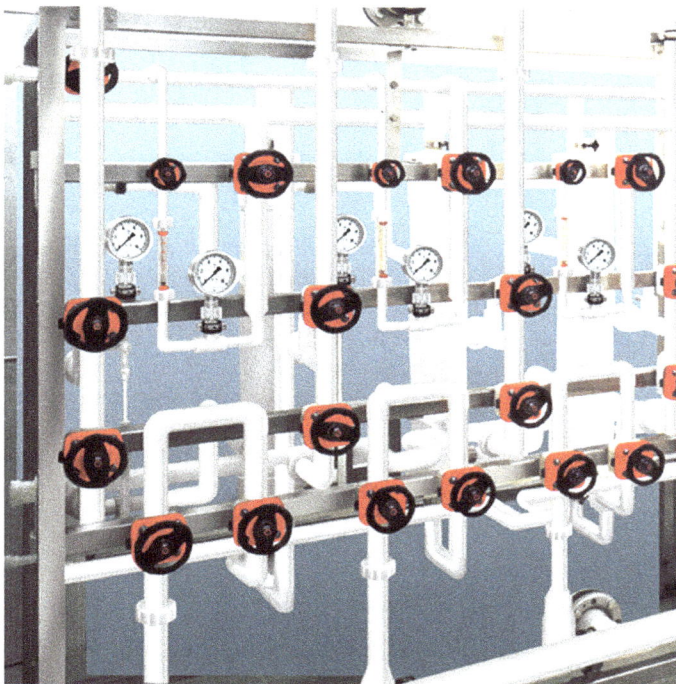

FIGURE 8.9   PVDF manifold. (Courtesy of Arkema)

8.6 Applications of PCTFE

A major application of polychlorotrifluoroethylene (PCTFE) is in specialty films for packaging in applications where there are high moisture barrier demands, such as pharmaceutical blister packaging and health-care markets. In electroluminescent (EL) lamps PCTFE film is used to encapsulate phosphor coatings, which provide an area of light when electrically excited. The film acts as a water vapor barrier,

FIGURE 8.10   Electroluminescent lamp with PCTFE film. (Courtesy of Honeywell)

protecting the moisture-sensitive phosphor chemicals. EL lamps are used in aircraft, military, aerospace, automotive, and business equipment applications, and buildings (Figure 8.10). Another use for PCTFE films is for the packaging of corrosion-sensitive military and electronic components. Because of their excellent electrical insulation properties, these films can be used to protect sensitive electronic components, which may be exposed to humid or harsh environments. They can be thermoformed to conform to any shape and detail. PCTFE films are also used to protect the moisture-sensitive liquid crystal display panels of portable computers [25].

PCTFE films can be laminated to a variety of substrates, such as PVC, polyethylene-terephtalate glycol (PETG), amorphous polyethylene terephthalate (APET), or polypropylene (PP). Metallized films are used for electronic dissipative and moisture barrier bags for sensitive electronic components (Figure 8.11), for the packaging of drugs (Figure 8.12), and for medical devices (Figure 8.13).

Other applications of PCTFE are in pump parts, transparent sight glasses, flow-meters, tubes, and linings in the chemical industry and for laboratory ware [26].

8.7 Applications of ECTFE

The single largest application of the copolymer of ethylene and chlorotrifluoroethylene (ECTFE) has been as a primary insulation and jacketing [27] for voice and copper cables used in building plenums [28]. In automotive applications, ECTFE is used for the jackets of cables inside fuel tanks for level sensors, for hookup wires, and in heating cables for car seats. Chemically foamed ECTFE is used in some cable constructions [28, p. 534]. In the chemical process industry, it is often used in chlorine/caustic environment in cell covers, outlet boxes, lined pipes (Figure 8.14), and tanks.

In the pulp and paper industries, pipes and scrubbers for bleaching agents are lined with ECTFE. Powder-coated tanks, ducts, and other components find use in the semiconductor and chemical process

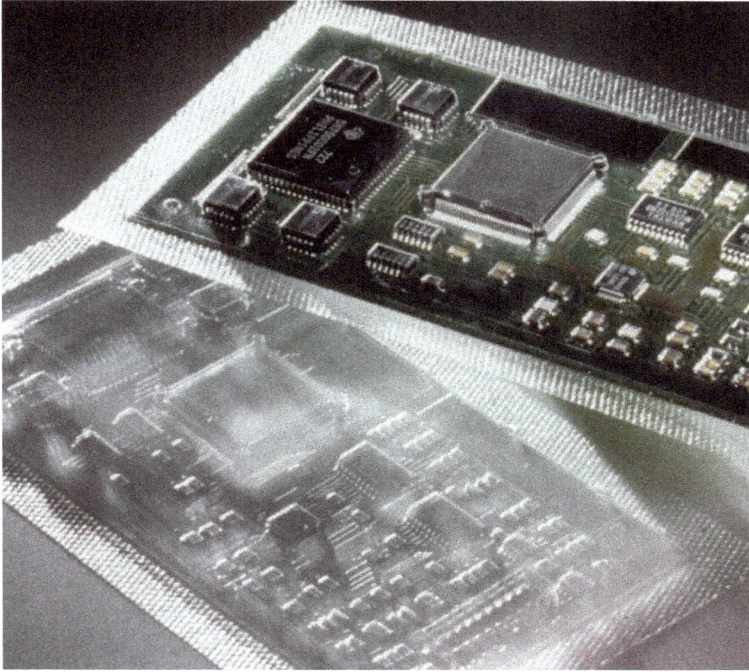

FIGURE 8.11 Packaging of electronic components with a barrier film of PCTFE. (Courtesy of Honeywell)

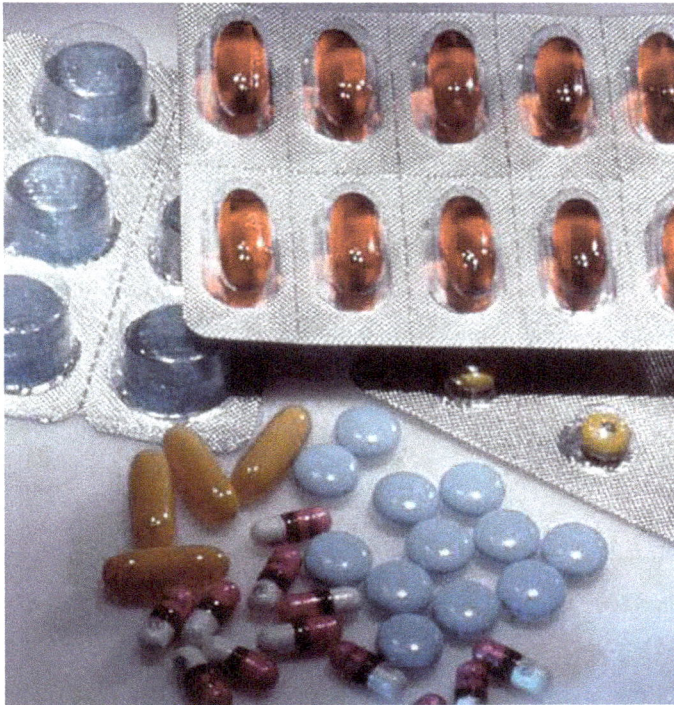

FIGURE 8.12 Packaging of drugs with a barrier film of PCTFE. (Courtesy of Honeywell)

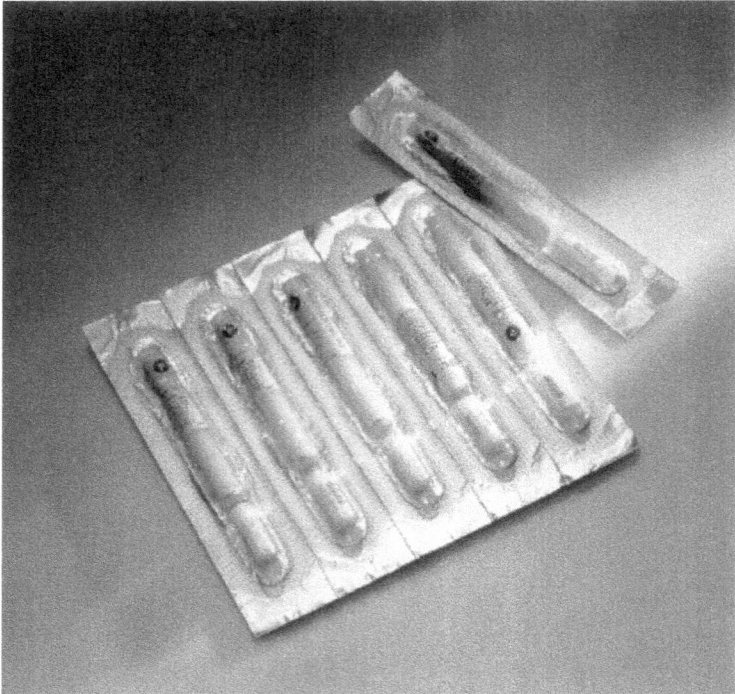

FIGURE 8.13 Packaging of medical devices with a barrier film of PCTFE. (Courtesy of Honeywell)

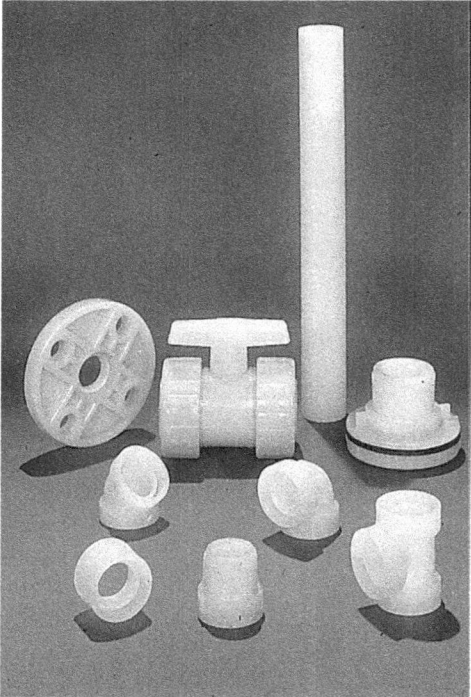

FIGURE 8.14 Parts using from PCTFE. (Courtesy of Solvay)

FIGURE 8.15   Reactor coated inside with an ECTFE layer applied by powder coating. (Courtesy of Fisher Co. and Moore)

industries (Figure 8.15). Monofilaments made from ECTFE are used for chemical-resistant filters and screens [12, p. 413].

Other applications include rotomolded tanks and containers for the storage of corrosive chemicals, such as nitric or hydrochloric acid. Extruded sheets can be thermoformed into various parts, such as battery cases for heart pacemakers [12, p. 413]. ECTFE film is used as a release sheet in the fabrication of high-temperature composites for aerospace applications. Braided cable jackets made from monofilament strands are used in military and commercial aircraft as a protective sleeve for cables [28, p. 538].

8.8 Applications of THV Fluoroplastics

Because the terpolymers of tetrafluoroethylene, hexafluoropropylene, and vinylidene fluoride (THV Fluoroplastics) are highly flexible, are resistant to chemicals and automotive fuels, and have good barrier properties, they are used as a permeation barrier in various types of flexible hoses in automotive applications and in the chemical process industry. Liners and tubing can be made from an electrostatic dissipative grade of THV, which is sometimes required for certain automotive applications [29].

THV is used for wire and cable jacketing, which is often cross-linked by electron beam to improve its strength and to increase its softening temperature. It is also used as primary insulation in less demanding applications, where high flexibility is required [29, p. 266].

The low refractive index of THV (typically 1.355) is utilized in light tubes and optical fiber communications applications where high flexibility is required. Its optical clarity and impact resistance make it suitable for laminated safety glass for vehicles and for windows and doors in psychiatric and correctional institutions. An additional advantage is that the film does not burn or support combustion, which may be a major concern in some applications [29, p. 267].

Other applications for THV are flexible liners (drop-in liners or bag liners), used in the chemical process and other industries, and blow-molded containers, where it enhances the resistance to permeation when combined with a less expensive plastic (e.g., high-density polyethylene, HDPE), which provides structural integrity [29, p. 366]. Optical clarity, excellent weatherability, and flexibility make THV suitable as a protection of solar cell surfaces in solar modules [29, p. 269].

8.9 Applications of EFEP

EFEP, terpolymer of ethylene, TFE, and hexafluoropropylene (HFP) have many of the mechanical properties of ETFE, but with a lower processing temperature, allowing it to be coextruded, overmolded, and laminated with conventional thermoplastics polymers and certain thermoplastic elastomers. Because of the nature of the polymer, this can be done without the use of adhesives or tie layers. EFEP also adheres to glass and metals and has: a higher degree of transparency than most fluoropolymers; low permeation; excellent resistance to heat aging, weathering, and chemicals; and stands up well to impacts at low temperatures [30, 31]. Because of these advantages, it can be used for numerous applications in the automotive, chemical process, electrical and electronic, and oil and gas industries, as well as in medical devices and construction. Examples include [32, 33]:

- Automotive fuel tubes.
- Transparent tubing in the chemical process industry.
- Multilayer tubing.
- Convoluted tubing.
- Wire insulations.
- Clear bottles (single-layer and multi-layer).
- Multilayer films and bags.
- Coextruded outer layers over coil- and braid-reinforced assemblies.
- Tubing for catheters.
- Dip tubes for perfume bottles.

8.10 Applications of PVF

As pointed out in Chapter 5, polyvinyl fluoride (PVF) is almost exclusively used as a film for lamination with a large variety of substrates. Its main function is as a protective and decorative coating. PVF films can be made transparent or pigmented to a variety of colors and can be laminated to hardboard, paper, flexible PVC, polystyrene, rubber, polyurethane, and other substrates [34]. These laminates are used for applications including wall coverings, aircraft cabin interiors, pipe covering, and duct liners. For covering metal and rigid PVC, the film is first laminated to flat, continuous metal or vinyl sheets using special adhesives; the laminate is then formed into the desired shapes. These laminates are used for the exterior sidings of industrial and residential buildings [34, p. 248]. Other applications are highway sound barriers [35], automobile truck and trailer sidings, vinyl awnings, and backlit signs. On metal or plastic, PVF surfaces serve as a primer coat for painting or adhesive joints [34, p. 248]. PVF films are used as a release sheet for the bag molding of composites from epoxide, polyester, and phenolic resins and in the manufacture of circuit boards [36]. Other uses of PVF films are in greenhouses, flat-plate solar collectors, and photovoltaic cells. Dispersions of PVF are used for coating the exterior of steel hydraulic brake tubing for corrosion protection [37]. More details about the common applications of PVF films are given in the sections below.

8.10.1 Aircraft Interiors

Typical application in aircraft interiors are: interior ceiling and sidewall decorative panels; window shades; stow bins; lavatories and galleys; ceiling panels; personal service units; bulkhead partitions;

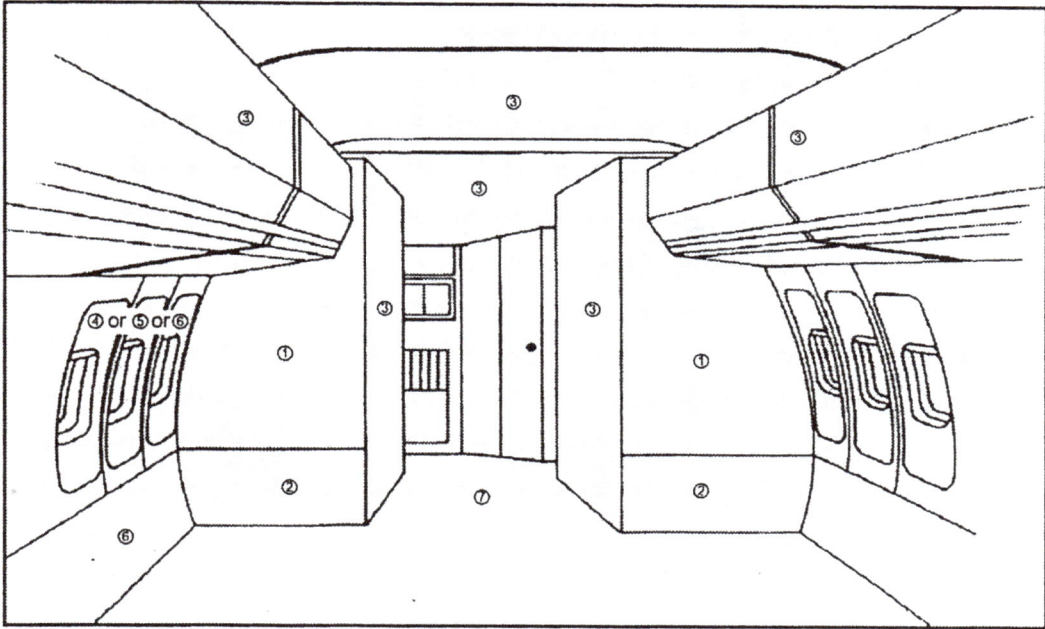

FIGURE 8.16 Schematic of an aircraft interior. (Courtesy of DPA) (1) Panel construction with tapestry cover; (2) panel construction with wainscot cove; (3) panel construction with decorated plastic laminate; (4) formed thermoplastic or laminate; (5) formed aluminum with decorative plastic laminate; (6) composite laminate with wainscot cover; (7) panel construction with carpet cover.

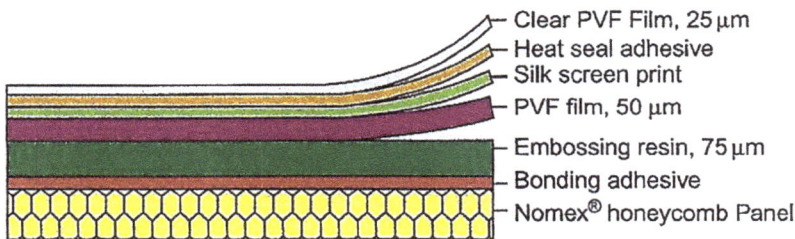

Clear PVF Film, 25 µm
Heat seal adhesive
Silk screen print
PVF film, 50 µm
Embossing resin, 75 µm
Bonding adhesive
Nomex® honeycomb Panel

FIGURE 8.17 Example of an aircraft interior laminate. (Courtesy of DPA)

insulation barriers; moisture barriers; cargo bin liners; aircraft wire markers; and composite noise panels (see Figure 8.16) [38]. An example of an aircraft interior laminate is shown in Figure 8.17 [38].

8.10.2 Architectural Applications

PVF films are easily laminated to an array of architectural substrates including metal, wood, cement, asphalt, vinyl, melamine composites, and fabrics. They provide superior protection that significantly prolongs their useful life and preserves their aesthetics for interior and exterior applications [38, Chapter 13]. They do not support bacterial growth and provide extra protection against contamination in hospitals, laboratories, clean rooms, and restaurants. The inertness of PVF films resists staining while their chemical stability lets the strongest cleaning agents be used without damaging the film or its substrate [34, p. 249].

Typical applications using PVF films in architectural applications include: wall coverings; ceiling and acoustic tiles; insulation jacketing; bagging film for thermal or acoustic materials; residential and

FIGURE 8.18   Example of solar panel with PVF film. (Courtesy of DPA)

commercial siding, trim, and accents; fiberglass-reinforced plastic (FRP) panels; formed or flat metal building panels; flexible laminates for air-inflated structures, canopies, awnings, and stadium domes; rigid composite fiberglass-reinforced polyester utility buildings, and skylights; and conformable building panels [34, p. 250].

8.10.3 Graphic Applications

PVF films provide a tough, cleanable surface with excellent resistance to weather and harsh chemicals. They are the best protection against graffiti and grime. Transparent versions shield UV radiation. Translucent and opaque pigmented films contain no plasticizers, and therefore resist fading, chalking, and color shifts for years [34, p. 250]. PVF films can be used in transfer, reverse, or direct-printed applications with many water-based, solvent-based, or ultraviolet-cured inks. For heavy-duty graphics protection, it is possible to print directly on white PVF film and cover it with a clear layer of PVF film [34, p. 250].

Graphics applications using PVF films include:

- Gas pump skirts.
- Flexible signs and awnings.
- Billboards.
- Highway signage.
- Labels.

8.10.4 Solar Applications

In solar applications PVF films are preferred as the backing sheet for photovoltaic modules due to their excellent strength, weather resistance, UV resistance, and moisture barrier properties.

Solar applications using PVF films include [34, p. 251]:

- Photovoltaic modules.
- Solar collectors.

An example of a solar panel containing PVF films is shown in Figure 8.18.

REFERENCES

1. Carlson, D. P., Schmiegel, W. in *Ullmann's Encyclopedia of Industrial Chemistry* (Gerhartz, W., Ed.), Vol. *A11*, VCH Publishers, Weinheim, Germany, p. 401, (1988).
2. Sperati, C. A. *Handbook of Plastic Materials and Technology* (Rubin, I. I., Ed.),. John Wiley & Sons, New York, p. 121, (1990).
3. Gore, R. W., U.S. Patent 3,962,153 (1976) to W.L. Gore & Associates.
4. Hishinuma, Y., Chikahisa, T., Yoshikawa, H. *1998 Fuel Cell Seminar*, Nov. 16–19, 1998, Palm Springs Convention Center, Palm Springs, CA, p. 655, (1998).
5. El-Sayed, A. Taher, N. H., Kamal, H., *J. Appl. Polym. Sci.* *38*:1229, (1989).
6. Hollahan, J. R., Stafford, B. B., Falb, R. D., Payne, S. T., *J. Appl. Polym. Sci.*, *13*:807, (1969).
7. Cirlin, E. H., Kaelble, D. H., *J. Polym. Sci., Polym. Phys. Ed.*, *11*:785, (1973).
8. Baumhard-Neto, R., Galembeck, S. E., Joekes, I., Galembeck, F., *J. Polym. Sci., Polym. Chem. Ed.*, *19*:819, (1981).
9. Rappaport, C. A., U.S. Patent 2,809,130 (1957) to General Motor Corporation.
10. Doban, R. C., U.S. Patent 2,871,144 (1959), to E.I. du Pont de Nemours & Co.
11. Gangal, S. V. *Encyclopedia of Polymer Science and Technology*, Vol. *16* (Mark, H. F. and Kroschwitz, J. I., Eds.), John Wiley & Sons, New York, p. 597, (1989).
12. Carlson, D. P., Schmiegel, W., *Ullmann's Encyclopedia of Industrial Chemistry* (Gerhartz, W., Ed.), Vol. *A11*, VCH Publishers, Weinheim, Germany, p. 408, (1988).
13. *Tefzel® Fluoropolymers, Product Information*, Du Pont Materials for Wire and Cable, Bulletin E-81467, E. I. Du Pont de Nemours & Co., Wilmington, DE, (1986).
14. Kerbow, D. L. *Modern Fluoropolymers* (Scheirs, J., Ed.), John Wiley & Sons, Ltd., Chichester, UK, p. 307, (1997).
15. Levy, S. B., U.S. Patent 4,510,301 (1985), to E.I. du Pont de Nemours & Co.
16. *Tefzel® Properties Handbook*, Publication No. E-31301-4, E. I. du Pont de Nemours, & Co., Wilmington, DE, (1993).
17. Humphrey, J. S., Dohany, J. E., Ziu, C., *Ultrapure Water*, First Annual High Purity Water Conference & Exposition, April 12–15, p. 136, (1987).
18. Benzinger, W. D., Robinson, D. N., U.S. Patent 4,384,047 (1983) to Pennwalt Corporation.
19. Sternberg, S., U.S. Patent 4,340,482 (1982) Millipore Corporation.
20. Lanza, V. L., Stivers, E. C., U.S. Patent 3,269,862 (1966) Raychem Corporation.
21. Sopory, U. K., U.S. Patent 4,318,881 (March 9, 1982); U.S. Patent 4,591,700 (May 27, 1986) Raychem Corporation.
22. Heaven, M. D., *Progr. Rubb. Plast. Tech.*, 2:16, (1986).
23. Bartell, F. E., U.S. Patent 3,582,457 (1971) Electronized Chemicals Corporation.
24. Eyrund, L., Eyrund, P., Bauer, F., *Adv. Ceram. Mater.*, *1*:233, (1986).
25. Aclar® Barrier Films, Allied Signal Plastics, Morristown, NJ.
26. Carlson, D. P., Schmiegel, W. in *Ullmann's Encyclopedia of Industrial Chemistry* (Gerhartz, W., Ed.), Vol. *A11*, VCH Publishers, Weinheim, Germany, p. 412, (1988).
27. Robertson, A. B., *Appl. Polym. Symp.*, *21*:89, (1973).
28. Stanitis, G. *Modern Fluoropolymers* (Scheirs, J., Ed.), John Wiley & Sons, Ltd, Chichester, UK, p. 533, (1997).
29. Hull, D. E., Johnson, B. V., Rodricks, I. P., Staley, J. B. *Modern Fluoropolymers* (Scheirs, J., Ed.), John Wiley & Sons, Ltd, Chichester, UK, p. 265, (1997).
30. Adhesive Neoflon™ EFEP, Bulletin May.2003 EG – 37(0005/AI) www.daikinchem.de
31. *Daikin America-Product Brochure*, daikin-america.com (2022).
32. *Zeus EFEP, Catalogue*, www.zeusinc.com/MATERIALS (2022).
33. *Teleflex Medical OEM Products*, teleflexmedicaloem.com (2022).
34. Drobny, J. G., *Applications of Fluoropolymer Films*, Elsevier, Oxford, UK, p. 247, (2020).
35. *Public Works* III, p. 78 (1980).
36. Schmutz, G. L., *Circuits Manufacturing*, 23:51, (1983).
37. Brasure, D. and Ebnesajjad, S. in *Encyclopedia of Polymer Science and Engineering*, Vol. *17* (Mark, H. F. and Kroschwitz, J. I., Eds.), John Wiley & Sons, New York, p. 488, (1989).
38. Ebnesajjad, S., *Polyvinyl Fluoride - Technology and Applications of PVF*, Chapter 12, Elsevier, Oxford, UK, (2013).

Part III

Fluoroelastomers

9

Fluorocarbon Elastomers

Jiri G. Drobny

Fluorocarbon elastomers represent the largest group of fluoroelastomers and have, as pointed out earlier, carbon-to-carbon linkages in the polymer backbone and a varied amount of fluorine in the molecule. They can be based on several types of monomers: vinylidene fluoride (VDF), tetrafluoroethylene (TFE), chlorotrifluoroethylene (CTFE), hexafluoropropylene (HFP), perfluoromethylvinyl ether (PMVE), 1-hydro-pentafluoropropane (HPFP), ethylene, and propylene (P). The proper combination of these monomers produces amorphous materials with elastomeric properties. A review of monomer combinations in commercially important fluorocarbon elastomers is given in [1]. VDF-based elastomers have been, and still are, commercially most successful among fluorocarbon elastomers. The first commercially available fluoroelastomer was KEL-F, developed by the M.W. Kellogg Co. in the late 1950s. Since then, a variety of fluorocarbon elastomers have been developed and made available commercially.

Currently, eight major manufacturers produce fluorocarbon elastomers, and these are listed in Table 9.1. The main commercially available fluorocarbon elastomers are listed in Table 9.2. In ASTM D1418, fluorocarbon elastomers have the designation FKM, and in ISO R1629 their designation is FPM.

Perfluoroelastomers represent a special subgroup of fluorocarbon elastomers. They are essentially rubbery derivatives of polytetrafluoroethylene (PTFE) and exhibit exceptional properties, such as unequaled chemical inertness and thermal stability. These have the ASTM designation FFKM.

An alternating copolymer of TFE and propylene (TFE/P) and a terpolymer TFE/P/VDF are fluorocarbon elastomers commercially available under the trademarks Aflas®, Viton™ Extreme™, and others. They are characterized by improved low-temperature and electrical properties and steam resistance when compared with FKM and are comparable to FFKM in chemical resistance at lower cost. TFE/P has the ASTM D1418 and the ISO 1629 designations FEPM, and in ASTM D2000/SAE J200 it is classified as Type/Class HK.

As pointed out earlier, the first commercial fluoroelastomer, KEL-F, was developed by M. W. Kellogg as a copolymer of vinylidene fluoride (VDF) and chlorotrifluoroethylene (CTFE). Another fluorocarbon elastomer, Viton™ A, is a copolymer of VDF and hexafluoropropylene (HFP), developed by DuPont and made available commercially in 1955. The products developed thereafter can be divided into two classes: VDF-based fluoroelastomers and tetrafluoroethylene (TFE)-based fluoroelastomers (perfluoroelastomers) [2]. Current products are mostly based on copolymers of VDF and HFP, or VDF and PMVE, or terpolymers of VDF with HFP and TFE. In the combination of VDF and HFP, the proportion of HFP has to be in the range from 19 to 20 mole % or higher to obtain an amorphous elastomeric product [2]. The ratio of VDF/HFP/TFE has also to be within a certain region to yield elastomers, as shown in the triangular diagram of Figure 9.1 [3].

Fluorocarbon elastomers are classified in ASTM D1418 as "M" class rubber, that is, a rubber having a saturated chain of the polymethylene type that utilizes vinylidene fluoride as a comonomer and has substituent fluoroalkyl, perfluoroalkyl, or perfluoroalkoxy groups on the polymer chain, with or without a cure site monomer (having a reactive pendant group). According to the monomer composition, they are classified as [4]:

Type 1: Dipolymer* of VDF and HFP; curable by bisphenol; general purpose, best balance of overall properties. Typically 66 wt. % of fluorine.

Type 2: Terpolymer of VDF, HFP, and TFE; curable by bisphenol or peroxide; higher heat resistance, best resistance to aromatic solvents of the VDF-containing FKMs. Typically 68–69.5 wt. % of fluorine.

DOI: 10.1201/9781003204275-12

TABLE 9.1

Major Manufacturers of Fluorocarbon Elastomers

Company	Trademarks	Web Page
3M Dyneon	3M™Dyneon™	www.3m.com
AGC Chemicals	Aflas®	www.agcchem.com
Chemours	Viton™	viton.com
Daikin	DAI-EL	daikinchemicals.com
DuPont	Kalrez®	dupont.com
Greene Tweed	Chemraz® Fluoraz®	www.gtweed.com
HaloPolymer OJSC	Elaftor SKF	www.halopolymer.com
Solvay	Tecnoflon®	www.solvay.com

TABLE 9.2

Main Commercially Available Fluorocarbon Elastomers and Their Composition

Monomer	HFP	PMVE	CTFE	P
VDF	Dyneon FC, FG, FPO		SKF-32	
	DAI-EL G7, G8			
	Viton A			
	SKF-26			
	Elaftor 2000 Series			
TFE		Dyneon PFE Series		Aflas 100 Series, 150 Series, 150 E, 400E
		Aflas PM		Fluoraz
		Daikin G-15, G-105, G-500		Viton Extreme TBR-605CS
		Viton Extreme ETP-600S*		
		Elaftor 1000		
		Chemraz		
		Tecnoflon PFR		
VDF + TFE	Dyneon FT, FX	Dyneon LTFE 6		Dyneon BRE 7231
	DAI-EL G-55, G-6, G-9, G-501	DAI-EL LT-2		Aflas 200P
	Viton B, GF, F, AL	Viton GLT, GFLT		
	Elaftor 3000, 7005	Tecnoflon PL,VPL		
	Tecnoflon P, T, FOR			
VDF + TFE + CSM		Elaftor 8000		

HFP = hexafluoropropylene; PMVE = perfluoromethyl vinyl ether; CFTFE = chlorotrifluoroethylene; P = propylene; VDF = vinylidene fluoride; TFE = tetrafluoroethylene; CSM = cure site monomer.
* Viton Extreme ETP-600S is a termonomer of ethylene, tetrafluoroethylene, and perfluoromethyl vinyl ether.

Type 3: Terpolymer of VDF, TFE, and fluorinated vinyl ether; curable by bisphenol or peroxide; improved low-temperature performance, higher cost. Typically 62 and 68 wt. % of fluorine.
Type 4: Terpolymer of TFE, propylene, and VDF; curable by bisphenol; improved base resistance, higher swelling in hydrocarbons; decreased low-temperature performance. Typically 67 wt. % of fluorine.
Type 5: Pentapolymer of TFE, HFP, VDF, ethylene, and fluorinated vinyl ether; curable by peroxide; improved base resistance; low swelling in hydrocarbons; improved low-temperature performance.

* "Dipolymer": This is how it is shown in the official ASTM document. An alternative, very commonly used term is "copolymer".

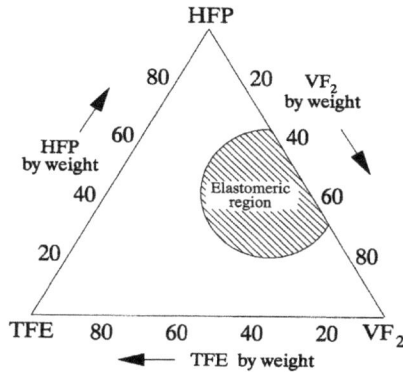

FIGURE 9.1 Compositions of VDF/HFP/TFE.

The "M" class also includes:

FFKM: Perfluorinated rubbers of the polymethylene type having all fluoroalkyl, perfluoroalkyl, or perfluoroalkoxy substituent groups on the polymer chain; a small fraction of these groups may contain functionality to facilitate vulcanization. They are essentially rubbery derivatives of PTFE and exhibit exceptional properties, such as unequaled chemical inertness and thermal stability.

FEPM: A fluoro rubber of the polymethylene type only containing one or more of the monomeric alkyl, perfluoroalkyl, and/or perfluoroalkoxy groups, with or without a cure site monomer (having a reactive pendant group).

ASTM D2000/SAE J200 [5] outlines the system used to classify the properties of *vulcanized* rubber materials, which are frequently used in automotive applications. The classification system helps to determine the properties of vulcanized rubber materials associated with (1) the type identified by heat aging resistance and (2) the class identified by resistance to swelling in oils. According to this system, FKM fluoroelastomers are classified as *HK polymers* and perfluoroelastomers as *KK polymers*, where

H indicates heat resistance to 250°C and K to 300°C and
K less than 10% volume swelling in IRM 903 oil.

9.1 Manufacturing Process for Fluorocarbon Elastomers

9.1.1 Industrial Synthesis of Monomers for Fluorocarbon Elastomers

The major fluorinated monomers for fluorocarbon elastomers are the same as those used for thermoplastic fluoropolymers. Production volumes of thermoplastic fluoropolymers, and thus of their monomers, are much higher than the volumes of fluoroelastomers. This allows the supply of modest amounts of monomers for fluoroelastomers at a reasonable cost. Most producers of fluoroelastomers are also producers of fluoroplastics, or allied with these suppliers. Vinylidene fluoride (VDF) accounts for about half the volume of monomers used for fluoroelastomers. Tetrafluoroethylene (TFE) and hexafluoropropylene (HFP) are the other main fluorinated monomers in fluoroelastomers. Perfluoromethyl vinyl ether (PMVE), used in specialty fluoroelastomers, is not commonly used in fluoroplastics, but can be made in facilities used for manufacturing the perfluoroalkyl vinyl ethers (PMVE and PPVE), monomers used in fluoroplastics. A number of cure-site monomers and fluorinated chain-transfer agents are also made in low volumes for specialty fluorocarbon elastomers. Safe handling of these monomers and a wide range of monomer mixtures is a major consideration for fluoroelastomer producers. The industrial synthesis of monomers for thermoplastic fluoropolymers was covered in detail in Chapter 3. More information specific to the monomers for fluorocarbon elastomers, including olefins (ethylene and propylene) and the main cure-site monomers, is available in [6].

9.1.2 Polymerization and Finishing of Fluorocarbon Elastomers

As pointed out earlier, most commercial fluorocarbon elastomers are copolymers of two or more monomers and are produced by high-pressure free-radical emulsion polymerization [6, p. 40]. Figure 9.2 is a schematic of the general process. The polymerization operation may be carried out in continuous or semi-batch mode. Numerous process variations are used to produce different products. The molecular structures of fluoroelastomers are determined by the polymerization and isolation process conditions.

As indicated in Figure 9.2, water and other liquid ingredients are added to the polymerization reactor. These include an initiator and soap as aqueous solutions and an optional chain-transfer agent and cure site monomer. Two or three major monomers are fed as gases by a compressor. The reactor is maintained at the temperature, pressure, and holdup time required for the particular product. Air and other impurities are carefully excluded from the feed and reactor systems. The polymer is formed in the reactor as a dispersion containing 15–30% solids, with particle size generally in the range 100–1,000 nm in diameter. At reactor conditions, much of the monomer present is dissolved in the particles at concentrations of 3–30%, depending on the polymer and monomer compositions, and on the prevailing temperature and pressure [4, p. 275].

Polymer dispersion is discharged from the reactor to a degassing vessel maintained at low pressure to allow removal of the residual gaseous monomer. In continuous reactor operation, the reaction vessel is maintained liquid-full and the dispersion is let down through a back-pressure control valve to the degasser. The recovered monomer is recycled continuously to the reactor through the monomer feed compressor. In semi-batch reactor operation, the dispersion is let down to the degasser at the end of the polymerization, and the recovered monomer is held for subsequent recharging of the reactor for succeeding batches of the same composition. Additional vessels may be provided for final monomer removal and dispersion blending prior to isolation.

Polymer isolation is effected by chemical coagulation of the dispersion, followed by separation of polymer crumb from the aqueous phase, removal of soluble soap and salt residues, and dewatering and drying of the polymer. The usual coagulants are soluble salts of aluminum, calcium, or magnesium. Various means of separating the polymer from the coagulated slurry are used commercially, including continuous centrifuges, filters, and dewatering extruders. Methods used for salt removal include washing by repeated reslurrying in fresh water and separation of the polymer; washing on a batch filter or continuous filter belt; or expelling most of the aqueous phase in a dewatering extruder. The purified polymer is

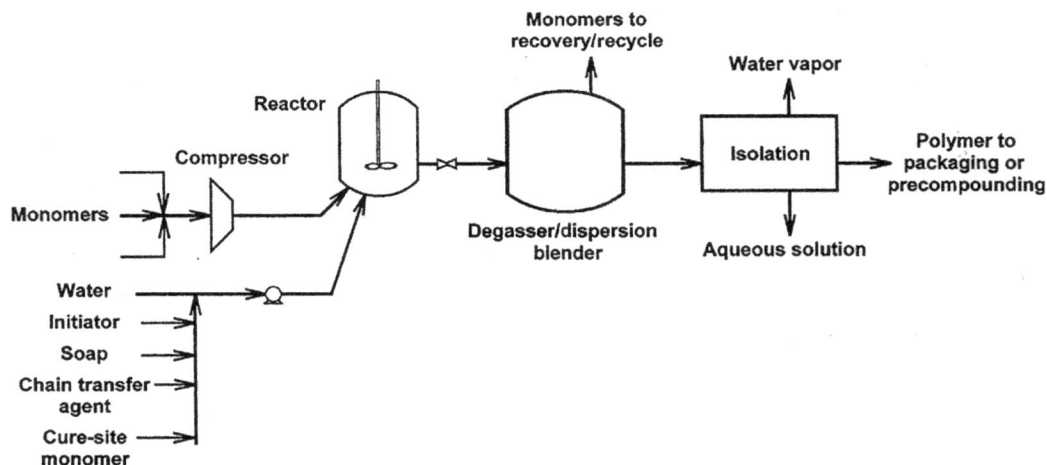

FIGURE 9.2 General fluorocarbon elastomer production process. From Drobny, J. G. *Fluoroelastomers Handbook*, Second Edition, Elsevier, p. 41 (2016). Reprinted with permission.

dried in a batch oven or continuous conveyor dryer, or in a drying extruder. The isolated fluoroelastomer is generally formed into pellets or a sheet for packaging and sale as gum polymer. Alternatively, the polymer may be pre-compounded by adding curatives and processing aids before forming and packaging [6, Chapter 5]. Some fluorocarbon elastomers are also available in latex form. More detailed description of currently used industrial methods are given in Sections 9.1.2.1 and 9.1.2.2 and in [4, p. 275].

9.1.2.1 Emulsion Polymerization

As previously described, polymerization occurs in monomer-swollen polymer particles some 100 to 1,000 nm in diameter, not in a liquid–liquid emulsion as implied by the name. Particles are stabilized by a surfactant, either added or made *in situ* by polymerization in the aqueous phase. A water-soluble initiator system generates free radicals, some of which grow and form or enter particles. In most fluoroelastomer polymerization systems, there is no sizeable reservoir of liquid monomer present. Much of the monomer is dissolved in the polymer particles, and is replenished by a continuous feed during polymerization. Even in semi-batch polymerization, the amount of monomer in the reactor vapor space is relatively small. The segregation of growing radicals in small particles under conditions of limited termination by incoming radicals allows attainment of the high molecular weights desired for good elastomeric properties. Especially for VDF-based elastomers, emulsion systems allow very high productivity in reactors of modest size [4, p. 276].

9.1.2.1.1 Continuous Emulsion Polymerization

DuPont pioneered VDF/HFP/TFE polymerization in continuous stirred tank reactors (CSTRs) in the late 1950s. An early version of a continuous fluoroelastomer production process, including isolation, is described in [7]. Recent versions of the continuous emulsion polymerization process, as it was run by DuPont Dow Elastomers, feature more feed components, monomer recovery with continuous recycling of unreacted monomers, and considerably more monitoring and control systems. A schematic diagram of such a continuous polymerization system, including monomer recovery and recycle, is shown in Figure 9.3.

Continuous polymerization has the advantage of allowing sustained production at a steady state. High rates are attained at moderately high dispersion solids (15–30%). Most or all of the heat of polymerization

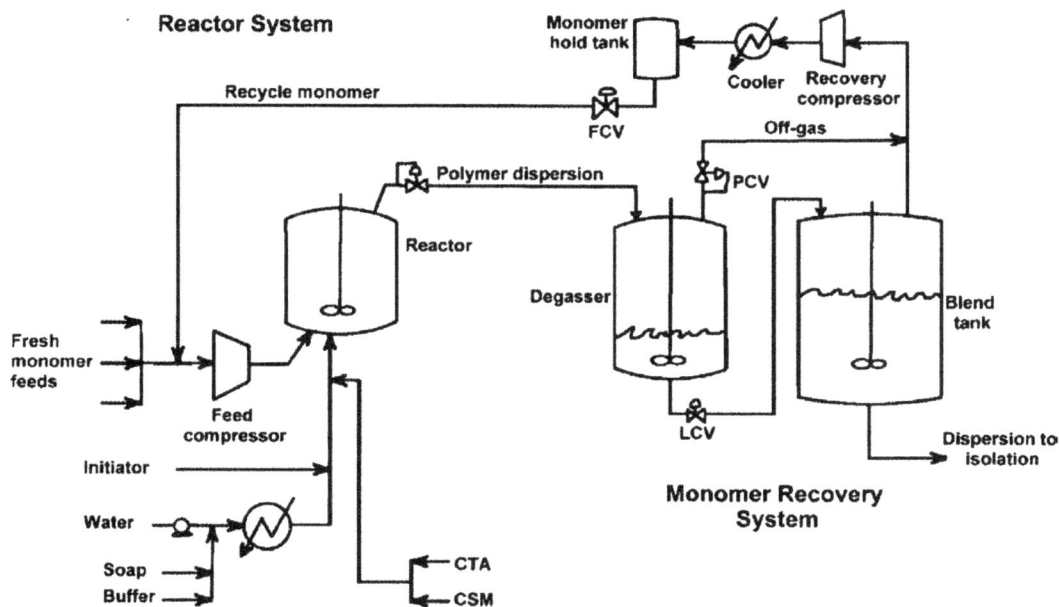

FIGURE 9.3 Continuous emulsion polymerization process. From Drobny, J. G. *Fluoroelastomers Handbook*, Second Edition, Elsevier, p. 54 (2016). Reprinted with permission.

is removed by the temperature rise of chilled feed water, so polymerization rates are not limited by relatively low rates of heat removal through a reactor cooling jacket. Continuous polymerization is particularly advantageous for the production of a few high-volume types, especially if individual product campaigns are two days or more in length. After initial adjustments are made, a uniform polymer can be produced under the same conditions for a considerable period.

Continuous polymerization is less attractive for a product line comprising many types requiring short campaigns with frequent reactor startups and shutdowns. Modern control systems allow rapid attainment of goal polymer characteristics and thus good quality even in this situation. However, semi-batch systems are better suited to making product lines with many low-volume specialty types [4, p. 276].

The range of products suitable for a continuous emulsion polymerization process is somewhat restricted. Monomer compositions must allow aqueous-phase oligomerization rates high enough that continuous generation of new particles occurs, and thus steady polymerization rates can be attained.

Reasonably high radical generation rates are required, with dispersion stabilization by ionic oligomers and added soap. Suitable compositions include most vinylidene fluoride copolymers, especially the commercially important VDF/HFP/(TFE) and VDF/PMVE/TFE products. For continuous emulsion polymerization of these VDF copolymers, low levels of highly water-soluble short-chain hydrocarbon alkyl sulfonates (e.g., sodium octyl sulfonate) are effective in place of fluorinated soaps [8]. TFE/PMVE perfluoroelastomers and ethylene/TFE/PMVE base-resistant elastomers can also be made in continuous reactors, though at much lower rates. Sustained particle nucleation is difficult to attain for TFE/propylene compositions; these do not appear suitable for production in continuous polymerization. Certain polymer designs that require initial formation of particles with little or no further initiation must be made in semi-batch reactors. An example is the Daikin family of polymers with almost all chain ends capped with iodine, made in a living radical polymerization [4, p. 277].

9.1.2.1.2 Semi-Batch Emulsion Polymerization

All fluoroelastomer producers use semi-batch emulsion polymerization systems. Detailed descriptions of commercial fluoroelastomer semi-batch systems are not available in the open literature, but smaller scale reactors are described in a number of patents. Figure 9.4 is a schematic representation of a fluoroelastomer semi-batch reactor with associated charging and feed systems, and monomer recovery system. Shown are components usually charged initially, and those that may be fed during the course of polymerization. Semi-batch polymerization is suitable for a wide range of compositions, including those having very slow polymerization rates. Semi-batch reactors are more versatile than continuous reactors for making specially designed polymers. Feeds of initiator, transfer agents, and cure-site monomers can be varied during the course of a batch to make polymers with different molecular weights and molecular weight distributions, end groups, and cure site distribution along chains. This allows control of rheology, processing, and curing behavior to an extent not attainable in CSTRs. Polymer composition and polymerization rates are readily controlled by setting monomer feeds during the reaction. Commercial semi-batch reactors are capable of making a considerable number of low volume specialty products. However, the necessity of keeping different products separate in downstream handling equipment limits the versatility of the reactor system [4, p. 277].

Semi-batch reactors have limitations compared to continuous reactors in the production of high-volume, fast-polymerizing types. The heat of polymerization must be removed by means of a cooling jacket. With this limited cooling capability, polymerization rates must be limited well below those possible in adiabatic CSTRs for many important high-volume products (e.g., VDF copolymers containing 60–80 mole % VDF). In campaigns of high-volume types, many batches with attendant shutdowns and startups are required, and batch-to-batch variability may be significant. For many types, reaction times may be too short to allow the monitoring of product characteristics, feedback, and adjustments within each batch. Adjustments can be made on subsequent batches, but large blend tanks may be required to reduce final product variability [4, p. 278].

The holdup of gaseous monomer mixtures in semi-batch reactors and feed systems is greater than that in CSTR systems. Considerable volumes of monomer mixtures under pressure in semi-batch reactor vapor spaces and in accumulators after compressors may present potential explosion hazards. The lower operating pressures of semi-batch reactors somewhat offsets this hazard, compared to the system depicted

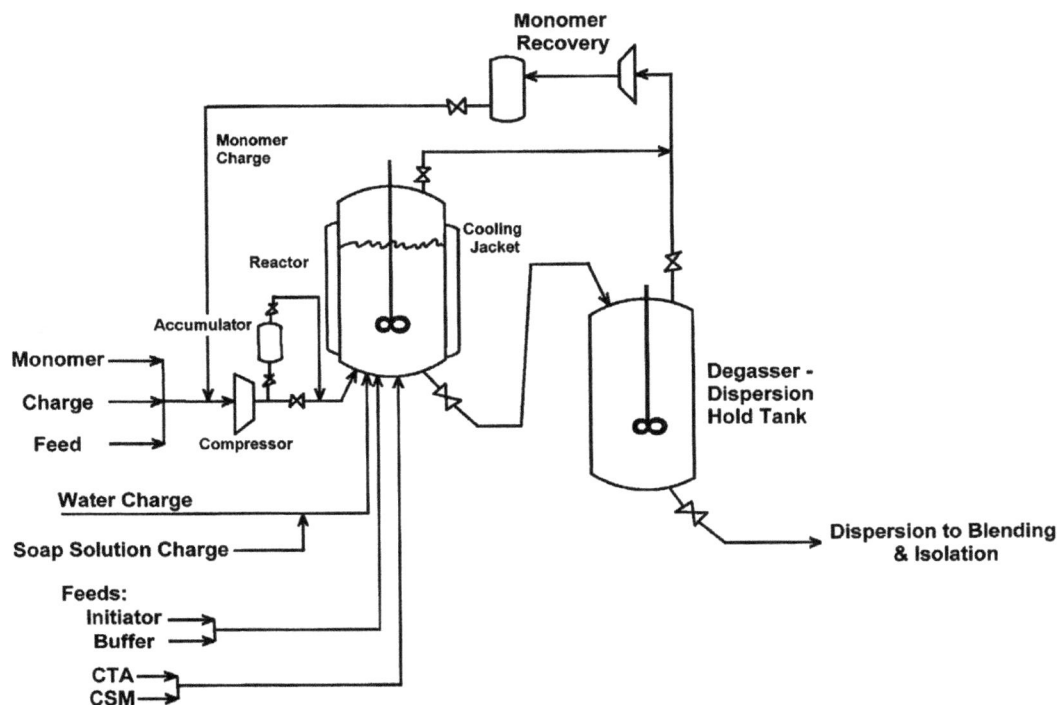

FIGURE 9.4   Semi-batch emulsion polymerization process. From Drobny, J. G. *Fluoroelastomers Handbook*, Second Edition, Elsevier, p. 58 (2016). Reprinted with permission.

by Figure 9.3. However, barricades around semi-batch reactors and feed facilities may be necessary to protect personnel [4, p. 278].

9.1.2.2 Suspension Polymerization

Suspension polymerization is used to make a number of thermoplastic polymers. In suspension polymerization, all reactions are carried out in relatively large droplets or in polymer particles stabilized by a small amount of water-soluble gum. Organic peroxide initiators are used to generate radicals within the droplets. A solvent may be used to dissolve a monomer at relatively high concentration. The main advantages of suspension polymerization over emulsion systems are that no surfactants, which are difficult to remove from the product, are used, and no ionic end groups are present which may be unstable during processing at high temperatures. What follows is a general introduction of suspension polymerization.

In one semi-batch suspension process for making VDF homopolymer [9], the reactor is charged with water containing a cellulose gum (about 0.03%) as the suspending agent, an initiator solution, and a VDF monomer. The initiator of choice is diisopropylperoxydicarbonate, which has a half-life of about two hours at 50°C. The jacketed reactor is heated with agitation to a temperature in the range 40 to 60°C, with a pressure in the range 6.5 to 7.0 MPa maintained by adding additional water or monomer during the polymerization period of about 3.5 hours. Chain-transfer agents may also be fed. Average particle diameter is typically about 0.1 mm for the dispersion obtained in suspension polymerization. At the end of polymerization, the reactor is cooled, the dispersion is degassed by letting off pressure from the reactor, the polymer is separated by filtering or centrifuging the dispersion, and washed to remove residual dispersion stabilizer. Major features of this process were adapted by Asahi Chemical Industry Co., Ltd. to make VDF/HFP/TFE fluorocarbon elastomers.

In the initial version of the Asahi Chemical suspension polymerization process [10], a relatively large amount of an inert solvent, trichlorotrifluoroethane (CFC-113, $CCl_2F–CClF_2$), is dispersed in water containing 0.01–0.1% of methyl cellulose suspending agent. The mixture is heated under agitation to the desired polymerization temperature (usually 50°C) and the proper composition of the VDF/HFP/TFE monomer mixture to make the desired copolymer is charged to the amount necessary to get the goal concentration in the monomer-solvent droplets. With the solvent used, the pressure is usually relatively low, about 1.2–1.6 MPa. Reaction is started by adding a diisopropyl peroxydicarbonate initiator solution and a monomer mixture, with composition essentially that of the polymer being made is fed to maintain the reactor pressure constant. Polymerization starts in the monomer-solvent droplets, with initial formation of a low molecular weight fraction. As polymerization proceeds, viscosity of the particles increases, long-lived radicals form, and both polymerization rate and molecular weight increase with reaction time. The resulting polymer has a bimodal molecular weight distribution, with the minor low molecular weight fraction acting as a plasticizer for the bulk high molecular weight polymer. Normally no chain-transfer agents are used for polymers cured with bisphenol. Polymer viscosity is set from the ratio of the total polymer formed to initiator charged. Since reaction times are fairly long (six hours or more) to attain high dispersion solids (30–40%), dispersion samples can be taken from the reactor during polymerization to monitor inherent viscosity and predict when to stop polymerization to attain goal viscosity.

After polymerization is stopped by turning off the monomer feed, monomers are removed by venting the reactor. Considerable care is necessary during this operation to reduce pressure in stages so that rapid release of the monomer from particles does not occur, and carryover of particles into vapor lines is avoided. Particle sizes after degassing are 0.1 to 1 mm in diameter, and are readily separated by filtering or centrifuging the dispersion. Fluorocarbon elastomers made by the suspension process have no ionic end groups and contain a significantly low molecular weight fraction. These copolymers can be made with high inherent viscosities for enhanced vulcanizate properties, while they still retain good processability because their compounds have relatively low viscosity at processing temperatures. Compared to the emulsion products of similar composition, bisphenol-curable suspension products exhibit better compression set resistance, faster cure, and better mold release characteristics. Asahi Chemical also developed peroxide-curable VDF/HFP/TFE fluorocarbon elastomers by charging methylene iodide along with the initiator to the suspension polymerization reactor. The resulting chain transfer reactions allow the incorporation of iodine on more than half the chain ends. Final polymer molecular weight is determined mainly by the ratio of total monomer fed during the polymerization to iodine incorporated. The suspension process has been adapted to make bimodal VDF/HFP/TFE polymers for extrusion applications, such as automobile fuel hoses, to get smooth extrudates with minimal die swell at high shear rates [11]. These polymers contain 50–70% very high molecular weight fractions (η_{inh} about 2.5 dL/g, M_n about 10^6) and 30–50% very low molecular weight fractions (η_{inh} about 0.15 dL/g, M_n about 17,000), with polymer bulk viscosity determined by the relative amounts of the two fractions. The low viscosity fraction has a molecular weight below the critical chain length for entanglement (M_e about 20,000 to 25,000), so it acts as a plasticizer to facilitate extrusion with low die swell. Similar bimodal polymers with low viscosity fractions having molecular weights greater than M_e would exhibit very high die swells. Synthesis of these polymers is carried out in two stages of suspension polymerization. A very small amount of initiator is used in the first stage to make the high molecular weight fraction. Then an additional initiator and a relatively large amount of methylene iodide are charged to attain the low viscosity fraction. The relative amounts of each fraction are estimated from the cumulative monomer feed in each stage. The amount of methylene iodide charged is that required to incorporate 1.5–2% iodine in the low viscosity fraction. The polymerization rate in the second stage is very low, so the total reaction time required for the bimodal polymer synthesis is some 40–45 hours. These bimodal polymers are ordinarily cured with bisphenol, but the iodine ends on the low viscosity fraction allow a mixed cure system with both bisphenol and radical components. The radical system links very short chains into longer moieties that can be incorporated into the bisphenol cross-linked network. Similar bimodal polymers made by emulsion polymerization with conventional chain-transfer agents are cured only with bisphenol. The resulting vulcanizates contain sizeable fractions of short chains that are not incorporated into the network and are thus susceptible to extraction when exposed to solvents.

The suspension process described above was used by Asahi Chemical for commercial production of Miraflon® fluoroelastomers during the early 1990s. However, it was recognized that the use of large amounts of the ozone-depleting solvent CFC-113 would need to be phased out. A second version of the suspension process uses a small amount of a hydrogen-containing solvent such as HCFC-141b,

CH_3-$CFCl_2$. Since only the minimum amount of solvent is used to dissolve the initiator, the reactor operating pressure must be increased to 1.5–3.0 MPa so that a fraction (10–30%) of the initial monomer charge condenses to form an adequate volume of droplets to serve as the polymerization medium. In a further improvement, the hydrochlorofluorocarbon solvent is replaced with a small amount of a water-soluble hydrocarbon ester, preferably methyl acetate or t-butyl acetate [12]. These polar hydrocarbon solvents are used mainly to feed the initiator to the reactor. The methyl or tertiary butyl groups are relatively inactive regarding transfer, and these solvents are so soluble in water that there is little in the polymer phase. After Asahi Chemical's suspension polymerization technology was acquired by DuPont in 1994, additional development was carried out to extend the technology to VDF/PMVE/TFE fluorocarbon elastomers with cure-site monomers incorporated along the chains [13]. Cure-site monomers can be incorporated evenly along chains by the careful feed in controlled ratio to polymerization rate of major monomers. In this way, bromine- or iodine-containing monomers can be incorporated, in addition to iodine on chain ends from methylene iodide transfer agent, to obtain polymers with improved characteristics in free radical cures. It should be noted that similar polymers can be made more readily by continuous emulsion polymerization [14]. Of more interest are bisphenol-curable VDF/PMVE/TFE compositions with 2H-pentafluoropropylene, CF_2=CH–CF_3, as a cure-site monomer. Bisphenol-cured parts from such polymers have better thermal stability than products made by radical curing.

The suspension polymerization process works well for VDF/HFP/TFE and VDF/PMVE/TFE compositions. These monomer mixtures exhibit high propagation rates at relatively low temperatures (45–60°C) and low monomer concentrations (less than 15% in monomer/polymer particles). Reasonably high polymerization rates are possible at temperatures below 60°C, so elastomer particle agglomeration is minimized. The amorphous polymers are insoluble in the monomer/solvent mixtures and also the monomer and solvent have low solubility in the polymer-rich phases. The high viscosity of the polymer-rich phase gives hindered termination, so that long-lived radicals can grow to high molecular weights. The initial monomer mixtures charged to the reactor can be partially condensed at about 50°C and moderate pressure to form droplets as the initial locus of polymerization, without the need for charging large amounts of solvent or for charging polymer seed particles [4, p. 280].

Slower propagating compositions like TFE/PMVE give a lower molecular weight and a less useful polymer when made by suspension polymerization than a polymer obtained by emulsion polymerization. For these perfluoroelastomers, monomer solubility in the polymer is high, so particle viscosity remains too low for hindered termination and the formation of long-lived radicals. A considerable initiator must be fed during the polymerization to sustain reasonable reaction rates. Several other TFE copolymer compositions give similar results [4, p. 280].

In 2001, DuPont Performance Elastomers developed Advanced Polymer Architecture (APA) technology [6, p. 78]. The fluoroelastomers produced by this technology have optimized structure (polymer branching), improved molecular weight distribution control, and an innovative cure site monomer. The materials exhibit significantly improved processing characteristics, including improved flow, cure, and mold release. They exhibit low die swell and a good dimensional stability during extrusion. Another important improvement is the rapid cure of the compounds made from these elastomers, which results in good physical properties and a low compression set with little or no post-curing [15]. An additional improvement listed is better long-term sealing force retention. In addition, APA technology elastomers can be cured without a metal oxide, and this improves their water and acid resistance [16]. FKM elastomers produced by APA technology are available commercially in a wide range of viscosities, with improved low-temperature resistance and expanded fluid resistance [15].

9.2 Properties of Fluorocarbon Elastomers

9.2.1 Properties Related to the Polymer Structure

Essentially, the high thermal and chemical stability of fluorocarbon elastomers, as of any fluoropolymers, is related to the high bond energy of the C–F bond, and to the high bond energy of the C–C and C–H links, caused by the presence of fluorine [17].

TABLE 9.3

Effect of Fluorine Content on Solvent Volume Swelling

FKM Type	Fluorine, %	Volume Swell, %	
		Benzene, at 21°C	Skydrol D[a], at 21°C
VDF/HFP	65	20	171 (at 100°C)
VDF/HFP/TFE	67	15	127
VDF/HFP//TFE/CSM[b]	67	15	127
TFE/PMVE/CSM[b]	71	3	10

Source: Data from [19].
[a] Aviation hydraulic fluid;
[b] cure site monomer.

The copolymers of VDF and HFP, completely amorphous polymers, are obtained when the amount of HFP is higher than 19 to 20% on the molar base [18]. The elastomeric region of terpolymers based on VDF/HFP/TFE is defined by the monomer ratios. Commercially, VDF-based fluorocarbon elastomers have been, and still are, the most successful among these elastomers [3]. The chemistry involved in the preparation of fluorocarbon elastomers has been discussed in some detail in Section 9.1.

The swelling resistance of fluorocarbon elastomers is directly related to the fluorine content in the molecule. This is demonstrated by the data given in Table 9.3 [19]. For example, when the fluorine content is increased by a mere 6% (from 65 to 71%), the volume swelling in benzene drops from 20 to 3%. The copolymers of VDF and HFP have excellent resistance to oils, fuels, and aliphatic and aromatic hydrocarbons, but they exhibit a relatively high swelling in low-molecular-weight esters, ketones, and amines, which is due to the presence of VDF in their structure [20]. VDF-based fluorocarbon elastomers (e.g., Viton™) have very good resistance to strong acids. For example, they remain tough and elastic even after prolonged exposure to anhydrous hydrofluoric acid or chlorosulfonic acid at 150°C (302°F) [20]. The general chemical resistances of fluorocarbon elastomers are shown in Table 9.4.

Perfluoroelastomers – that is, elastomers based on perfluoromethylvinyl ether and TFE – have a fully fluorinated backbone. The macromolecule includes oxygen atoms in the pending ether groups, which provide elasticity. Depending on the length of the ether group (marked by the length of the side chain), the fluorine content varies. FFKM elastomers contain a small amount of a cross-linkable monomer, referred to as a cure site monomer (CSM), which is necessary for their vulcanization, which is typically proprietary, though it is known that it is a cyano-functional vinyl ether with a single iodine or bromine substitution. Perfluoroelastomers exhibit a virtually unmatched resistance to a broad class of chemicals apart from fluorinated solvents. On the other hand, they are adversely affected by hydraulic fluid, diethyl amine, and fumed nitric acid, which cause swelling of the elastomer by 41, 61, and 90%, respectively [20, p. 29].

Fluorocarbon elastomers based on TFE, propylene, and FEPM (e.g., Aflas®) swell to a high extent in aromatic hydrocarbons because of the relatively low fluorine content (54%). However, because of the

TABLE 9.4

General Chemical Resistance of FKM Elastomers

Outstanding Resistance	Good to Excellent Resistance	Poor Resistance
Hydrocarbon solvents	Low-polarity solvents	Strong caustic
Automotive fuels[a]	Oxidative environments	NaOH, KOH
Engine oils[b]	Dilute alkaline solutions	Ammonia and amines
Apolar chlorinated solvents	Aqueous acids	Polar solvents
Hydraulic fuels	Highly aromatic solvents	Ketones
Aircraft fuels and oils	Water and salt solutions	Methyl alcohol

Source: Data from [19].
[a] Unleaded fuels give some problems due to the presence of methyl alcohol;
[b] certain amine additives in engine oils can be detrimental.

absence of VDF in their structure, they exhibit a high resistance to highly polar solvents such as ketones, which swell greatly all fluorocarbon elastomers containing VDF. In addition, elastomers based on copolymers of TFE and propylene exhibit a high resistance to dehydrofluorination and embrittlement by organic amines. This class of fluorocarbon elastomers has a high resistance to steam and hot acids but shows extensive swelling in chlorinated solvents such as carbon tetrachloride, trichloroethylene, and chloroform (86, 95, and 112%, respectively, after seven days at 25°C, or 77°F). Surprisingly, they have a high swelling (71%) in acetic acid [20, p. 30]. FEPM has a volume resistivity equivalent to that of EPDM. The high glass transition temperature of the pure polymer is −3°C. When correctly compounded, it is alone among elastomers in having steam resistance up to 200°C (392°F) [21].

The low-temperature flexibility of fluorocarbon elastomers depends on their glass transition temperature (T_g), which, in turn, depends on the freedom of motion of segments of the polymeric chain. If the chain segments are flexible and rotate easily, the elastomer will have a correspondingly low T_g and exhibit good low-temperature properties. Copolymers of VDF and HFP represent the largest segment of the fluorocarbon elastomer industry but exhibit a T_g of only −20°C (−4°F), which results in very poor low-temperature properties of parts made from them. Terpolymers of VDF, TFE, and perfluoroalkoxy vinyl ethers (e.g., PMVE) have much better low-temperature properties but are considerably more expensive. The importance of flexibility of vulcanizates from fluorocarbon elastomers at low temperatures is demonstrated by the well-known disaster of the space shuttle *Challenger*. The O-rings on its solid rocket boosters stiffened in the cold and consequently lost their ability to form an effective seal. Useful ranges of service temperature of some commercially available fluorocarbon elastomers are shown in Figure 9.5 [20, p. 35].

The thermal stability of fluorocarbon elastomers also depends on their molecular structure. Fully fluorinated copolymers, such as copolymers of TFE and PMVE (e.g., KALREZ®), are thermally stable at temperatures exceeding 300°C (572°F). Moreover, with heat aging these perfluoroelastomers become more elastic than embrittled. Fluorocarbon elastomers containing hydrogen in their structures (e.g., Viton™, Dyneon™, and DAI-EL FKM) exhibit a considerably lower thermal stability than the perfluorinated elastomer. For example, the long-term maximum service temperature for FKM is 215°C (419°F) compared with 315°C (599°F) for FFKM. In addition, it has been shown that heating Viton™ A at 150°C (302°F) results in unsaturation and that metal oxides promote this dehydrofluorination at even lower temperatures [20, p. 36]. Copolymers of VDF and CTFE (e.g., KEL®-F) with an upper long-term use temperature of about 200°C (392°F) are less heat resistant than copolymers of VDF and HFP [22]. Fluorocarbon elastomers based on hydropentafluoropropylene (HPFP), such as Tecnoflon® SL

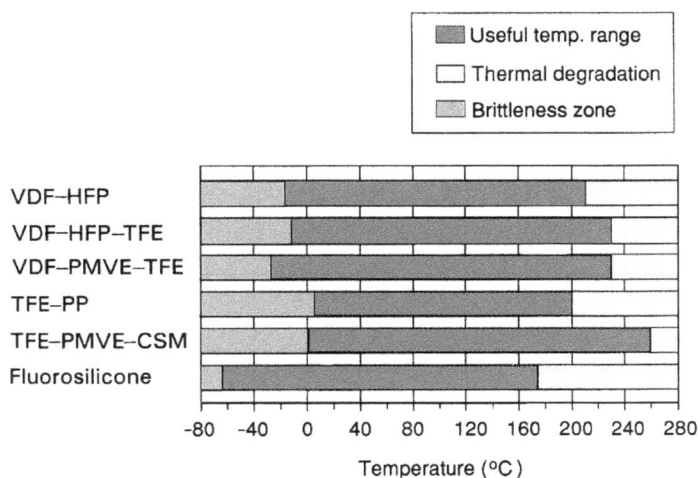

FIGURE 9.5 Useful service temperature ranges for commercial fluoroelastomers. From Scheirs, J. in *Modern Fluoropolymers* (Scheirs, J., Ed.) John Wiley & Sons Ltd., p. 35 (1997). Reprinted with permission.

(a copolymer of HPFP and VDF) and Tecnoflon® T (a terpolymer of VDF/HPFP/TFE), because of a lower fluorine content than that of their analogs with HFP, also exhibit lower thermal stability when compared with them [20, p. 36].

Another factor affecting the thermal stability of compounds based on fluorocarbon elastomers is the curing (cross-linking) system used. This subject is discussed at some length in the section on compounding.

9.2.2 Other Properties

Raw-gum fluorocarbon elastomers are transparent to translucent with molecular weights from approximately 5,000 (e.g., Viton™ LM with a waxy consistency) to over 200,000. The most common range of molecular weights for commercial products is 100,000 to 200,000. Polymers with molecular weights over 200,000 (e.g., Kel-F® products) are very tough and difficult to process. Elastomers prepared with vinylidene fluoride as a comonomer are soluble in certain ketones and esters, copolymers of TFE, and propylene in halogenated solvents; perfluorinated elastomers are practically insoluble [30, p. 431]. The same types can be cross-linked by ionizing radiation (mainly an electron beam or gamma rays) [23].

9.2.3 Properties of Currently Available Commercial Fluorocarbon Elastomers

In general, currently available commercial fluorocarbon elastomers can be classified in two groups, namely standard and specialty grades. Specialty fluorocarbon elastomers exhibit unique properties such as improved low-temperature flexibility or resistance to bases [16]. Standard polymers are usually available in gums or precompounds in which cure chemicals are already incorporated. Specialty fluorocarbon elastomers and peroxide cured standard types are usually supplied in the gum form only [16].

Suppliers provide a large number of grades with specific performance characteristics, though this often is confusing. For the purpose of this publication, we will use the framework of ASTM D1418. But this is just a beginning and the user should do some research and contact the supplier for details to find the proper material. We will briefly introduce the main points for each group and then show the brands and grade prefix for each supplier in a simple table. Each grade is identified by the prefix and a specific number assigned by the supplier. In most cases we just show the prefix that normally identifies the composition of the material and "*X*" that indicates additional numbers and/or letters to identify a specific grade. Where applicable, we identify the curing system specific to the given grade. It should be noted that we have chosen to include only several well-established suppliers, which have published sufficient data to be used for guiding authors as well as potential users.

9.2.3.1 FKM Type 1: Copolymers of HFP and VDF

This group consists of general purpose elastomers with the best overall balance of properties. Some suppliers have grades with an incorporated cure system while others require the user to add it to the compound. Details regarding the grades of this type of FKM are provided in Table 9.5.

TABLE 9.5

FKM Type 1, Suppliers, and Their Grades

Supplier	Brand	Grade Prefix	Cure System
3M™ Dyneon™	Dyneon™	FC *XXXX*	Bisphenol
		FG *XXXX*	Bisphenol
		FPO	Peroxide
Chemours	Viton	A	Bisphenol
Daikin	DAI-EL™	G-7*XX*	Bisphenol
		G-8*XX*	Peroxide
Solvay	Tecnoflon®	N *XXX*	Bisphenol
		FOR *XXX*	Bisphenol (incorporated)

TABLE 9.6

FKM Type 2, Suppliers, and Their Grades

Supplier	Brand	Grade Prefix	Cure System
3M™ Dyneon™	Dyneon™	FT *XXXX*	Bisphenol
		FX *XXXX*	Bisphenol
Chemours	Viton™	B	Bisphenol
		GF	Bisphenol
		F	Bisphenol
Daikin	DAI-EL™	G-55*X*	Bisphenol
		G-6*XX*	Peroxide
		G-9*XX*	Bisphenol
		G-501	Diamine
Solvay	Tecnoflon®	P *XXX*	Peroxide
		T *XXX*	Bisphenol
		FOR*XXXX*	Bisphenol (incorporated)

9.2.3.2 FKM Type 2: Terpolymers of TFE, VDF, and HFP

The Type 2 group of elastomers listed here contain fluorine in a range from 67 to 69% by weight and exhibit TR-10 values near −5°C [24]. Because of the increased fluorine content, they exhibit considerably improved resistance to aromatic hydrocarbons, alcohols, and engine lubes. The downside of that is a decrease of flexibility at low temperatures. Another reported benefit of a Type 2 FKM is better heat resistance [24]. Details regarding the grades of this type of FKM are given in Table 9.6.

9.2.3.3 FKM Type 3: Terpolymers of TFE, PMVE, and VDF

The main feature of these fluorocarbon elastomers is their improved performance in service at low temperatures with TR-10 values in a range from −20 to −40°C [24, p. 9]. The fluorine content of these polymers is in the range 60 to 65%. Details regarding the grades of this type of FKM are given in Table 9.7.

9.2.3.4 FKM Type 4: Terpolymers of TFE, P, and VDF

These products contain about 60 wt. % of fluorine and exhibit a TR-10 value ranging between −8 and −10°C. High temperature performance is somewhat reduced due to the presence of propylene in the backbone [24, p. 10]. These products are referred to as base resistant elastomers because they reportedly exhibit an improved resistance to oils, lubes, coolants, and transmission fluids containing high amounts

TABLE 9.7

FKM Type 3 Suppliers and Their Grades

Supplier	Brand	Grade Prefix	Cure System
3M™ Dyneon™	Dyneon™	LTFE 6*XXX*	Peroxide
Chemours	Viton™	GLT	Peroxide
		GFLT	Peroxide
Daikin	DAI-EL™	LT-2*XX*	Peroxide
Solvay	Tecnoflon®	PL*XXX*	Peroxide
		VPL *XXX**	Peroxide

* The Solvay VPL product is reported to be based on perfluoromethylvinyl ethers and prepared by their MOVE technology [24, p. 9].

TABLE 9.8

FKM Type 4 Suppliers and Their Grades

Supplier	Brand	Grade	Cure System
3M™ Dyneon™	Dyneon™	BRE 7231	Bisphenol (incorporated)
AGC Chemicals	Aflas	200P (SPL-FKM)	Peroxide-TAIC*

* TAIC = triallyl cyanurate (cross-linking coagent).

TABLE 9.9

FKM Type 5 Supplier and the Currently Available Grade

Supplier	Brand	Grade	Cure System
Solvay	Tecnoflon®	BR 9151	Peroxide

of amines and can operate at lower temperatures than TFE/P type FEPM elastomers [24, p. 10]. Details regarding the grades of this type of FKM are given in Table 9.8.

9.2.3.5 FKM Type 5: Pentapolymers of TFE, HFP, VDF, Ethylene, and PMVE

Solvay appears to be the only supplier which discloses this composition. The material contains reportedly 65 wt. % of fluorine and exhibits a TR-10 value of −7°C. This product exhibits excellent resistance to aggressive oils, amine containing fluids, bases, and steam [25]. Details regarding the grades of this type of FKM are given in Table 9.9.

9.2.3.6 FFKM Perfluoroelastomers

As pointed out earlier (in Section 9.2.1) perfluoroelastomers are copolymers of tetrafluoroethylene (TFE) and a perfluoroalkyl vinyl ether (PAVE) with a small amount of cure site monomer (CSM). FFKM elastomers fill an important niche for applications at temperatures up to 325°C (617°F) [26]. They are the most chemically resistant elastomers available on the market and also have an outstanding resistance to steam, ozone, and weathering and very low gas permeability. Additionally, they exhibit high reliability regarding static and dynamic loads due to their low compression set. Their disadvantages are poor abrasion resistance and only moderate mechanical properties that deteriorate rapidly at elevated temperatures and below 0°C (32°F). Another drawback is their very high price [26, 27].

The first perfluoroelastomers were developed by DuPont, whose marketing approach was not to sell the polymer or uncured compounds, but only cured products, such as O-rings and other molded and cured products designed for specific applications with the tradename Kalrez® They continue this practice to this day. Other elastomer manufacturers could not produce and market perfluoroelastomers for several decades until the expiration of the applicable DuPont patents in the late 1980s. Currently, there are several manufacturers that offer FFKM elastomers. These companies and some of their current products are listed below. It should be noted that Daikin offers both elastomers and cured products, offering the latter through their TOHO KASEI Co. Ltd [24, p. 16]. We do not include any cured products and their properties, because of the large number of manufacturers and products involved. The currently available grades of FFKM elastomers are supplied by several companies. Given the large number of different products, often developed for specific industries or even for specific users, we limit the number of products and include these major manufacturers: 3M Dyneon, AGC Chemicals, Daikin, Greene Tweed, and Solvay. It should be noted that Greene Tweed manufactures and is at the time of writing marketing FFKM and FEPM compounds with specific properties for certain applications. No uncompounded elastomers are marketed. The amount, depth, and style of the information provided by the manufacturers and suppliers varies greatly and we advise readers to contact the manufacturers to obtain more details from them. Details regarding the commercial grades of FFKM from different suppliers are given in Tables 9.10 to 9.17.

TABLE 9.10

FFKM Elastomer Grades from 3M Dyneon, Peroxide Cured

Grade	TR-10, °C	Mooney ML(1+10) @ 121°C	Compression Set 70 h @ 200°C
PFE 40	−6	40	19
PFE 60	−2	60	49
PFE 80Z	−2	80	49
PFE 90	−2	98	40

Note: Engineered for both reliable performance in harsh environments and ease of processing.

TABLE 9.11

FFKM Elastomer Grades from 3M Dyneon, High-Temperature Grades

Grade	TR-10, °C	Mooney ML(1+10) @ 121°C	Compression Set 70 h @ 2,032°C	Compression Set 70 h @ 300°C
PFE 81T	−2	80	27	50
PFE 131T	−2	80	20	43
PFE 191T	−2	80	15	33
PFE 194T	−2	90	15	31
PFE 132TB	−2	100	36	73
PFE 133BT	−2	110	26	60

Note: Designed to resist high temperatures with an upper continuous temperature of 315°C (599°F) and a high compression set resistance.

TABLE 9.12

FFKM Elastomer from 3M Dyneon, High-Temperature Grade, Cure Incorporated

Grade	TR-10, °C	Mooney ML(1+10) @ 121°C	Compression Set 70 h @ 2032°C	Compression Set 70 h @ 300°C
PFE 4131	−2	100	17	39

TABLE 9.13

FFKM Elastomer Grades from 3M Dyneon, Clear Perfluoroelastomer Systems

Grade	TR-10, °C	Mooney ML(1+10) @ 121°C	Compression Set 70 h @ 250°C	Compression Set 70 h @ 275°C
PFE 300Z	−2	80	19	33
PFE 301Z	−2	110	26	32

Source: Technical Data 9040HB 1/13 and 9041HB 1/13 (2013), www.3M.com/fluoropolymers
Note: Proprietary polymers and catalyst systems enable the manufacture of optically clear finished parts. The catalyst for PFE 300Z is PFE300C and that for PFE 301Z is PFE 301C.
Service temperature for the above systems is 300°C (continuous) and 350°C (peak).

9.2.3.7 FEPM

FEPM, the copolymer of tetrafluoroethylene (TFE) and propylene (P) or TFE/P, is a partially fluorinated fluorocarbon elastomer that can be cross-linked using a variety of curatives, such as peroxides or specialized proprietary systems [21]. TFE/P provides a unique combination of chemical, heat, and electrical resistance. Chemically, it resists both acids and bases, as well as steam, amine-based corrosion inhibitors, hydraulic fluids, alcohols, and petroleum fluids. It is also resistant to ozone and weather. The typical service temperature of commercial FEPM resins is between −55°C (−65°F) and + 230°C (+445°F).

TABLE 9.14

FFKM Elastomer Grades from AGC Chemicals (Brand: Aflas®)

Grade*	TR-10, °C	Mooney ML(1+4) @ 121°C	Compression Set 70 h @ 200°C	Service Temperature, °C Continuous/Peak
PM-1100	N.A	80	9.6	230/250
PM-3000	N.A	80	6.0	250/270
Premium PM-3000	−5	80	6.0	230/270
CP-4000 (R&D)**	N.A	85	N.A	280/300

Source: AGC Chemicals, Document CA011E Catalogue/2018.10, agc-chemicals.com/products.
* All products are peroxide cured;
** peroxide cure precompound.

TABLE 9.15

FFKM Elastomer Grades from Daikin (Brand: DAI-EL)

Grade	Mooney ML(1+10) @ 100°C	Service Temperature, °C Continuous/Peak	Defining Characteristics
G-15	15	200/220	Peroxide-cured base polymer, with glass transition temperature of −20°C
G-105	25	200/220	Super clean peroxide-cured base polymer with minimal metal ion contamination
G-500	141	300/320	Super clean imidazole or triazine-cured base polymer with minimal metal ion contamination and best chemical resistance

Source: Daikin Fluoroelastomers, daikin-america.com/fluoroelastomers (2022).

TABLE 9.16

FFKM Elastomer Grades from Greene Tweed (Brand: Chemraz®), Examples

Grade	TR-10/50, ASTM D1329, °C	Service Temperature Range, °C (°F)	Compression Set, % ASTM D1414 70h @ 204°C**	Defining Characteristics
615	6	−18 to +324 (0 to +615)	14	Product with highest temperature resistance; mainly for oil, gas, and chemical process industries
G20	N.A.	−14 to + 293 (7 to +559)	25*	Plasma resistant, superior fluorine resistance at high temperatures; high purity compound (particulation and metals); zero outgassing up to 200°C; broad chemical resistance; specifically for semiconductor applications
SD625	6	−20 to +260 (−4 to +500)	25	FDA registered compound

Source: Greene Tweed, www.gtweed.com/resources (2022).
* 70h @ 200°C;
** 25% deflection, 214 O-rings.

FEPM typically retains its exceptional chemical resistance even at high temperatures (short exposures up to 232°C, or 450°F), and tests have shown that electrical resistance actually improves with heat exposure [28]. FEPM is not compatible with aromatic and chlorinated hydrocarbons or some fairly common organic solvents such as MEK, toluene, and acetic acid. Some other major disadvantages include a high compression set, high glass transition temperature, and rather poor performance at low temperatures. Furthermore, it is rather expensive, thus it can be used when only for components that have to withstand harsh environment [27]. Currently available grades of FEPM elastomers from major manufacturers are listed in Table 9.18.

TABLE 9.17

FFKM Elastomer Grades from Solvay (Brand: Tecnoflon®) Peroxide Cured

Grade	TR-10, °C	Mooney ML(1+10) @ 121°C	Service Temperature Range, °C	Compression Set** 70 h @ 200°C
PFR 94*	−2	35	−10 to 230	25
PFR 95*	−1	35	−10 to 280	18
PFR 95HT*	−1	75	−10 to 300	18
PFR LT	−30	25	−40 to 230	27
PFR 06HC	−2	75	−10 to 230	20

Sources: 1. Tecnoflon® FKM/FFKM Fluoroelastomers and Perfluoroelastomers-Material Guide, Solvay (2016) 2. Omnexus FFKM Data, omnexus.special.chem.com/products 2021.

* FDA registered for food contact;
** ASTM D395 Method B on O-ring, 25% deformation.

TABLE 9.18

FEPM Grades

Manufacturer	Brand	Grade	Characteristics/Use	Cure
AGC	Aflas®	100 Series*	High strength	Peroxide
	Aflas®	150 Series*	Standard grade	Peroxide
	Aflas®	150E	Grade for insulating high-voltage cables	Peroxide
	Aflas®	400E	Designed for multi-layer hose used in high-pressure, high-temperature areas around engines	Peroxide
Chemours	Viton™ Extreme™	ETP-600S	Resistance to acids, bases, esters, ketones, amines, hydrocarbons; low-temperature flexibility; lower compression set; faster cure rates; improved steam resistance; FDA approved; made by APA technology	Peroxide
	Viton™ Extreme™	TBR-605CS	Resistance to acids, esters, ketones, hydrocarbons, bases; lower compression set; faster cure rates; higher state of cure; made by APA technology	Proprietary bisphenol cure system
Greene Tweed	Fluoraz®	790A	Service temperature range: 20 to −450°F (−7 to 232°C) TR-10/50, °F (°C): 40 (4) (ASTM D1329) Compression set**, 22 h at 200°C: 21%	NA
	Fluoraz®	797	Resistance to hot water, amines; very low permeability material. Service temperature range: 23 to −450°F (−5 to 232°C) TR-10/50, °F (°C): 42 (6), (ASTM D1329) Compression set**, 70 h at 200°C: 34%	N.A.
	Fluoraz®	799	Service temperature range: 23 to −450°F (−7 to 232°C) TR-10/50, °F (°C): 44 (7), (ASTM D1329) Compression set**, 70 h at 200°C: 40%	N.A.
	Fluoraz®	SD 890	FDA compliant; service temperature range: 20 to −450°F (−5 to 232°C) Compression set**, 22 h at 200°C: 30%	N.A.

Sources: 1. Aflas® Fluoroelastomers, High-Function fluoroelastomers, www.agcchem.com/products (2022)
2. Viton™ Extreme™ ETP-600CS, Viton™ Fluoroelastomer Products viton.com/en/products (2022)
3. Viton™ Extreme™ TBR 605CS, MatWeb, matweb.com/search/datasheettext.aspx (2022)
4. Fluoraz®, Greene Tweed, www.gtweed.com/resources (2022).
* Aflas® grades 100S, 100H, and 150P are FDA approved for food contact;
** ASTM D395, Method B, 25% deflection, O-rings.

9.3 Fabrication Methods for Fluorocarbon Elastomers

This section covers manufacturing methods used for producing useful commercial products from standard fluorocarbon elastomers. It is organized so that it starts with the mixing of solid rubber compounds and follows with subsequent processes, such as calandering, extrusion, and molding. It also includes an introduction to the processing of elastomers in liquid form such as volatile solvents and latexes.

9.3.1 Mixing and Processing of Compounds from Fluorocarbon Elastomers

Mixing and processing methods used for most other synthetic elastomers can be used for fluorocarbon elastomers, sometimes with considerable adjustments to take account of the special characteristics of the polymers and their compounds. The slow relaxation rates of fluorocarbon elastomers cause difficulties in mixing, extrusion, and injection processes normally run at high shear rates. Many curatives and additives are insoluble in them, so special procedures may be necessary to get adequate dispersion in compounds for reproducible curing. The relatively small volumes involved in producing fluorocarbon elastomer parts require that equipment and processes used normally for high-volume elastomers be adapted.

9.3.1.1 Mixing of Compounds of Fluorocarbon Elastomers

Compounding is a procedure to prepare a rubber material (compound) that has the desired properties and is suitable for specific methods of processing. Thermosetting rubbers require the addition of a cure system typical for the base polymer. A recipe used for the rubber compound is based on 100 parts of rubber (elastomer), which is written as "phr", meaning "parts per hundred parts of rubber". Thus the amount of each component in the recipe is shown in phr.

Compounds from solid fluorocarbon elastomers are mixed in equipment common in the rubber industry. However, the mixing procedures typical for standard types of elastomers are often modified to be suitable for mixing fluorocarbon elastomers.

Open-mill mixing is used mainly for special compounds prepared in small volumes. The advantages of mill mixing are its simplicity, the fact that the operator can control the temperature of the material on the rolls, and an easy cleanup. However, mill mixing, especially at the production scale, is rather difficult, especially for a number of gum fluorocarbon elastomers. Polymers with narrow molecular weight distribution and low levels of ionic end group levels may not have adequate cohesive strength to form a smooth band without holes on the mill rolls. Very high-molecular weight fluorocarbon elastomers undergo significant breakdown during initial passes through a tight nip of a cold mill, which leads to reduction of the physical properties of the resulting vulcanizate. On the other hand, bimodal blends (formed by latex mixing before isolation) have excellent milling characteristics with negligible breakdown of high-molecular weight fraction [29]. High-viscosity elastomers with considerable long-chain branching and gel content may also break down during milling, possibly improving subsequent processing characteristics (e.g., extrusion) [6, p. 109]. In many production operations, two-roll mills are used for sheeting off stock from internal mixers or for the warm-up of compounds for sheet feed to extruders or calanders. A typical two-roll mixing mill is shown in Figure 9.6.

Mixing in internal mixers is considerably more productive; however, because of the high intensity of mixing in an enclosed chamber, there is a relatively high risk of the premature onset of cross-linking ("scorch"). Compounds tending to scorch are most commonly mixed in two steps ("passes"). In the first pass, the elastomer is mixed with processing aids, fillers, pigments, activators, and acid acceptors. The cross-linking agents are almost always added in the second pass. An internal mixer is illustrated by Figure 9.7.

9.3.1.2 Processing of Fluorocarbon Elastomers

Mixed rubber compounds are almost always transformed into products with required shapes and dimensions. There are several methods to accomplish this. Tubes, solid round profiles, and profiles with

FIGURE 9.6 Two-roll mixing mill. (Courtesy of DPA)

irregular, often complex shapes are prepared by extrusion. Sheets, slabs, and rubber-coated fabrics are made mainly by calandering.

9.3.1.2.1 Calandering

Calandering, mentioned earlier, is used to produce sheets, slabs, and certain types of coated fabrics. In this operation, a stack of three or four rolls (see Figure 9.8) turn at the same surface speed to squeeze the elastomer stock through two or three nips to produce a sheet of about 1 mm (0.040 in.) thickness per pass.

The grades most suitable for calandering are those with low viscosity. Processing aids are necessary to improve surface smoothness, good flow in the processing equipment, and a good release of sheets from the rolls.

Mixed stocks should be used promptly or stored at temperatures below 18°C (65°F) to prevent scorching, and great care should be taken to exclude moisture. Typical roll temperatures for calendering using a three-roll calander recommended for Viton™ E-60C or related Dyneon™ types are [19, p. 421]:

Top roll: 85 ± 3.5°C (185 ± 5°F).
Middle roll: 74 ± 3.5°C (165 ± 5°F).
Bottom roll: cool (ambient temperature).
Speed: 7 to 10 m/min or 7.6 to 11 yd. /min.

The setting of roll temperatures depends on the cure systems used. Typically, stocks with a diamine curative (e.g., Diak #3) are calendered at temperatures where the top and middle rolls are set 15–20°C (59–68°F) lower than stocks with bisphenol and peroxide curing systems [30].

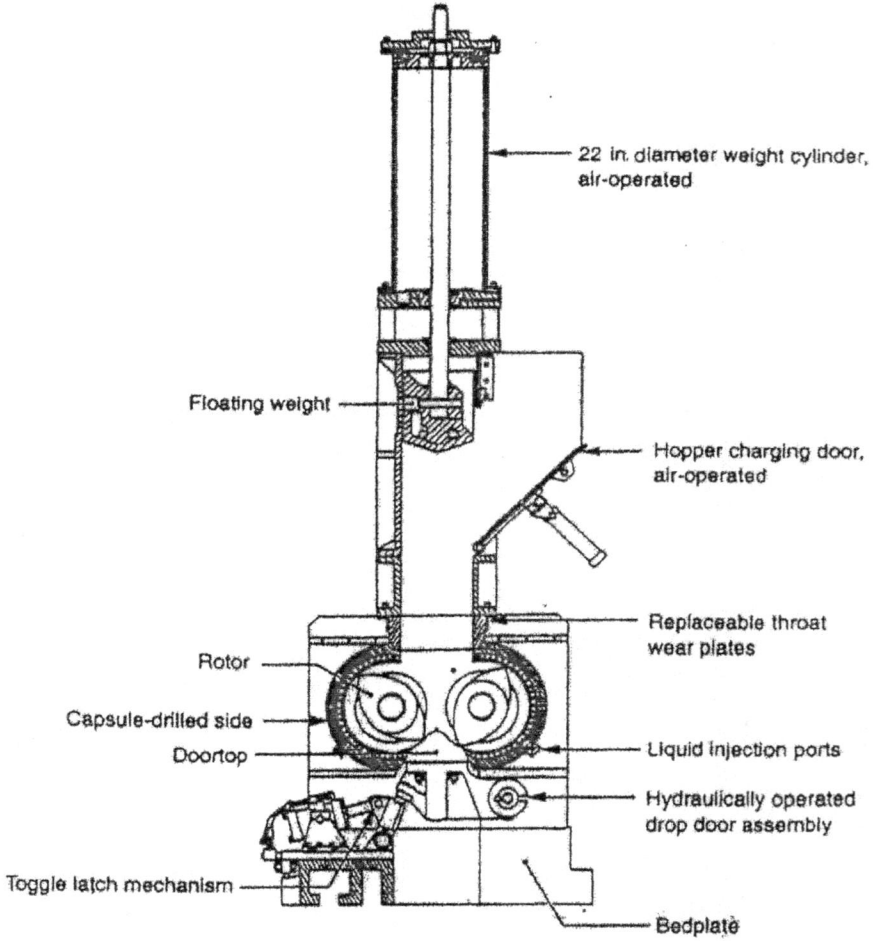

FIGURE 9.7   Internal mixer. (Courtesy of Farrel Corporation)

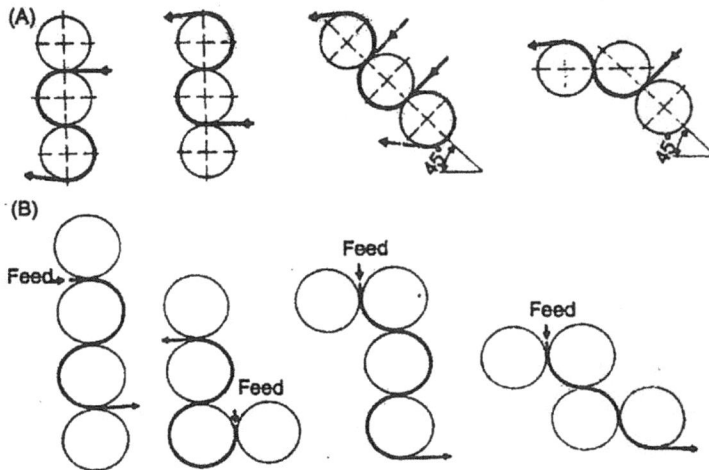

FIGURE 9.8   Calander. (Courtesy of DPA)

9.3.1.2.2 Extrusion

Extrusion of tubes, hose, and profiles is done on standard extruders for rubber. The usual temperature pattern is a gradual increase of temperature from the feed zone to the die. The die temperature is typically 100°C (212°F) and the screw temperature is approximately the same as the temperature of the feed zone [19, p. 419]. Processing aids are almost always required to improve the surface appearance and to increase the extrusion rate. Extrusion represents only a small proportion (less than 10%) of the total consumption of fluorocarbon elastomers [3, p. 86]. There are essentially two types of extruders used for rubber, namely the *hot feed extruder* and the *cold feed extruder*. Hot feed extruders are considerably shorter than cold feed extruders and they are fed by strips preheated on a two-roll mill. The cold feed extruders are fed by cold strips, chunks, or pellets. Often the barrel of the cold feed extruder is heated by steam, circulating oil, or by electric heaters clamped to it [31]. Currently, cold feed extruders are used more often for the processing of rubber because of their operating flexibility, better control of the stock temperature, and reduced labor cost [31].

Cold feed extruders typically used for rubber stocks are also suitable for fluorocarbon elastomer stocks. Extrusion should be carried out at temperatures below 120°C (250°F) to avoid scorch. In most cases a breaker plate and screen pack are used to generate backpressure on the screw and to remove foreign particles from the stock. A straight head is used for the extrusion of profiles or tubing, whereas a cross-head die is used for coating wires or the extrusion of veneer on a mandrel as the inner layer of a fuel hose. An example of a cold feed rubber extruder is shown in Figure 9.9.

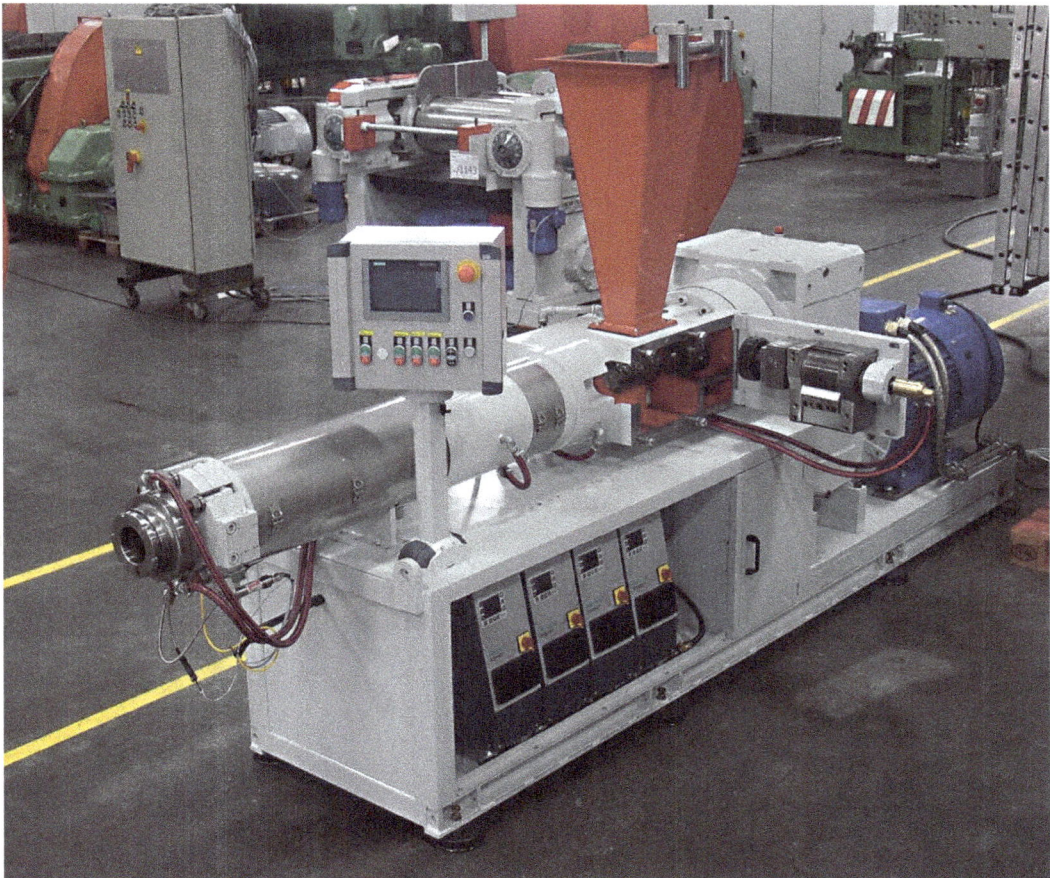

FIGURE 9.9 Cold feed extruder. (Courtesy of Deguma Schütz GmbH)

FIGURE 9.10   Barwell Preformer. (Courtesy of Barwell, Global Ltd.)

Ram extruders, such as a Barwell Precision Preformer [32], are widely used for the production of blanks for compression molding. Typical ram pressures are up to 35 MPa (5,070 psi). Various dies, such as dies for rods, tubing, and strips, are available for extrudate diameters up to 190 mm (7.5 in.). Barwell Preformers are useful for processing high-cost specialty fluorocarbon elastomers used in limited volumes in precision molded parts [29, p. 110]. An example of a Barwell Precision Preformer is shown in Figure 9.10.

9.3.1.2.3 Solution and Latex Coating

Certain substrates (woven and nonwoven fabrics, foils, and films) are coated by dipping, spreading, or spraying with fluorocarbon elastomers in liquid form. The older method using a solution of fluorocarbon elastomers in volatile solvents (e.g., methyl ethyl ketone, toluene) is gradually being replaced by the use of water-based latexes. Fluorocarbon elastomer latexes can also be used for chemically resistant and heat-resistant coatings. Some fluoropolymer producers offer latex in limited quantities to processors skilled in latex applications. Such products are typically based on VDF/HFP/TFE terpolymers with a 68% fluorine content [9, p. 119]. These terpolymers are polymerized into relatively stable dispersions (latexes) containing 30 to 40% solids [33]. The dispersions are then stabilized by pH adjustment and the addition of anionic or nonionic hydrocarbon surfactants. A water-soluble gum (e.g., sodium alginate) is then added to increase particle size, allowing creaming (actually settling) to concentrate the latex (about 70% solids). The supernatant serum is discarded. Small amounts of biocides are usually added to prevent growth of microorganisms. Formulations used by processors for particular applications are proprietary. An example of a currently available high viscosity and high solids content (~70%) product is Tecnoflon® TN Latex [33].

9.3.2 Curing of Fluorocarbon Elastomers

Semi-finished products made from fluorocarbon elastomers are cured (vulcanized) at elevated temperatures, typically from 170 to 220°C (338 to 428°F) [34]. During this process, the elastomers are cross-linked and the material becomes a thermoset. There are several methods to accomplish this which are discussed in the next sections.

9.3.2.1 Cross-Linking Chemistry

The cross-linking method for fluorocarbon elastomers depends on their type. Elastomers based on vinylidene fluoride can be cross-linked by ionic mechanism. However, if the polymer has been prepared in the presence of a cure site monomer (CSM) it can be cross-linked (cured) by a free radical mechanism. Moreover, some fluorocarbon elastomers can be cross-linked by ionizing radiation (see Section 9.3.2.1.3).

9.3.2.1.1 Cross-Linking by Ionic Mechanism

Fluorocarbon elastomers based on VDF/HFP and VDF/HFP/TFE can be mainly cured by bisnucleophiles, such as bisphenols and diamines. The mechanism, proposed in [3, p. 78], is:

1. Formation of a $-C(CF_3)=CH-$ double bond by elimination of "tertiary" fluorine.
2. Double bond shift catalyzed by fluoride ion and formation of $-CH=CF-$ double bond.
3. Nucleophilic addition of the $-CH=CF-$ double bond with:
   a. allylic displacement of fluoride affording the new $-C(CF_3)=CH-$ double bond;
   b. addition/fluoride elimination from the same double bond.

A detailed depiction is shown in Figure 9.11 [3, p. 79], where the bis-nucleophile Nu–R–Nu represents a bisphenol or diamine cross-linking agent.

The general disadvantage of curing fluorocarbon elastomers by an ionic mechanism is that the dehydrofluorination required for this reaction produces considerably more double bonds than required for the cross-linking itself. This excess of unsaturation represents weak points in the polymeric chain, which can be attacked by basic substances contained in a contact fluid. This has actually been found when parts cured by this method were exposed to new oil and fuels containing basic additives [35, 36]. The diamine cure system is used very little now. The exception is in latex compounding. Its major deficiency is a

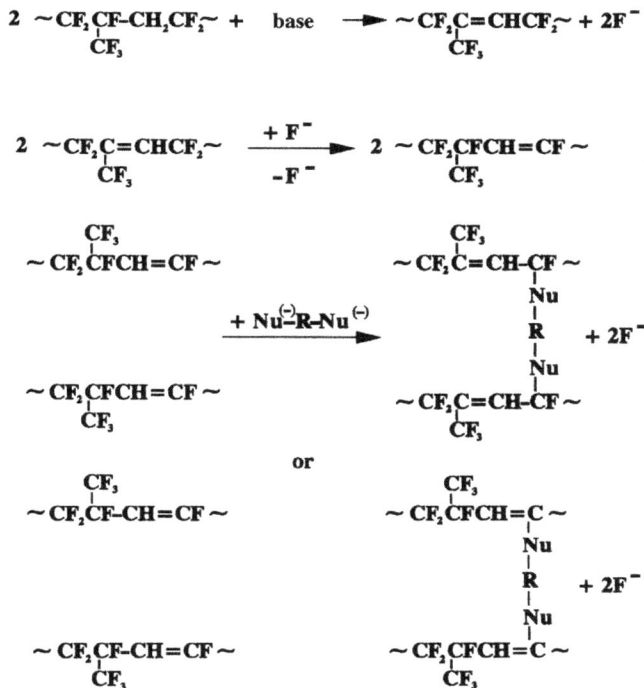

FIGURE 9.11 Reaction mechanism of ionic curing. From Arcella, V. and Ferro, R. in *Modern Fluoropolymers* (Scheirs, J., Ed.) John Wiley & Sons Ltd., p.79 (1997). Reprinted with permission.

tendency to the premature onset of cross-linking ("scorch") typically at processing temperatures in the range 100 to 140°C (212 to 285°F) and relatively slow cure rates at temperatures used in molding, that is, 160 to 180°C (320 to 356°F). Moreover, the retention of physical properties on exposure to temperatures above 200°C (392°F) is relatively poor [29, p. 78].

The bisphenol cure system has gradually displaced the diamine system for curing VDF/HFP and VDF/HFP/TFE fluorocarbon elastomers. Curing with bisphenols has the advantage of excellent processing safety, fast cures to high states of cure, excellent final properties, and especially high resistance to compression set at high temperatures [65]. Several bisphenols are suitable for curing these elastomers, but the preferred one is bisphenol AF, chemically 2,2-bis(4-hydroxyphenyl) hexafluoropropane. An accelerator such as benzyltriphenylphosphonium chloride (BTPPC) is necessary, along with inorganic bases, such as calcium hydroxide and magnesium hydroxide with small particle sizes. Typical amounts are 2 phr of bisphenol AF, 0.5–0.6 phr of the accelerator, 3 phr of MgO, and 6 phr of $Ca(OH)_2$ [29, p. 78].

9.3.2.1.2 Cross-Linking by Free Radical Mechanism (Peroxide Cure)

The reaction is activated by an organic peroxide that decomposes thermally during the cure. The fluoroelastomer has to contain reaction sites to produce a sufficiently high cross-link density. Bromine-containing fluoroelastomers form a stable network in the presence of peroxide. However, bromine-based fluoroelastomers were found to cause processing problems, mainly mold fouling. Iodine-based fluoroelastomers were found to be much better since they produce much less mold fouling and are suitable for more sophisticated molding techniques, such as injection molding [3, p. 79]. They also exhibit excellent sealing properties; however, their thermal stability is lower than that of bromine-based fluoroelastomers [3, p. 80]. The use of peroxides for cross-linking requires the addition of a coagent (radical trap), for example, triallyl isocyanurate (TAIC) or triallyl cyanurate (TAC).

Perfluoroelastomers (FFKM) contain a cure site monomer (CSM) that is essential for their cross-linking. Examples of CSMs are:

- Perfluoro(8-cyano-5-methyl-3,6-dioxa-1-octene) (8-CNVE) [37].
- Perfluoro(2-phenoxypropyl vinyl ether).
- VDF.
- Bromine- or iodine- or nitrile-containing monomers.

Each of these has a specific curing behavior and provides vulcanizates with different characteristics. To attain sufficient heat resistance, the compounds require long post-cures at high temperatures, such as 288°C (550°F), in some cases under nitrogen [29, p. 93]. Details of the different systems are in [29, p. 93].

9.3.2.1.3 Cross-Linking by Ionizing Radiation

Fluorocarbon elastomers with ASTM designation FKM are predominantly copolymers or terpolymers of different fluorinated or perfluorinated monomers with vinylidene fluoride, as pointed out earlier. The presence of vinylidene fluoride in their molecules is responsible for their propensity to cross-link by responding to ionizing radiation, that is an electron beam (EB) and gamma rays [4, p. 280]. The final result depends on the ratio of cross-linking to chain scission, which occur simultaneously. Prorads (radiation promoters), such as TAC, TAIC, trimethylolpropane trimethacrylate (TMPTM), trimethylolpropane triacrylate (TMPTA), and N, N'-(m-phenylene) bismaleimide (MPBM), reduce the damage to the elastomeric chain by the radiation [28]. It appears that each fluorocarbon elastomer has the best cross-link yield with a specific prorad. In general, optimized compounds from fluorocarbon elastomers irradiated at optimum conditions attain considerably better thermal stability and mechanical properties than chemical curing systems [39–41]. A typical radiation dose for a sufficient cross-linking of most fluorocarbon elastomers is in the range 10 to 100 kGy.

Perfluoroelastomers (ASTM designation FFKM) are essentially copolymers of two perfluorinated monomers, TFE and PMVE with a CSM, which is essential for their cross-linking. Perfluoroelastomers can be cured by ionizing radiation without any additives. The advantage of radiation-cured FFKM is the absence of any additives, so that the product is very pure. The disadvantage is the relatively low upper-use temperature of the cured material, typically 150°C, which limits the material to special sealing applications only [42].

9.3.2.2 Molding Processes

Molding is a process used to transform a material to a desired shape, dimensions, and surface. In general, elastomeric materials are fabricated by compression, transfer, or injection molding. All these methods are used commercially for fluorocarbon elastomers with a number of factors determining the choice of equipment. In this section, the process of molding also includes curing (vulcanization) of the compounds, so that finished products are thermoset materials. Fluorinated thermoplastic elastomers (see Chapter 10) are also molded but in this case molding methods without curing typical for thermoplastic melt-processible fluoropolymers (see Chapter 7) are used. The compounds used for thermoset curing have to be designed so that they include a delay in the onset of cross-linking to allow sufficient time for the stock to fill the mold cavity and allow a good mold release for the removal of the cured part. Sections 9.3.2.2.1 to 9.3.2.2.3 describe the molding used for curing rubber compounds in general and Section 9.3.2.2.4 focusses on the molding of fluorocarbon elastomer compounds in particular. Additional details on the thermoset molding of fluorocarbon elastomer compounds are in [29, p. 110].

9.3.2.2.1 Compression Molding

Compression molding is the simplest and most widely used molding method. A piece of preformed material is placed directly in the mold cavity and compressed under hydraulic clamp pressure (see Figures 9.12A and 9.12B). At the completion of the required cure cycle, the hydraulic clamp is released, the mold

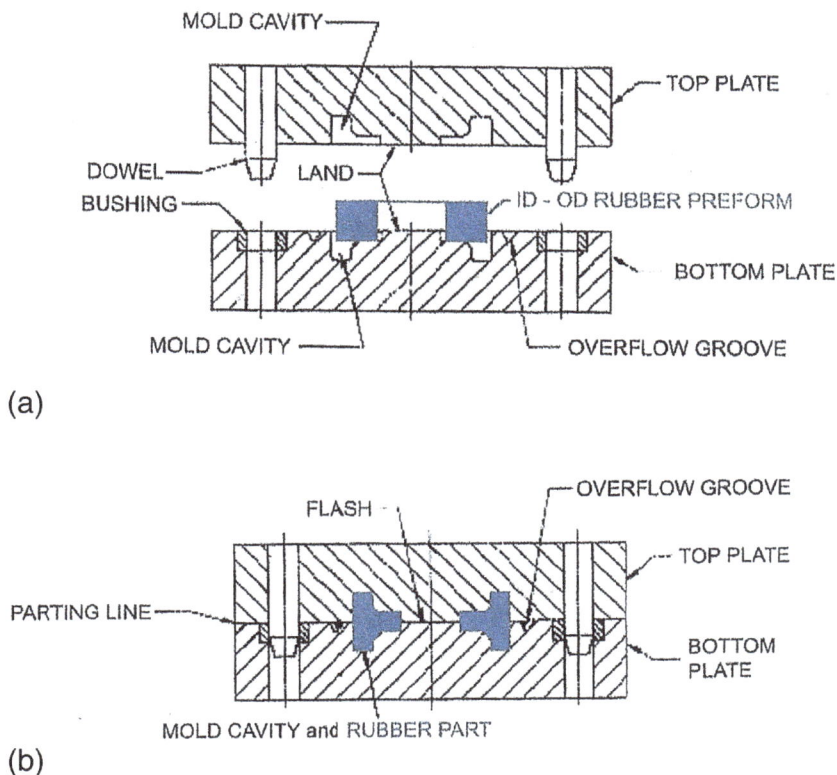

FIGURE 9.12   A. Compression mold open; B. Compression mold closed. (Courtesy of Vanderbilt Chemicals, LLC, copyright successor of R.T. Vanderbilt Company, reprinted with permission)

is opened, and the part is stripped from the mold. Compression molding has several advantages for the fabrication of parts from expensive compounds. Loss of expensive material may be minimized by careful control of preform size. The process is advantageous for the relatively small production volumes of parts of any size.

Compression molding works best with stocks of medium to high viscosity. The disadvantage of compression molding is the high labor cost since the process requires operator attention is given to loading preforms, closing and opening the molds, and removing the cured parts.

9.3.2.2.2 Transfer Molding

Transfer molding involves using a piston and cylinder device to force rubber through small holes into the mold cavity. A piece of uncured compound is put into a part of the mold called the pot, and a plunger then pushes the stock into the closed mold through a sprue. The mold is kept closed while the stock is curing. The plunger is then raised, and the "transfer pad" material is removed and discarded. The mold is then opened for removal of the part; then the flash and sprue material are trimmed off and discarded.

When compared to compression molding, transfer molding provides better product consistency, shorter cycle times, and better bonding of rubber to metal [43]. However, a considerable amount of material is lost as scrap in the transfer pad, sprues, and flash. The stock must have relatively low viscosity and adequate scorch safety for sufficient flow into the mold [44]. The basic three-plate multiple cavity mold is more complex and expensive than a comparable compression mold, but is suited better to the intricate parts or the securing inserts [45]. Several transfer molding variants are described by Ebnesajjad [46]. A schematic of transfer molding is shown in Figures 9.13A and 9.13B.

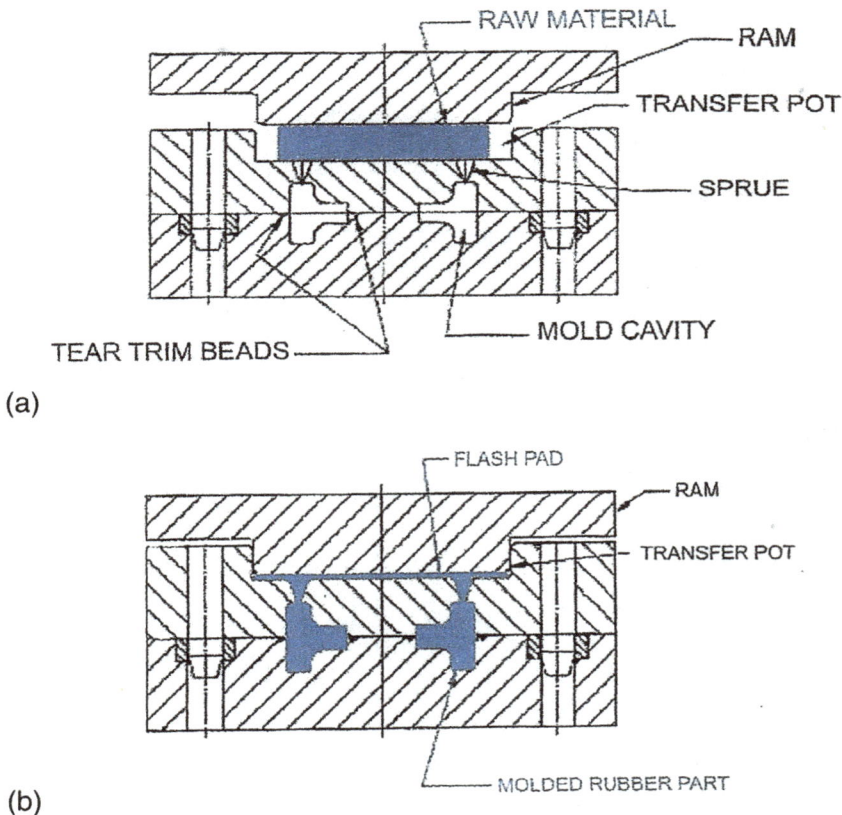

(a)

(b)

FIGURE 9.13 A. Transfer mold open. B. Transfer mold closed. (Courtesy of Vanderbilt Chemicals, LLC, copyright successor of R.T. Vanderbilt Company, reprinted with permission)

9.3.2.2.3 Injection Molding

Injection molding is the most advanced method of molding rubber products [43]. The most widely used equipment is the reciprocating screw machine. The compound is usually fed to the screw as a continuous strip or as pellets from a hopper. As the stock accumulates at the front of the screw, the screw is forced backward a specified distance in preparation for the shot. Then the rotation of the screw is stopped and it is pushed forward to inject the specified amount into the closed mold through the sprue bushing. The stock flows through the runner system into the parts via the sprues. While the rubber cures, the screw is initially held in the injection position to maintain the predetermined pressure to consolidate the stock. Then after a preset time, the screw rotates again to refill the barrel. Then the mold is opened for the removal of the cured part and subsequently closed for the next shot. A schematic of the injection molding process is shown in Figures 9.14A and 9.14B.

Ram or piston injection molding machines are also used for rubber processing [44]. These are somewhat similar to the transfer molding process (see Section 9.3.1.2.2). In this case, the rubber stock is fed to a cylinder, which is heated to the required temperature, and from there it is forced by a hydraulic ram through a nozzle, mold runners, and restrictive gates to the mold. Ram injection molding machines are somewhat lower in cost than the reciprocating screw unit, but are less efficient, especially for high-viscosity stocks.

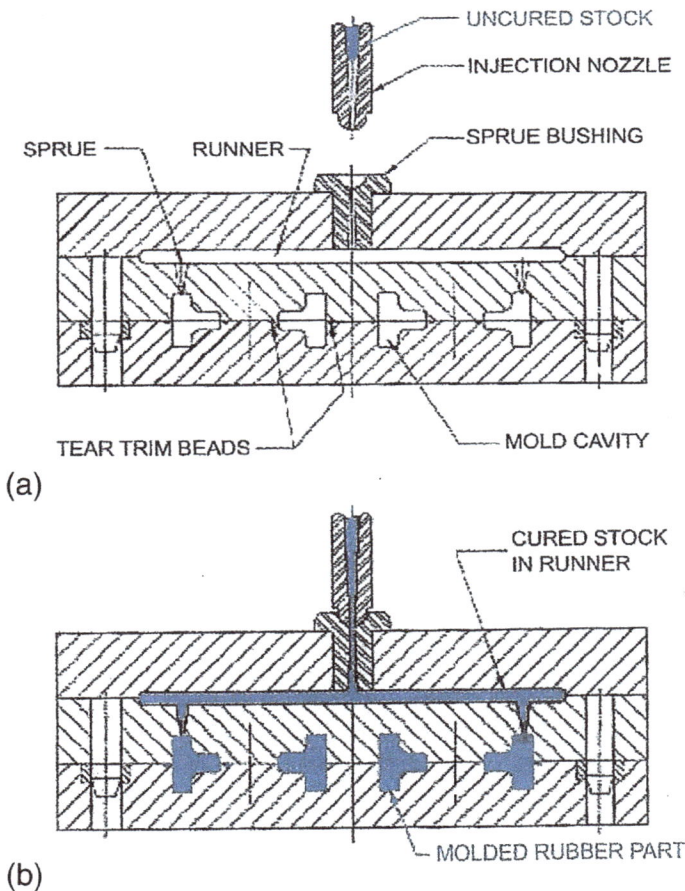

FIGURE 9.14 A. Injection mold before shot. B. Injection mold filled. (Courtesy of Vanderbilt Chemicals, LLC, copyright successor of R.T. Vanderbilt Company, reprinted with permission)

Of all the molding processes, injection molding provides the maximum product consistency, shortest cycle times, and minimum flash. The main disadvantage is the highest investment cost of the machine, molds, and auxiliary equipment. The process is most suitable for high volume production. Fluorocarbon elastomers with low viscosity (less than 30 as measured on the Mooney viscometer at 121°C, or 250°F) are required for injection molding, The use of one or more process aids is essential to enhance mold flow during injection and the easy release of parts from the mold after curing.

Most molds for injection molding are unique in design, which depends on the application, type of elastomer compound, and feed system (hot or cold runners). Standard systems are distinguished as two-plate, three-plate, or stack molds [46].

9.3.2.2.4 Molding of Fluorocarbon Elastomers

The largest volume of fluorocarbon elastomers (about 60% of the total) is processed by compression molding. In the mold design for fluorocarbon elastomer stocks, it is necessary to take into consideration the fact that fluorocarbon elastomers shrink considerably more during cure than standard elastomers (3.0 to 3.5% vs. 1.5 to 2.0%) [46]; the use of vacuum devices improves quality and reduces scrap.

Injection molding is another method to produce parts from fluorocarbon elastomers. It is particularly suitable for small parts such as O-rings, seals, and gaskets produced in large volumes. The nozzle temperature is usually set at 70 to 100°C (158 to 212°F) and the mold temperature at 180 to 220°C (356 to 428°F) [3, p. 85]. The best results are achieved by applying a vacuum during the injection step to avoid air trapping, splitting, and porosity. Typical process conditions for the injection molding of fluorocarbon elastomers are given in Table 9.19 [44].

TABLE 9.19

Typical Process Conditions for Injection Molding of FKM Elastomer Compounds

Machine	Ram Type	Screw Type
Temperature Settings, °C		
Barrel		
Feed zone	80–90	25–40
Middle zone	80–90	70–80
Front zone	80–90	80–100
Nozzle		
Nozzle extrudate	165–170	165–170
Mold	205–220	205–220
Stock in mold	165–170	165–170
Pressure Settings, MPa		
Injection	14–115	14–115
Hold pressure	—	1/2 injection pressure
Back pressure	—	0.3–1
Clamping pressure	Maximum	Maximum
Screw speed, rpm	—	40–60
Time Setting, Seconds (for thin parts)		
Total cycle	58–75	43–60
Clamp	48–65	33–50
Injection	3–5	3–5
Hold	—	10–15
Cure (includes hold)	45–60	30–45
Open-ejection of parts	10	10

Source: Processing Guide Viton® Fluoroelastomer, Technical Information Bulletin VTE-H90171-00-A0703, DuPont Dow Elastomers, 2003.

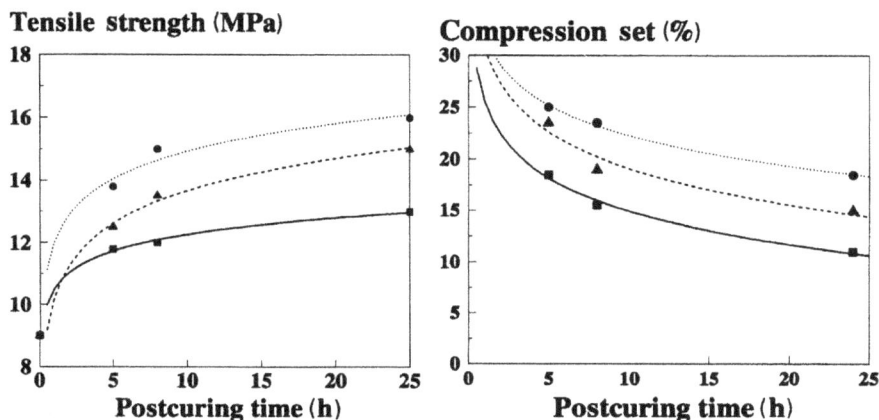

FIGURE 9.15 Effect of post-curing time on tensile strength and compression set From Arcella, V. and Ferro, R. in *Modern Fluoropolymers* (Scheirs, J., Ed.) John Wiley & Sons Ltd., p. 87 (1997). Reprinted with permission. Post-curing temperatures: ■ 200°C, ▲ 225°C, ● 250°C

9.3.3 Post-Curing Process

As pointed out in Section 9.3.2 products made from fluorocarbon thermoset elastomers are cured (vulcanized) typically at temperatures from 170 to 220°C (338 to 428°F). However, to achieve optimum properties, post-curing in a circulating air oven is required to complete the cross-linking reaction and to remove volatile byproducts, including water. Standard post-cure conditions are 18 to 24 hours at 220 to 250°C (428 to 482°F) [3, p. 87]. Figure 9.15 illustrates the effects of post-cure at different temperatures on tensile strength and compression set of a carbon black–filled fluorocarbon elastomer compound [3, p.87].

9.4 Physical and Mechanical Properties of Cured Fluorocarbon Elastomers

As mentioned previously, fluorocarbon elastomers are chemically very stable. They exhibit a unique combination of properties (e.g., resistance to heat, aggressive chemicals, solvents, ozone, UV light) in which they excel over other elastomeric materials. Moreover, they have a very good high-temperature compression set and flexibility at low temperatures. A comparison of the heat aging and oil resistance of typical FKMs and several other elastomeric materials is given in Table 9.20 [19, p. 423].

TABLE 9.20

Heat Aging and Oil Resistance, Comparison of Different Cured Elastomers (ASTM D2000-SAEJ200 Classification)

Type	Heat Aging Temperature, °C (70 h)[a]	Volume Swell, % 10 h/130°C in IRM 903 Oil
Nitrile[b]	100	10, 40, or 60
Polyacrylic[b]	130	0 or 60
Silicone	200 or 225	120 or 80
Fluorosilicone	200	10
Fluorocarbon (FKM)	250	10

Source: Data from [19].

[a] Tensile change + 30%, elongation change −50%, hardness change + 15 points;

[b] varying acrylonitrile content or acrylate content.

TABLE 9.21

Service Life Dependence on Temperature

Limit, Hours of Service	Temperature, °C (°F)
>3,000	230 (356)
1,000	260 (410)
240	290 (464)
48	315 (509)

Source: Data from [19].

TABLE 9.22

Heat Resistance of Cured FKM Elastomers

Formulation	A (Parts by Weight)	B (Parts by Weight)
Viton A	100	—
Viton B	—	100
Diak #3	2	3
Cure, minutes at 163°C (325°F)	30	30
Oven post-cure, h at 204°C (400°F)	24	24

Property	Original		100 Days at 232°C (450°F)		20 Days at 260°C (500°F)		2 Days at 316°C (600°F)		1 Day at 343°C (650°F)	
	A	B	A	B	A	B	A	B	A	B
Tensile strength, MPa psi	15 2,175	15.5 2,250	6.90 1,000	4.31 625	8.62 1,250	3.79 550	7.24 1,050	3.45 500	– Brittle	3.97 575
Elongation at break, %	470	410	360	480	100	400	60	240	—	15
Hardness, Shore A	68	74	87	75	94	83	91	83	99	91
Weight loss, %	—	—	—	—	—	—	18	11	36	22

Source: Data from [19].

9.4.1 Heat Resistance

Vulcanizates from fluorocarbon elastomers can be exposed continuously to temperatures up to 200°C (396°F) almost indefinitely without appreciable deterioration of their mechanical properties. With increasing temperature the time of service is reduced, as illustrated in Table 9.21 [19, p. 427]. An example of the heat resistance of two compounds is shown in Table 9.22 [19, p. 428].

9.4.2 Compression Set Resistance

The largest volume of fluorocarbon elastomers is used for O-rings and seals. In these applications, the compression set is the most important property affecting the performance of the seal. The lowest values of the compression set are achieved when using a phosphonium chloride accelerator system with bisphenol AF or other phenol cures with certain grades of FKM (e.g., Viton™ E-60C or Dyneon™ 2170). Peroxide cures give generally a poorer compression set than bisphenol cures. Coagents for the peroxide curing system have an effect on the compression set: TAIC gives, for instance, a lower compression set than TAC [47].

9.4.3 Low-Temperature Flexibility

Most commercial fluorocarbon elastomers have brittle points between −25°C (−13°F) and −40°C (−40°F). The low-temperature flexibility depends on the chemical structure of the polymer and cannot be improved markedly by compounding. The use of plasticizers may help somewhat, but at a cost of reduced heat

TABLE 9.23

Low-Temperature Properties of Several Different FKM Elastomer Grades

Viton Grade	A-401C	B-50	B-70	GLT-200S
Dyneon grade	FC2144	FT2350	—	—
Tecnoflon grade	FOR532	T636	—	PL455
DAI-EL grade	G7451	G-551	G-671	LT-304
Fluorine, %	66	69	66	65
Brittle point,[a] °C (°F)	−25 to −30 (−13 to −22)	−35 to −40 (−30 to −40)	−35 to −40 (−30 to −40)	−51 (−59)
Clash-Berg[a] @ 69MPa				
°C	−16	−13	−19	−31
°F	(+2)	(+9)	(−3)	(−24)
TR-10, °C	−18	−14	−20	−30

Source: Data from [16, 19].

[a] These values are often difficult to reproduce.

stability and worsened aging. Peroxide-curable polymers may be blended with fluorosilicones, but such blends exhibit considerably lower high-temperature stability and solvent resistance and are considerably more expensive than the pure fluorocarbon polymer. Viton™ GLT is a product with a low brittle point of −51°C (−59°F) [48]. Tecnoflon® PL455 containing a stable fluorinated amide plasticizer reportedly exhibits improved low-temperature hardness, brittle point, and compression set without sacrificing physical properties [48]. The low-temperature characteristics of selected fluorocarbon elastomers are listed in Table 9.23 [19, p. 430].

9.4.4 Resistance to Automotive Fuels

The use of aromatic compounds in automotive fuels, higher under-the-hood temperatures, combined with automotive regulations, presents a challenge for the rubber parts (e.g., hose, seals, diaphragms) used in vehicles. Traditional elastomers do not have a high enough resistance to meet all these requirements, but fluorocarbon elastomers do. They are being used successfully, for example, in automotive hoses for gasoline/alcohol mixtures and "sour" gasoline (containing peroxides), where epichlorohydrin copolymer depolymerizes and nitrile butadiene rubber (NBR) materials embrittle [49]. Moreover, studies of permeation have shown that FKM hose has superior resistance to permeation in comparison with other fuel-resistant elastomeric materials, with permeation rates often over 100 times lower [19, p. 432]. The swelling of selected fuel-resistant elastomeric materials is shown in Table 9.24 [19, p. 431].

TABLE 9.24

Swelling of Different Cured Elastomers in Fuel Blends

Fuel	Fuel Composition (Volume %)				
Gasoline (42% aromatic)	100	85	75	85	75
Methanol	—	15	25	—	—
Ethanol	—	—	—	15	25
Rubber	**Equilibrium Volume Increase, % at 54°C (129°F)**				
Viton AHV	10.2	28.6	34.9	19.7	21.1
Viton B	10.2	22.9	25.6	17.3	18.3
Viton GF	7.5	13.6	14.6	12.6	13.1
Fluorosilicone (FMQ)	16.4	25.3	26.6	23.1	23.0
Nitrile rubber (NBR)	40.8	90.5	95.8	62.4	66.6
Epichlorohydrin (ECO)	42.4	92.6	98.1	75.6	78.5

Source: Data from [19].

9.4.5 Resistance to Solvents and Chemicals

As pointed out earlier, fluorocarbon elastomers are highly resistant to hydrocarbons, chlorinated solvents, and mineral acids. Vulcanizates from them swell excessively in ketones and in some esters and ethers. They also are attacked by amines, alkalis, and some acids, such as hot anhydrous hydrofluoric acid and chlorosulfonic acid [19, p. 430]. Generally, stability and solvent resistance increase with increasing fluorine content, as shown previously in Table 9.3.

Other than the type of fluorocarbon elastomer used, the main determinant of resistance to acids is the metal oxide used in the compound. Compounds of FKM contain litharge swell markedly less than those containing magnesium oxide or zinc oxide [19, p. 430]; the current tendency is to remove and replace the litharge because of its toxicity. Compounds based on KALREZ® and AFLAS® are considerably more resistant to strong alkalis and amines than are compounds based on FKM [19, p. 430]. FKM terpolymers cured with peroxides exhibit exceptional resistance to wet acidic exhaust gases in desulfurization systems in coal-fired plants [19, p. 427].

9.4.6 Steam Resistance

Resistance to the steam of FKM-based vulcanizates increases with fluorine content. Peroxide cures are superior to diphenol and diamine cures. Compounds based on AFLAS® and particularly on Kalrez® surpass FKM in this respect [19, p. 431].

9.5 Formulation of Compounds of Fluorocarbon Elastomers

When compounding fluorocarbon elastomers, the basic principles are the same as for other elastomers. The selection of the elastomer grade and of the remaining compounding ingredients depends on the required physical and chemical properties of the vulcanizate (cured compound) as well as on the desired behavior of the compound during processing and curing.

A typical FKM compound usually contains the following ingredients: one or more fillers, an acid scavenger, and a curing system (cross-linker). Inorganic or organic colorants are used for colored compounds. The development of a compound requires a great deal of experience and understanding of the chemistry involved and of the interactions among the individual ingredients. However, the compounding of fluorocarbon elastomers is relatively simpler than that of other types of elastomers [3, p. 81].

9.5.1 Fillers

The type and amount of filler affect not only the final properties of the vulcanizate but also the processing behavior of the compound. Since the compounds stiffen very soon after mixing, only relatively small amounts of fillers, typically 10 to 30 phr, can be used [50].

Various carbon blacks are used for black compounds. Medium thermal black (N990) is the most widely used grade, because it offers the best compromise between physical properties and cost. More reinforcing grades of carbon blacks, such as N774 or N750, produce a higher hardness and better physical properties at the expense of somewhat higher compression set and cost. Lowest compression set values are obtained with Austin black [50].

White (silica) fillers, often surface treated, are sometimes used to improve flow, moisture resistance, and tensile properties [51–53]. Fillers commonly used in fluorocarbon elastomers are listed in Table 9.25.

9.5.2 Acid Acceptor Systems

Acid acceptors serve the purpose of neutralizing the hydrogen fluoride generated during the cure or on prolonged aging at high temperatures. The compounds used for that purpose are listed in Table 9.26. Low-activity magnesium oxide is used in diamine cures and *not* in bisphenol cures. High-activity magnesium oxide is used in bisphenol cures and *not* in diamine cures. Lead oxide (PbO) is optimum, where

TABLE 9.25

Common Fillers for Fluorocarbon Elastomer Compounds

Filler	Comments
Austin Black (coal fines)*	Better high-temperature compression set resistance than MT carbon black, but less reinforcing and poorer in processing; lower tensile strength and elongation at break
Blanc Fixe (barium sulfate)	Best compression set resistance of nonblack fillers; neutral filler good for colored compounds; poorer for tensile strength than MT carbon black
Conductive carbon black (e.g., Vulcan XC-72)	Used for conductive compounds and for compounds with static dissipation
Diatomaceous silica (calcium metasilicate)	General-purpose neutral filler that provides good tensile strength
Fibrous $CaSiO_3$ (e.g., Nyad 400)	General purpose mineral filler; neutral and good for control stocks; tensile strength comparable with MT carbon black
Graphite powder	Used combined at 10 to 15 phr with other fillers to improve wear resistance
N762 (SRF) carbon black	High strength, high modulus compounds; aggravates mold sticking in peroxide cures; compression set not too good
N990 (MT) carbon black	Best general purpose filler; excellent compression set and heat aging
Precipitated calcium carbonate	Good, general-purpose mineral filler that exhibits the least amount of change in physical properties due to ambient moisture
PTFE micropowders	Adding up to 5% of PTFE micropowder can improve the friction properties, wear resistance, tear strength, and demolding behavior of the cured material
Red iron oxide	Used with other mineral fillers for red-brown stocks
Titanium oxide	Goof filler for light colored compounds; provides good tensile strength but features poorer heat aging than other fillers

Sources: [16, 50] and author's own working experience.
* Proper post-cure oven loading and ventilation must be used when working with Austin Black to prevent an oven fire.

TABLE 9.26

Common Acid Acceptors for Fluorocarbon Elastomer Compounds

Acid Acceptor	Recommended Usage	Caution Information
Magnesium oxide, low activity	General purpose for diamine cure	Sensitive to water and acids
Magnesium oxide, high activity plus calcium hydroxide	General purpose for bisphenol cure	Sensitive to water and acids
Zinc oxide	To improve heat aging, tear resistance, and safety for peroxide cures	Results in a lower rate and state of cure; reducing water and acid resistance
Calcium oxide	To minimize fissuring, improve adhesion, and reduce mold shrinkage	Difficult to disperse; is hygroscopic; causes slow cure rate
No metal oxide (APA peroxide cured Viton™)	To improve water, steam, and acid resistance and replace old lead oxide formulations	Lack of metal oxide reduces heat resistance above 200°C (392°F)

Source: [16, 50].

the vulcanizate is exposed to hot acids, and dibasic Pb-phosphite with ZnO for exposure to steam or hot water. However, lead-based curing systems have largely been abandoned due to environmental and health concerns. Superior performance in dry heat is achieved with CaO and MgO [50].

9.5.3 Curatives

Generally, as discussed previously, the mechanism involved in the cross-linking of fluorocarbon elastomers is the removal of hydrogen fluoride to generate a cure site that then reacts with diamine [54], bisphenol [55], or organic peroxides [56] that promote a radical cure by hydrogen or bromine extraction. Preferred amines have been blocked diamines such as hexamethylene carbamate (Diak™ #1) or

TABLE 9.27

Relative Performance of Various Fluorocarbon Elastomer Cure Systems

Process and Physical Property Characteristics	Diamine Cure (Diak #1, #3, #4)	Bisphenol Cure	Peroxide Cure
Scorch safety	Poor to fair	Good to excellent	Good to excellent
Balance of fast cure and good scorch safety	Poor	Excellent	Excellent
Mold release	Good	Excellent	Good to excellent
Ability to single pass Banbury mix	#1 No; #3 Risky; #4 OK	Yes	Yes
Adhesion to metal	Excellent	Good	Good
Tensile strength	Good to excellent	Fair to excellent	Good to excellent
Compression set	Fair	Excellent	Good
Resistance to water, steam, acids	Fair	Good	Excellent

Source: [16, 50].

bis(cinnamylidene) hexamethylene diamine (Diak™ #3). Preferred phenols are hydroquinone and bisphe-nols such as 4,4′-isopropylidene bisphenol or the corresponding hexafluoro-derivative bisphenol AF.

The nucleophilic curing system is most common and is used in about 80% of all applications. It is based on the cross-linker (bisphenol AF) and accelerator (phase transfer catalyst, such as phosphonium or amino-phosphonium salt). Both diaminic and bisphenol type cure systems are permitted by U.S. Food and Drug Administration (FDA) regulations governing rubber articles in contact with food. The diaminic curing system is also used in some coating and extrusion applications [3, p. 81].

Peroxidic cure systems are applicable only to fluorocarbon elastomers with cure sites that can generate new stable bonds. Although peroxide-cured fluorocarbon elastomers have inferior heat resistance and compression set, compared with bisphenol cured types they develop excellent physical properties with little or no post-curing. Peroxide cured fluorocarbon elastomers also provide superior resistance to steam, acids, and other aqueous solvents because they do not require metal oxide activators used in bisphenol cure systems. Their difficult processing was an obstacle to their wider use for years, but recent improve-ments in chemistry and polymerization offer more opportunities for this class of elastomers [3, p. 81].

Solid fluorocarbon elastomers are commercially available as pure gum polymers or precompounded grades with a bisphenol-type curing system included. Some precompounded stocks include processing aids, adhesion promoters, or other application-specific additives. The relative strengths and weaknesses of commonly used curing systems are listed in Table 9.27.

Precompounded grades are optimized by the supplier to provide the best combination of accelerator and cross-linker for a given application [19, p. 418]. Then, the final compounding consists of only the addition of fillers, activators, and other ingredients needed to achieve the required physical properties and processing characteristics.

Although development of a formulation for a specific product and process requires a great deal of knowledge and experience, there are some basic rules typical of FKM compounding. The levels of acid acceptor (MgO) and activator (Ca(OH)$_2$) in the bisphenol cure system strongly affect not only the cross-link network as reflected by the physical properties of the material but also the behavior of the compound during vulcanization. Therefore, the curing system must be optimized to achieve the best balance of properties. Examples of formulations for different curing systems are given in Table 9.28.

9.5.4 Plasticizers and Processing Aids

The processing behavior of fluorocarbon elastomers can be improved by the addition of small amounts of plasticizers and processing aids. High-molecular-weight hydrocarbon esters, such as dioctyl phtha-late (DOP) and pentaerythritol stearate, are effective plasticizers in fluoroelastomer compounds. Lower-molecular-weight esters also soften such compounds, but they reduce their high-temperature stability because they are less stable than fluorocarbons and highly volatile at the usual service temperatures. Carnauba wax, low-molecular-weight polyethylene (e.g., AC-617), and sulfones act as good processing

TABLE 9.28

Examples of Formulations for Different Curing Systems

Ingredients	Curing System Parts by Weight)				
	Diamine	Bisphenol		Peroxide	
FKM	100	100	100*	—	—
FKM with CSM[a]	—	—	—	—	100**
Magnesium oxide, low activity	15	—	—	—	—
Magnesium oxide, high activity	—	3–6	3	—	—
Calcium hydroxide	—	6	6	—	—
Zinc oxide	—	—	—	0–3	2–6
Filler(s)	5–50	5–60	5–60	5–60	5–60
Process aids	1–4	0–2	0.5–2.0	0.5–1.5	0.5–1.5
Diamine curative	1–3	—	—	—	—
Bisphenol curative	—	3–8	*	—	—
Peroxide	—	—	—	1.25–3.00	2–6

Source: Data from [16, 50].
[a] Cure site monomer;
* bisphenol curatives incorporated into fluoroelastomer;
** APA peroxide G-types like GF-S, GBL-S, GLT-S, and GFLT-S.

aids. These additives ensure improved calendering, smoother extrusion, and an improved flow in molds. Low-molecular-weight polyethylene should not be used in compounds with peroxide-curing systems because it aggravates mold sticking. Other commercially available processing aids are low-viscosity fluorocarbon elastomers that improve processing without having an adverse effect on the physical properties of the vulcanizate [19, p. 419].

9.6 Applications of Fluorocarbon Elastomers

9.6.1 Applications of FKMs

The estimated market shares of fluoroelastomers is [57]:

    Automotive 65%.
    Aerospace 4%.
    Oil/gas/chemical 14%.
    Industrial 10%.
    Other 7%.

Annual growth is estimated at approximately 5 to 8%, mainly for new applications or replacement of parts made previously from inferior elastomers. O-rings and gaskets consume about 30 to 40%, shaft seals and oil seals about 30%, and hoses and profiles 10 to 15% [3, p. 88]. Because of their high price, fluorocarbon elastomers are used in special applications with very high demands on high-temperature resistance and resistance to corrosive chemicals and hot oils. They are most widely used in molded and extruded products, mainly gaskets, used in: the aircraft, aerospace, and automotive industries; hoses; membranes; rubber covered rolls; fabrics; and in flame resistant coatings on flammable substrates. Because of their dielectric properties, they are used in electrical insulations for low voltages and frequencies, when resistance to heat and aggressive chemicals is required.

Other applications include the food industry, binders for solid rocket fuels, and expanded (foamed) rubber. In the latex form, FKM can be used for coated fabrics and as a binder for fibrous materials. A comprehensive review of the current applications of FKM elastomers is given in [6, Chapter 11]. Typical current applications are listed in the following sections.

FIGURE 9.16   Valve stems and valve seats. (Courtesy of Daikin)

9.6.1.1 *Typical Automotive Applications*

- Valve stems and valve seals (Figure 9.16).
- Shaft seals.
- Transmission seals.
- Engine head gaskets.
- Water pump gaskets.
- Seals for exhaust gas and pollution control equipment.
- Bellows for turbo-charger lubricating circuits.
- Fuel-handling systems including diaphragms for fuel pumps (see Figures 9.17), fuel hose or fuel hose liner (Figure 9.18), injection or nozzle seals, needle valves, filter casing gaskets, fuel shutoff valves, and carburetor parts.

9.6.1.2 *Typical Aerospace and Military Applications*

- Shaft seals.
- O-ring seals in jet engines (Figure 9.19).
- Hydraulic hose.
- O-ring seals in fuel, lubricant, and hydraulic systems.
- Fuel tanks and fuel tank bladders.
- Manifold gaskets.
- Lubricating systems.
- Electrical connectors.
- Gaskets for firewalls.

FIGURE 9.17   FKM diaphragms. (Courtesy of Diacom Corporation)

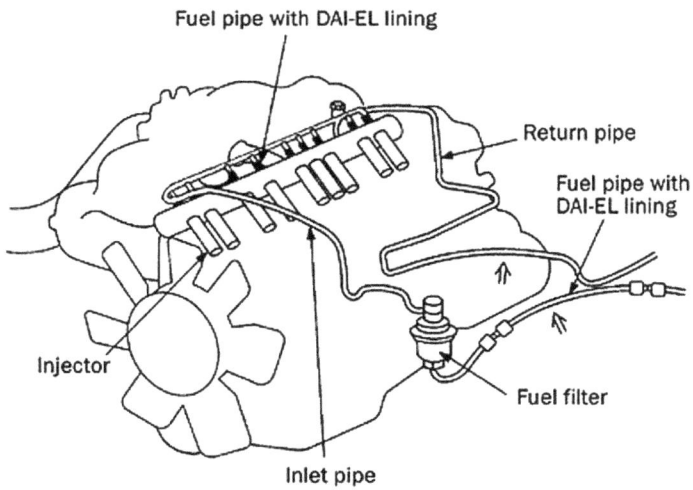

Fuel pipe with DAI-EL lining

Return pipe

Fuel pipe with DAI-EL lining

Injector

Fuel filter

Inlet pipe

FIGURE 9.18   Fuel pipe with FKM lining. (Courtesy of Daikin)

- Traps for hot engine lubricants.
- Heat-sealable tubing for wire insulation.
- Tire valve stem seals.
- Flares.

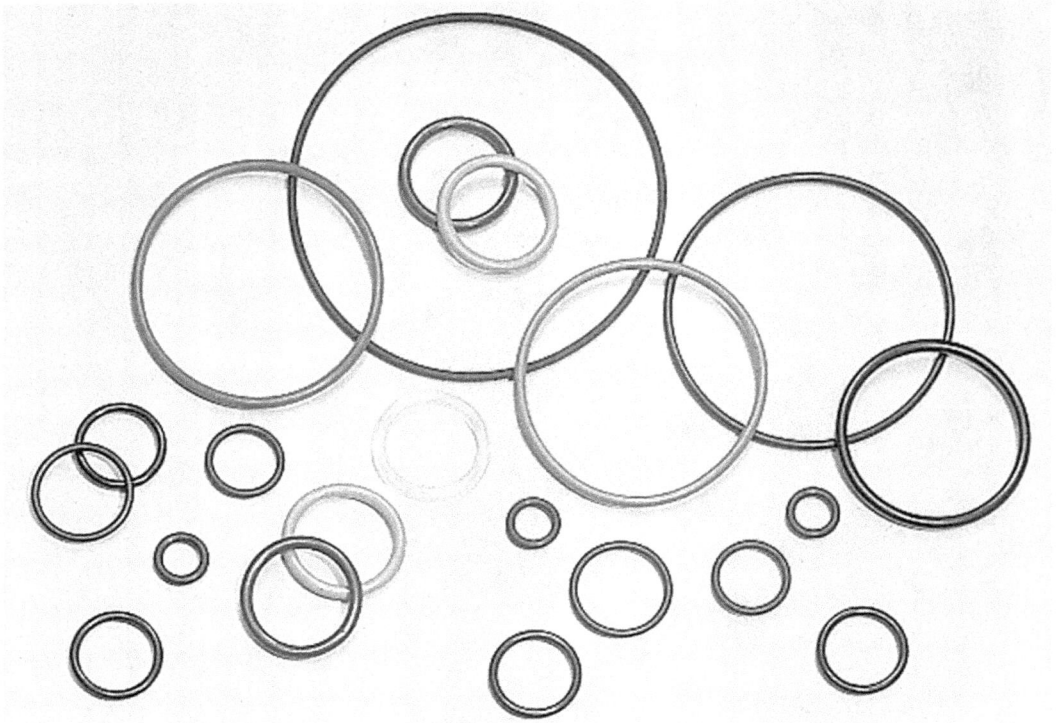

FIGURE 9.19   FKM O-rings for different applications. (Courtesy of Daikin)

9.6.1.3 *Typical Chemical and Petrochemical Applications*

- O-rings (as shown in Figure 9.19).
- Expansion joints.
- Diaphragms (as shown in Figure 9.17).
- Blow-out preventers.
- Valve seats.
- Gaskets.
- Hose.
- Safety clothing and gloves.
- Stack and duct coatings.
- Tank linings.
- Drill bit seals.
- V-ring packers.

9.6.1.4 *Other Industrial Applications*

- Valve seals.
- Hose (rubber-lined or rubber-covered).
- Wire and cable covers (in steel mills and nuclear power plants).
- Diaphragms (as shown in Figure 9.17).
- Valve and pump linings.

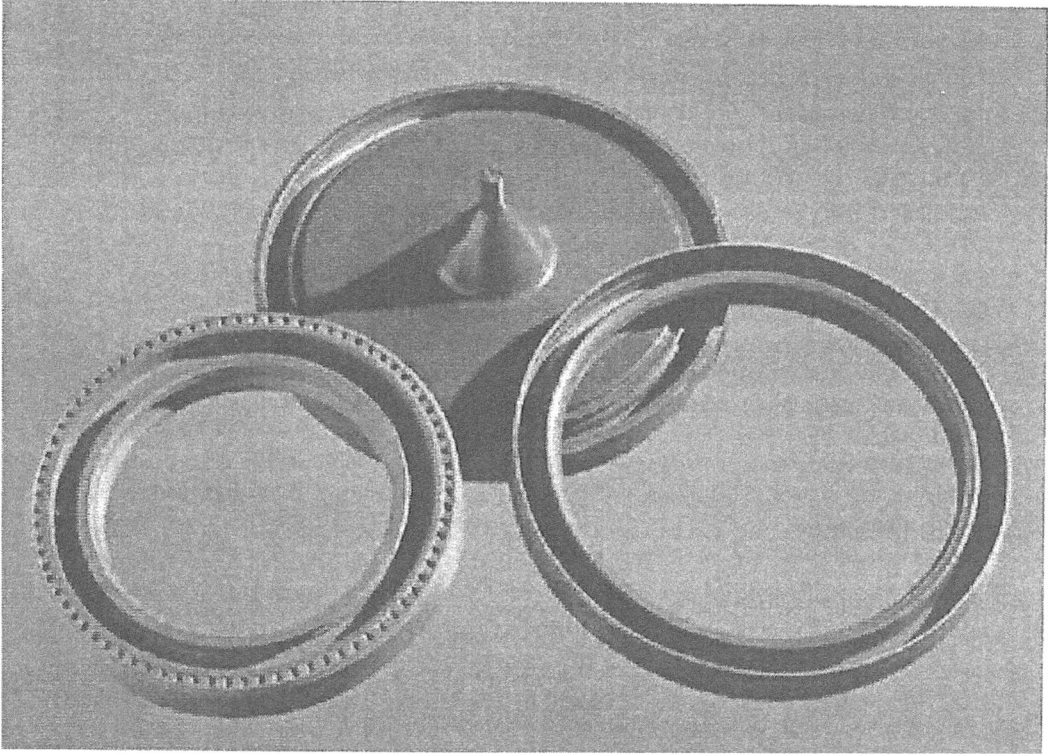

FIGURE 9.20   Molded parts from FKM. (Courtesy of Solvay Solexis)

- Reed valves.
- Rubber-covered rolls (100% fluorocarbon elastomer or laminated to other elastomers).
- Electrical connectors.
- Pump lining and seals.
- Different molded parts (Figure 9.20).
- Seals in food-handling processes approved by Food and Drug Administration.

9.6.2 Applications of FFKMs

Perfluoroelastomers (FFKMs), such as Kalrez™, Chemraz®, Perfluor®, and Tecnoflon®, are particularly suited to extreme service conditions. They are resistant to more than 1,800 chemical substances, including ethers, ketones, esters, aromatic and chlorinated solvents, oxidizers, oils, fuels, acids, and alkali and are capable of service at temperatures up to 327°C (620°F) [58, 59]. Because of the retention of resilience, low compression set, and good creep resistance, they perform extremely well as static or dynamic seals under conditions where other materials, such as metals, FKM, PTFE, and other elastomers, fail. Parts from FFKMs have very low outgassing characteristics and can be made from formulations which comply with FDA regulations [58]. Primary areas of application of FFKMs are paint and coating operations, oil and gas recovery, semiconductor manufacture, the pharmaceutical industry, the chemical process industry, and the aircraft and aerospace industry [60]. Examples of FFKM applications are [58, 59]:

- O-ring agitator shaft seals in an oxidation reactor operating at temperatures above 220°C (428°F) and in contact with 70% acetic acid.

- Mechanical seals of a process pump in a chemical plant pumping alternately acetone, dichloromethane, and methyl isocyanate at elevated temperatures.
- Pipeline seals exposed to chloromethyl ether at elevated temperatures.
- Pipeline seals exposed to dichlorophenyl isocyanate at elevated temperatures.
- Seals for outlet valves exposed to a 50/50 mixture of methylene chloride/ethanol at ambient temperature.
- Mechanical seals of a pump handling a mixture of ethylene oxide and strong acids at 200°C (390°F) and high pressures.
- Static and dynamic seals in a pump for hot asphalt at 293 to 315°C (560 to 600°F).
- O-ring seals in a pump handling 99% propylene at –45°C (–50°F).
- O-ring seals in a pump pumping chromate inhibited water at 196°C (385°F).

Since FFKM parts are primarily used in fluid sealing environments, it is essential to pay attention to seal design parameters, especially as they relate to the mechanical properties of the elastomeric material being used. The sealing performance depends on the stability of the material in the fluid, its mechanical properties, mechanical design, and installation of the seal [60, 61]. A variety of O-rings made from FFKMs are shown in Figure 9.21.

9.6.3 Applications of FEPM

The FEPM designation was originally directed at the copolymers of TFE and P. TFE/P provides a unique combination of chemical, heat, and electrical resistance. Chemically, TFE/P resists both acids and bases, as well as steam, amine-based corrosion inhibitors, hydraulic fluids, alcohol, and petroleum fluids. TFE/P is also resistant to ozone and weather. It typically retains its remarkable chemical resistance even at high temperatures (short exposures up to 232°C, or 450°F), and tests have shown that electrical resistance

FIGURE 9.21   O-rings from FFKM. (Courtesy of DPA)

actually improves with heat exposure (see Section 9.2.3.7). The first TFE/P compound to be commercially marketed was Aflas® (a product of Asahi Glass). In a sense, Aflas defined the initial boundaries for base-resistant materials.

The basic elastomer is available in molecular weights ranging from 40,000 to 135,000. The higher molecular weight products are normally used for mechanically demanding applications. Lower molecular weight polymers are used for calendered and extruded products where resistance to chemical attack is the primary requirement [21]. Because of their unique compatibility with the complete range of common hydraulic fluids such as esters (di- and poly-), polyglycols, and hydrocarbon-based oils, they are suitable for applications such as downhole oilfield service, as seals and gaskets for high-temperature engine lubricants and engine coolants stabilized by amines and amine-type inhibitors [21], and for multilayer automotive hoses. Their high volume resistivity [21] makes them useful for heat resistant and chemically resistant insulation of high-voltage wires and cables. Additional applications of FEPM elastomers are the aerospace and chemical process industries.

9.6.4 Applications of FKMs in Coatings and Sealants

Liquid FKM-based systems were developed to satisfy the need for products that combine the physical properties of solid fluorocarbon elastomers, such as excellent chemical and heat resistance, in a form that is easy to apply and versatile to use. There are two main types of liquid fluorocarbon elastomer system, namely the solvent-borne and water-borne. Both types are based mainly on VDF-HFP or VDF-HFP-TFE and contain only relatively small amounts of filler to obtain soft flexible coatings. Solvent-borne liquid systems are made by dissolving compounds of low-viscosity FKM elastomers, such Viton™ A-35 or Dyneon™ 2145, in methyl ethyl ketone, ethyl acetate, methyl isobutyl ketone, amyl acetate, or other related ketones [19, p. 432]. These systems are cured with amines, bisphenol A, or peroxides depending mainly on the end-use properties required. The products have a typical useful storage life of seven days at 24°C (75°F) and cure within two weeks [62]. Daikin offers a solvent-borne DAI-EL™ DPA-382 system and several DAI-EL™ GL and GLS water-borne systems. One-part and two-part water-borne FKM coatings are commercially available [63–66]. They are used in the following applications [67]:

- Flue gas desulfurization units in coal-fired power plants.
- Wind turbine blades.
- Fuel cells.
- Mesh covering.

One-part and two-part adhesives and sealants based on 3M and Chemours FKMs are used for the following applications [68, 69]:

- Sealing of flue duct expansion joints in power plants.
- Door gaskets on industrial ovens.
- Joint sealants for steel and concrete secondary containment areas.
- Acid resistant joint sealants for industrial floorings.
- Bonding FKM gaskets to metal.
- Adhesives for splicing and bonding fluorocarbon elastomers and O-rings.
- Adhesives used in jet engine maintenance.
- Coating of fuel injector hoses in cars.
- Coatings of rubber and metal rollers in the printing industry.

9.6.5 Applications of FKMs as Polymeric Processing Additives

There has been long standing interest in fluorinated polymeric processing additives because of the improvements that adding small concentrations of these compounds impart to processing polymers.

Technically, the effect is an increase in the critical shear rate of the major phase polymer. Generally, thermoplastic materials must be processed below the velocity at which melt fracture occurs, referred to as the "critical shear rate". Melt fracture in molten plastics takes place when the velocity of the resin in flow exceeds the critical velocity, the point where the melt strength of the polymer is surpassed by internal stresses. A melt-fractured extrudate is no longer smooth, often with a washboard appearance, and has a milky appearance [70]. Fluorocarbon elastomers are known to be effective polymeric processing aids. There is a multitude of fluoroelastomeric polymeric processing aids (PPAs). The manufacturers retain the most important proprietary information, and manufacturing details are protected by patents. This section covers this subject to a limited extent.

The earliest reports of this application is in [71]. A small amount (0.005–2.000 weight %) of a copolymer of vinylidene fluoride and hexafluoropropylene (Viton™ A) was added to poly (ethylene-butene-1) and high density polyethylene. There was a drastic reduction in the tendency of hydrocarbon polymers to undergo melt fracture as a result of adding Viton™ A. This effect was measured by comparing the transparency of extrudates with and without Viton™ A. Adding as little as 500 ppm of Viton™ A increased the transparency by over one order of magnitude from 5.7 to 69.0%.

In [72] the development of free flowing powders of perfluoroelastomers which could be easily added to other polymer powders as a processing aid is reported. An antiblocking agent such as silica, calcium silicate, or calcium carbonate was added before or during the spray-drying of the polymer powder. The free flow powder containing an antitack or antiblocking agent is relatively easy to handle during the process because it does not cake or bridge easily.

Another development reported in [73] is that extrusion of thermoplastic polyolefins, especially linear low density polyethylene, was improved by the use of a processing additive comprising a blend of a thermoplastic acrylic polymer and a fluorocarbon elastomer. The processing additive is a homogeneous blend of a thermoplastic styrene/methyl methacrylate polymer and a copolymer of vinylidene fluoride and hexafluoropropylene containing less than 15% by weight of hexafluoropropylene. The polyolefin extrudate produced according to this invention has reduced levels of melt fracture when extruded in conventional polymer extrusion equipment.

Dyneon™ [74] has suggested that the use of multimodal fluorinated polymers is an effective PPA. The term "multimodal terpolymer" means a terpolymer having two or more discrete molecular weight ranges. The PPA compositions contain a fluoroelastomer that is a copolymer of an ethylenically unsaturated fluoromonomer, tetrafluoroethylene, and at least one ethylenically copolymerizable monomer like ethylene or propylene. For example, useful PPAs are made by copolymerizing tetrafluoroethylene, hexafluoropropylene, and ethylene or propylene.

Useful multimodal fluoropolymers can be prepared in a number of ways. For example, the polymer can be produced by means of a suitable polymerization process ("step polymerization"). This process employs the use of specific initiators and chain transfer agents such as short-chain alkanes and halogen alkanes plus hydrogen.

In current commercial applications, amounts of fluoropolymer (elastomer or plastic) dispersed in hydrocarbon thermoplastics can greatly improve their extrusion characteristics, reducing melt fracture and die buildup. These improvements are especially important in the film extrusion of high density polyethylene (HDPE) and linear low density polyethylene (LLDPE) resins. To serve this market, Chemours™ offers FreeFlow™ additives, Dyneon offers Dynamar™, Solvay offers Technoflon® NM FKM, and AGC Chemicals offers Fluon® Polymer Processing Additives (PPAs). These processing aids form a non-stick fluoropolymer coating on the inside of the die, reducing friction so that the resin flows freely and more rapidly through the die to produce an extrudate with smooth surfaces.

9.7 Examples of Fluorocarbon Elastomer Formulations

These examples were collected in good faith from the available literature based on experimental work and in many cases on industrial practice. However, we cannot verify any of the details and/or guarantee the results when used. Readers are advised to contact the manufacturers of the ingredients to be used.

9.7.1 Typical Formulations of Fluoroelastomer Compounds

Ingredients	Diamines	Bisphenol		Peroxides**
FKM elastomer	100	100	100*	100
Magnesium oxide	15	3	3	—
Calcium hydroxide	—	6	6	—
Zinc oxide	—	—	—	0–3
Filler(s)	5–50	5–60	5–60	5–60
Process aid(s)	1–4	0–2	0.5–2	0.5–1.5
Diamine curatives	1–3	—	—	—
Bisphenol curatives	—	3–8	*	—
Coagent(s)	—	—	—	1–4
Peroxide	—	—	—	1.25–3

Source: Data from [15, 16].
* Bisphenol curatives are incorporated into the elastomer;
** APA peroxide G-types like GF-S, GBL-S, GLT-S, GFLT-S.

9.7.2 Rotary Seal for Airborne Applications

Formulation

Ingredient	Amounts (Parts by Weight)
Viton™ A-HV	100
Magnesium oxide	15
N990 (medium thermal) carbon black	60
Copper Inhibitor 55	0.50
Diak #1	0.25
Total	**175.75**

Physical Properties from Samples Press, Cured for 30 min at 140°C (284°F)

Property	Value
Tensile strength, psi (MPa)	3.000 (20.7)
Elongation at break, %	120
Hardness, Shore A	85–90

Source: Data from [75].

9.7.3 Compounds for Compression Molded Seals

Ingredient	Amounts, Parts by Weight	
	Compound I	Compound II
Viton A	100	100
N990 (medium thermal) carbon black	15	15
Magnesia	20	15
Calcium oxide	—	3
Diak #3	3	—
HMDAC[a]	—	1.2
Total	**138.0**	**134.2**

[a] Hexamethylene diamine carbamate (curing agent).

Typical Properties

Scorch	Compound I	Compound II
Mooney Scorch, MS at 250°F, min.	25+	44
Minimum reading, units	44	50
Physical properties: Press-cured for 30 min at 300°F (149°C); post-cured in oven for 24 h at 400°F (204°C)		
Original physical properties		
Tensile strength, psi (MPa)	2,640 (18.2)	1,750 (12.0)
Elongation at break, %	305	195
Hardness, Durometer A	65	73
Aged in oven for 16 h at 400°F (204°C)		
Tensile strength, psi (MPa)	1,500 (10.3)	1,460 (10.1)
Elongation at break, %	160	170
Hardness, Durometer A	79	80
Aged in oven for 2 days at 600°F (316°C)		
Tensile strength, psi (MPa)	Brittle	1,160 (8.0)
Elongation at break, %	—	—
Hardness, Durometer A	98	87
Compression set (ASTM D395, Method B), 22 h at 450°F (232°C)		
Compression set, %	47	38

Source: Data from [76]

9.7.4 Basic O-Ring Compounds

Formulation

Ingredient	Compound				
	I	II	III	IV	V
Dyneon™ FE 5640Q*	100	100	100	100	100
N990 (MT) carbon black	2	9	20	35	60
Ca(OH)$_2$	4	6	6	6	6
MgO (Elastomag® 170)	3	3	3	3	3
Total Parts	**109**	**118**	**129**	**144**	**169**

* Incorporated cure dipolymer.

Physical Properties, Cured in Press, Post-Cured for 16 h at 232°C (450°F)

Property	I	II	II	IV	V
Hardness Shore A (specified)	**55**	**60**	**70**	**80**	**90**
Tensile strength, psi (MPa)	1100 (7.6)	1490 (10.1)	1945 (13.4)	1840 (12.7)	1830 (12.6)
Elongation at break, %	240	240	240	185	150
Modulus, 100%, psi	220	340	565	930	1325
Hardness, Shore A (measured)	55	60	70	81	91
Specific gravity	1.85	1.85	1.86	1.85	1.85
Air aging, 70 h at 270°C (518°F), ASTM D573					
Tensile strength, % change	−12	−9	−26	−22	−17
Elongation at break, % change	19	6	−4	−8	−10
Hardness, Shore A, points change	−4	−3	−3	0	−4
Compression set, 70 h at 200°C (392°F), ASTM D395					
Compression set, O-ring 0.139 in., %	13	14	15	16	22
Molded 0.50 in. button, set, %	16	11	13	14	18

Source: Data from [77].

9.7.5 Compound for Closed Cell Sponge

Formulation

Ingredient	Amounts (Parts by Weight)
FKM	100
Magnesia (low activity)	15
N990 (medium thermal) carbon black	25
Petrolatum	3
Diak #1 (curing agent)	1.25
Cellogen AZ (blowing agent)	5
Diethylene glycol	2
Total	**151.25**

Properties, Cured in Beveled Compression Mold for 30 min. at 163°C (325°F)

Property	Value
Density, lb/cu.ft. (kg/m³)	22 (352)
Compression set (ASTM D395, Method B), 50% deflection, 22 h at 70°C (158°F)	
Set value, %	48

Source: Data from [78].

9.7.6 Example of Compounds Based on FEPM (TFE/P) Elastomers

Formulation

Ingredient	Amounts, Parts by Weight	
	Low Set	General Purpose
Aflas® FA 100S	100	—
Aflas® FA 150P	—	100
N990 (medium thermal) carbon black	15	35
Austin black	15	—
Carnauba wax	1.0	1.0
Sorbitan monostearate	1.0	1.0
VAROX 802- 40 KE[a]	4.0	3.0
TAIC	5.0	4.0
Total	**141.0**	**144.0**

[a] α,α'-bis(t-butylperoxy)diisopropyl benzene (40% active).

Properties, Press Cured for 10 min. at 177°C (350°F), Oven Post-Cured for 4 h at 204°C (400°F)

Property	Values	
Hardness Shore A	**75**	**75**
Tensile strength, MPa (psi)	14.9 (2,160)	13.9 (2,030)
Elongation at break, %	235	280
Compression set, ASTM D395, Method B		
Set after 70 h at 200°C (392°F), %	28	45

Source: Data from [21].

9.7.7 Example of Steam Resistant Formulations

Formulation

Ingredient	Amounts (Parts by Weight)	
FKM (APA technology, branched polymer)[a]	100	100
Varox DBPH-50 (peroxide)	2.5	2.5
TAIC DLC-A (coagent)	3	3
Litharge[b]	0	5
N990 MT (medium thermal) carbon black	40	40
Struktol WS 280 Paste (processing aid)	0.5	0.5
Total	**146.0**	**151.0**

a For example, Tecnoflon P459;
b litharge is being phased out due to its toxicity.

Properties from Specimens Cured for 10 h at 177°C (350°F), Post-Cured for 4 h at 230°C (496°F)

Property	Value	
Hardness, Durometer A	83	85
Tensile strength, MPa	19.3	19.4
Elongation at break, %	223	225
Modulus at 100% elongation, MPa	7.3	8.0
Fluid resistance (ASTM D471), water, 70 h at 200°C; O-rings		
Change in Durometer A, points	−12	−12
Change in tensile strength, %	−42	−45
Change in elongation at break, %	7	37
Change in modulus @ 100% elongation, %	−37	−44
Volume change, %	15	7
Fluid resistance (ASTM D471), steam, 22 h at 200°C; O-rings		
Change in Durometer A, points	−9	−9
Change in tensile strength, %	−54	−53
Change in elongation at break, %	38	71
Change in modulus at 100% elongation, %	−32	−38
Volume change, %	4	2

Source: Data from [79].

9.7.8 Basic Extrusion Compounds

Formulation

Ingredient	Amounts (Parts by Weight)	
	O-Ring Cord	**Hose/Tubing**
3M™ Dyneon™ FC or FE Type FKM*	100	100
N990 (MT) carbon black	10–50	—
N762 (SRF) carbon black	—	10–20
BaSO$_4$, CaCO$_3$, or talc	—	20–40
MgO	3	3
Ca(OH)$_2$	6	6
Carnauba wax	0.5–1.0	0.5–1.5
Polyethylene wax	—	0–0.05
Processing conditions		
Barrel and die temperatures	65 to 104°C (150 to 220°F)	

Source: Data from [77].
* Incorporated cure polymers.

9.7.9 Example of FKM Compounds for Aerospace AMS 3216 and AMS 7276H

Formulation

Ingredient	Amounts (Parts by Weight)	
	Compound I	Compound II
3M™ Dyneon™ FC 2174[a]	100	—
3M™ Dyneon™ FE 5640Q[b]	—	100
N990 carbon black	30	30
Ca(OH)$_2$	6	6
Magnesium oxide	3	3
Struktol® WS-280	0.5	—
Carnauba wax	—	0.5
Total	139.5	139.5

a Incorporated cure dipolymer;
b incorporated cure terpolyme.

Properties from Specimens Cured in Press for 10 min. at 177°C (350°F); Post-Cured for 16 h at 232°C (450°F)

Property	Compound I	Compound II
Tensile strength, psi (MPa)	2280 (15.72)	1750 (12.06)
Elongation at break, %	194	170
Hardness, Shore A	78	77
Specific gravity	1.83	1.83
ASTM fuel resistance: reference fuel B, 70 h at 23°C (75°F)		
Tensile strength, % change	− 8.0	1.5
Elongation at break, % change	3.1	0
Hardness Shore A, points change	0	−1
Volume, % change	1.4	2.9
ASTM fuel resistance: AMS oil 3023, 70 h at 200°C (392°F)		
Tensile strength, % change	−23.4	−17.2
Elongation at break, % change	−3.1	10.2
Hardness Shore A, points change	−10	−9
Volume, % change	16.0	19.9
Compression set,%, 0.139 in. O-rings	8	3
Dry heat resistance: 70 h at 270°C (518°F)		
Tensile strength, % change	−26.8	−20.8
Elongation at break, % change	−9.8	21.3
Hardness Shore A, points change	2	−4
Weight loss, %	−3.9	−2.4
Compression set resistance: plied disks, 22 h at 200°C (392°F)		
Compression set, %	12	9
Compression set resistance: plied disks, 336 h at 200°C (392°F)		
Compression set, %	34	29
Low temperature resistance: TR-10		
TR-10, °C	−18	−18

Source: Data from [77].

9.7.10 No Post-Cure Fluoroelastomer Articles

Formulation

Ingredient	Amounts (Parts by Weight)
Tecnoflon® P757	100
N990 (MT carbon black)	30
Zinc oxide	5
Luperco 101 XL*	3
TAIC,** 75% dispersion	4
Total	**142**

* 2,5-Dimethyl-2,5-di(tetr-butylperoxy)hexane (peroxide);
** triallyl isocyanurate (coagent).

Properties from Samples Press Cured for 10 min. at 170°C (338°F)

Hardness, Shore A	67
Tensile strength, psi (MPa)	2,354 (16.2)
Elongation at break, %	320
Compression set, 70 h at 200°C*,%	33

* O-rings

Chemical Resistance

Immersion for 168 h at 150°C (302°F)	
SH motor oil	
Hardness change, points	+2.5
Tensile strength change, %	+8
Elongation at break change, %	+2
Volume change, %	−0.7
Engine coolants	
Hardness, change, points	+0.1
Tensile strength, change, %	−11
Elongation at break, change, %	−9
Volume change, %	+0.1
Synthetic gear oil	
Hardness change, points	+0.7
Tensile strength change, %	+1
Elongation at break, change, %	−3
Volume change, %	+1.5

Source: Data from [80]

9.7.11 Low Temperature Service Seals

Formulation

Ingredient	Amounts (Parts by Weight)
Tecnoflon® P710	100
N990 (MT) carbon black	30
Zinc oxide	5
Luperco® 101 XL*	3
TAIC, 75% dispersion**	4
Total	142

* 2,5-Dimethyl-2,5-di(tetr-butylperoxy)hexane, (peroxide)
** Triallyl sisocyanurate (coagent)

Properties from Specimens Cured in Press for 10 min. at 177°C (350°F); Post-Cured for 8 + 16 h at 230°C (445°F)

Physical Properties	Values
Hardness, Shore A	69
Tensile strength, psi (MPa)	2,760 (19.0)
Elongation at break, %	200
Compression set, 70 h at 200°C*,%	35
Low temperature properties	
TR-10, °C	−30
TR-20, °C	−26
TR-50, °C	−24

Source: Data from [80]

9.7.12 FFKM Compound for Aerospace-AMS 7257E

Formulation

Ingredient	I	II
Dyneon™ PFE 131T	65	93
Dyneon™ PFE 81T	28	—
Dyneon™ PFE 02CZ*	2.5	2.5
Dyneon™ PFE 01CZ*	6	6
N550 (Fast Extrusion Furnace) carbon black	15	15
Aerosil® R972 (silica reinforcing filler)	1.5	1.5
Total	**118.0**	**118.0**

* Curative

Properties from Samples Cured in Press for 15 min. at 188°C (370°F); Post-Cured for 16 h at 250°C (482°F)

Original Properties	I	II
Tensile strength, psi (MPa)	1,865 (12.9)	2,115 (14.6)
Elongation at break, %	130	120
Hardness, Shore A	80	80
Compression set, 70 h at 230°C (446°F), ASTM D1414		
Compression set (O-rings), %	21	21
TR10, °C	−1.9	−1.8
Thermal air aging, 70 h at 290°C (554°F), ASTM D573		
Tensile strength, % change	−8	2
Elongation at break, % change	31	50
Hardness, points change	−3	−2
Weight loss, %	1	1
Fluid aging, ASTM Reference Fuel B, 70 h at 25°C, ASTM D471		
Tensile strength, % change	7	−3
Elongation at break, % change	8	8
Hardness, pts. change	−1	0
Volume change, %	0	1
Fluid aging, block oven, 70 h at 125°C, AS 1241, Type IV, ASTM D471		
Tensile strength, % change	−25	−23
Elongation at break, % change	−8	0
Hardness, points change	−6	−5
Volume change, %	6	6

Original Properties	I	II
Fluid aging, block oven, 70 h at 200°C, AMS 3085 (Mobil Jet 254), ASTM D471		
Tensile strength, % change	2	0
Elongation at break, % change	4	13
Hardness, points change	−3	−2
Volume change, %	1	1

Source: Data from [77]
Note: For more on fluoroelastomer compounding see [6], Chapters 12 to 15.

9.8 Fluoroelastomer Safety, Disposal, and Sustainability

Different safety issues predominate in the various stages of the fluoroelastomer life cycle. The production of fluoroelastomers involves the handling of a number of hazardous raw materials under conditions that must be closely controlled. Processors have to carry out compounding and curing operations involving additional components with reactions at high temperatures that may generate hazardous by-products. Fabricated fluoroelastomer products are often used in severe environments where failures may have dangerous consequences. Disposal of these products at the end of their useful service life may be complicated by the possible presence of contaminants and hazardous components.

9.8.1 Safety in Production

In the production of fluoroelastomers, safe handling of monomers is a major concern, largely because of the potential for explosion of various mixtures. Polymerization process safety has been discussed in some detail in Section 9.1.2, and hazards of handling various monomers have been covered in Chapter 4. Explosion hazards are minimized by eliminating possible ignition sources (e.g., electrical arcs, trace amounts of oxygen, and hot spots in equipment). The consequences of a deflagration are lessened by putting limits on the compositions of monomer mixtures, and on the temperatures and pressures in the process, so that explosion containment or relief is possible without harm to personnel or serious damage to plant equipment. Most major monomers are not highly toxic, but minor components such as cure-site monomers, modifiers, surfactants, and initiators often require special handling procedures to protect personnel. Monitoring systems are often necessary to detect low levels of hazardous materials that may come from small leaks. Adequate ventilation should be provided to protect workers from exposure to airborne contaminants.

The processing of fluoroelastomers involves compounding, forming, and curing operations. Equipment used for mixing, extrusion, and molding of elastomer compounds requires adequately trained operators safely following procedures designed to produce high-quality finished parts. Some of the curing ingredients are quite reactive and may be toxic. All these must be well dispersed in the elastomer matrix for good results, including avoiding hot spots from excessive local reactions. Care must be taken to insure addition of proper amounts of curatives according to well-designed recipes. Wrong ratios of some components may lead to runaway reactions or the production of excessive amounts of toxic by-products. Adequate ventilation should be provided to protect operators from toxic fumes, especially where hot stock is present (e.g., around mills, at the discharge of internal mixers and extruders, and in the vicinity of the openings of hot presses).

Material safety data sheets are available from the suppliers of fluoroelastomer gums, masterbatches, curatives, and processing aids; these cover potential hazards and handling precautions for particular compositions. Several general precautions for handling fluoroelastomers are listed by all suppliers: store and use fluoroelastomers only in well-ventilated areas; avoid eye contact; do not smoke in areas where fluoroelastomers are present; and after skin contact, wash with soap and water. Also note that potential hazards, including evolution of toxic vapors, may exist during compounding, processing, and curing of fluoroelastomers, especially at high temperatures [81]. More detailed handling precautions are provided in a DuPont bulletin [82]. Measurements of volatile products evolved during curing of fluoroelastomers by bisphenol and amines have been reported [83]. For a bisphenol-cured polymer, Viton® E-60C, the weight loss during press cure at 193°C (379°F) was about 0.3%, with about 1.5% additional weight loss after post-curing for 24 hours at 232°C (450°F) in an air oven. Most of the weight loss (about 95%) was water, with minor amounts of carbon dioxide and fragments from curatives. Very small amounts of hydrogen fluoride were detected, amounting to about 80 ppm based on fluoroelastomer compounds. Volatiles are generated from peroxide curing. Most of the volatiles are water and hydrocarbon fragments from peroxide decomposition, but small amounts of methyl bromide are evolved (methyl iodide would be present when the polymer contains iodine cure sites). Small amounts of hydrogen iodide or hydrogen bromide may be given off during the peroxide curing of fluoroelastomers [84].

Finely divided metals should not be used in fluoroelastomer compounds, since stocks containing them may undergo vigorous exothermic decomposition at high temperature [82]. Aluminum and magnesium powder are particularly sensitive. Some metal oxides such as litharge, dispersed at high levels in fluoroelastomers, may undergo exothermic decomposition at about 200°C (392 °F). However, litharge is no longer recommended for compounding because of toxicity problems.

Fires have occurred in air ovens during fluoroelastomer post-curing for various reasons [85]. Fluoroelastomer parts should not be cured in the same oven as other elastomers. Silicones are a particular problem because of the chemical interactions between silicone rubber and the small amounts of hydrogen fluoride (HF) generated by fluoroelastomer compounds. Adequate fresh air should be supplied to allow removal of volatiles from the mainly recirculating flow. Parts to be post-cured should be placed evenly around the oven, and not piled too deeply, to allow adequate air flow around them. Small pieces of flash and accumulated residues of processing aids may also serve as ignition sources for oven fires. Combustion products from fluoroelastomer compounds burned in a deficiency of oxygen (as is likely in an oven or building fire, or from fluoroelastomer dust on a cigarette) have been determined [82]. Besides major amounts of water and carbon dioxide, combustion products include carbon monoxide, hydrogen fluoride, carbonyl fluoride, fluoroform, and traces of fluorocarbon monomers.

9.8.2 Safety in Applications

Applications of fluoroelastomer parts often involve contact with hazardous fluids at elevated temperatures. Failure of parts such as seals may result in personal injury in some cases. Care should be taken by users of parts to assure that the proper fluoroelastomer composition is chosen for the application. This is not always easy to determine, since the suppliers of many parts may not disclose the type of fluoroelastomer used. More detailed information can be obtained from fluoroelastomer suppliers.

9.8.3 Disposal

Disposal of fluoroelastomers can be carried out by recycling, incineration with energy recovery, or by burying in a landfill [85]. Recycling is generally possible with uncured stock. Incineration is preferable for most material, including parts contaminated by absorbed fluids. However, the incinerator must be capable of scrubbing out acidic combustion products [82]. Fluoroelastomer compounds burned in excess oxygen give off water, carbon dioxide, and hydrogen fluoride as volatile products. The landfill is an option for most solid fluoroelastomers and parts if they are not contaminated by toxic fluids. It is suggested that government and local regulations are reviewed as to a proper procedure. Production scrap and post-consumer recycled goods from fluorinated thermoplastic elastomers can be reprocessed by techniques commonly used for other thermoplastic polymers.

9.8.4 Sustainability

Environmental sustainability is concerned with human interaction with nature and technology. Often it relates the use of renewable, natural materials to the use of extracted, refined materials. The current concept of environmental sustainability is represented as the goal of using technology and resources to meet our current needs without preventing future generations from meeting their needs. Development is then sustainable if it involves no decrease of average quality of life.

The efforts to replace fossil feedstocks by bio-renewable ones provide greater sustainability and eliminate greatly their supply/price volatility. The subject of bio-based polymers is covered in an article in *Chemical & Engineering News* [86].

At the time of writing, out of all monomers for fluorocarbon elastomers, only ethylene is available from bio-based raw materials. It can be produced by the dehydration of bioethanol or by the cracking of bio-naphtha. Bio-naphtha is produced during the processing of renewable feedstocks in processes such as Fischer-Tropsch fuel production [6, p. 501].

REFERENCES

1. Logothetis, A. L., *Prog. Polym. Sci.*, *14*:251, (1989).
2. Ajroldi, G., Pianca, M., Fumagalli, M., Moggi, G., *Polymer*, *30*:2180, (1989).
3. Arcella, V., Ferro, R. *Modern Fluoropolymers* (Scheirs, J., Ed.), John Wiley & Sons, Ltd, Chichester, UK, p. 73, (1997).
4. Drobny, J. G. *Introduction to Fluoropolymers, Second Edition*, (Ebnesajjad, S., Ed.), Elsevier, Oxford, UK, p. 274, (2021).
5. *ASTM D200/SAE J200 Classification of properties of vulcanized rubber materials which are frequently used in automotive applications*, Smithers, (www.smithers.com), (2021).
6. Drobny, J. G., Fluoroelastomers Handbook, Second Edition, Chapter 4, Elsevier, Oxford, UK, (2016).
7. Bailer, F. V., Cooper J. R., US Patent 3,536,683 (1970), to DuPont Co.
8. Tang, P. L., US Patent 6,512063 (2003), to DuPont Dow Elastomers.
9. Dumoulin, J. US Patent 4,524,194 (1985), to Solvay & Cie.
10. Hayashi, K., Matsuoka, Y., US Patent 4,985,520 (1998), to Asahi Chemical Industries Co. Ltd.
11. Hayashi, K., Saito, H., Toda, K., US Patent 5,218,026 (1993), to Asahi Chemical Industries Co. Ltd.
12. Hayashi, K., Hashimura, K., Kasahara, M, Ikeda Y., US Patent 5,824,755 (1998), to DuPont Co.
13. Duvailsaint, F., Moore, A. L., US Patent 6,348,552 B2 (2000), to DuPont Dow Elastomers, LLC.
14. Moore, A. L., US Patent 5,032,655 (1991), to DuPont Co.
15. Viton™ made with Advanced Polymer Architecture expand fluoroelastomer performance in a variety of applications; Brochure VTS-H90160-00-D0606, DuPont Performance Elastomers (2009).
16. Stevens, R. D., Ferrandez, P. A., Fluorocarbon Elastomers, *The Vanderbilt Rubber Handbook*, 14th ed. (Sheridan, M. F., Editor), p. 365, R.T. Vanderbilt Company, Norwalk, CT, (2010).
17. Banks, R. E., *Fluorocarbons and Their Derivatives*, Macdonald, London, 17, (1970).
18. Ajroldi, G., Pianca, M., Fumagalli, M., Moggi, G., *Polymer*, *30*:2180, (1989).
19. Schroeder, H. *Rubber Technology* (Morton, M., Ed.), 3rd ed., Chapt. 14. Van Nostrand Reinhold Co., New York, (1987).
20. Scheirs, J. *Modern Fluoropolymers* (Scheirs, J., Ed.), John Wiley & Sons, Ltd, Chichester, UK, p. 25, (1997).
21. Hertz, D. L., Jr., Tetrafluoroethylene-Propylene Elastomers, in *The Vanderbilt Rubber Handbook*, 14th ed. (Sheridan, M. F., Editor), p. 405, R.T. Vanderbilt Company, Norwalk, CT, (2010).
22. Novikov, A. S., Galil, F., Slovokhotova, N. A., Dyumaeva, T. N., *Vysokomol. Soed.*, *4*:423, (1962).
23. Drobny, J. G., *Radiation Technology for Polymers, Third Edition*, CRC Press, Boca Raton, FL, p. 134, (2021).
24. Hertz, D. L., III, *Fluoroelastomer Compendium for the Non-metallic Practitioner*, Paper presented at the EPG Educational Symposium, Galveston, TX, September 19–20, (2017), p. 8.
25. *Solvay Tecnoflon® BR 9151, Technical Data Sheet*, solway.com/en/products (2021).
26. *Omnexus FFKM Technical Data Sheet* omnexusspecialchem.com/products (2021).
27. *Polymer Properties Database*, FFKM-Perfluoroelastomers, FEPM-Tetrafluoroethylene Propylene, polymerdatabase.com, Crow, (2021).
28. Hudson, R. L., rlhudson.com/materials, (2022).
29. Moore, A. L., *Fluoroelastomers Handbook*, William Andrew Publishing, Norwich, NY, p. 104, (2006).
30. *Processing Guide Viton® Fluoroelastomer*, Technical Information Bulletin VTE-H 90171-00-A 0703, DuPont Dow Elastomers (2003).
31. Rauwendaal, C., *Polymer Extrusion*, Chapter 2, Hanser Publisherss, Munich, (1986).
32. Barwell Preformer, www.barwell.com (2020).

33. Tecnoflon® TN Latex, Technical Data Sheet, Solvay Specialty Polymers, www.solvay, (2015).
34. Arcella, V., Chiodini, G., Del Fanti, N., Pianca, M., Paper #57, the 140th ACS Rubber Division Meeting, Detroit, MI (1991).
35. Arcella, V., Geri, S., Tommasi, G., Dardani, P., *Kautsch. Gummi Kunstst.*, *39*:407, (1986).
36. Arcella, V., Ferro, R., Albano, M., *Kautsch. Gummi Kunstst.*, 44:833, (1991).
37. Brezeale, A. F., U.S. Patent 4,281,092 (1981) to E.I. du Pont de Nemours & Co.
38. Lyons, B. J., Weir, F. E., *Radiation Chemistry of Macromolecules* (Dole, M., Ed.), Vol. *II*, Chapter 14, Academic Press, New York, (1974), p. 294.
39. Vokal, A. et al., *Radioisotopy 29*(5–6):426 (1988).
40. McGinnis, V. D. *Encyclopedia of Polymer Science and Engineering*, Vol. *4* (Mark, H. F. and Kroschwitz, J. I., Eds.), John Wiley & Sons, New York, 1986, p. 438.
41. Banik, I. Bhowmick, A. K., Raghavan, S. V., Majali, A. B., Tikku, V. K., *Polymer Degradation and Stability*, **63**(3):413, (1999).
42. Marshall, J. B. in *Modern Fluoropolymers* (Scheirs, J., Ed.), John Wiley & Sons Ltd., Chichester, UK, 1997, p. 352.
43. Molding Solutions www.molders.com (2015).
44. Viton™ Fluoroelastomer, Processing Guide, Technical Information Bulletin VTE-H90171-00-A0703, DuPont Dow Elastomers (2003).
45. Rubber Molding www.hawthornerubber.com (2021).
46. Ebnesajjad, S., *Fluoroplastics, Volume 2, Melt-Processible Fluoropolymers, Second Edition* Chapter 10, Elsevier, Oxford, UK, (2015).
47. Apotheker, D. et al., *Rubber Chem. Technol.*, *55*:1004, (1982).
48. Geri, S., Lagana, C., Grossman, R., "New Easy Processing Fluoroelastomer", Paper #30, 124th Meeting of ACS Rubber Division, Houston, TX, October 25–28, 1983.
49. Nersasian, A., Paper No. 790659 and MacLachlan, J. D., Paper No. 790657, SAE Passenger Car Meeting, Detroit (1979).
50. Hoffmann, W., *Rubber Technology Handbook*, Carl Hanser Verlag, Munich, p. 122, (1989).
51. Skudelny, D., *Kunststoffe/German Plastics*, *77*(11), 17, (1987).
52. Ferro, R., Giunchi, G., Lagana, C., *Rubber Plastics News*, 19 Feb. (1990).
53. Struckmeyer, H., *Kautsch. Gummi Kunstst. 43*(9), (1989).
54. Smith, J. F., Perkins, G. T., *Proc. Inst. Rubb. Conf.*, Preprint 575 (1959).
55. Moran, A. L., Pattison, D. B., *Rubber World*, *103*:37, (1971); Schmiegel, W. W., *Kautsch. Gummi Kunstst.* 31, 137 (1978).
56. Apotheker, D., Krustic, P. J., U.S. Patent 4,214,060 (1980) to E.I. du Pont de Nemours & Co).
57. Stahl, W., *Compounding Fluoroelastomers*, Educational Symposium, International Elastomer Conference, Rubber Division ACS, Pittsburgh, PA, October 4–7, 2021.
58. *Key Uses and* Advantages of *Kalrez® Sealing Solutions*, DuPont™ Kalrez®, dupont.com/kalrez, (2021).
59. *Chemraz® FFKM Overview*, Greene Tweed & Co, Publication DS-US-GE-203 7/21 GT, www.gtweed. com, (2021).
60. Schnell, R. W., *Using KALREZ® Parts, Seal Design Guide*, Publication E-3308-3 7/92, DuPont Polymers (1992).
61. *Greene Tweed Sealing Solutions*, Greene Tweed & Co. www.gtweed.com (2021).
62. VITON® Technical Bulletin VT 240C10, E.I. DuPont de Nemours & Company (2010).
63. Kirochko, P., Kreiner, J. G., *Surf. Coat. Int. B Coat. Trans.*, *84*(5):161 (2001).
64. Kirochko, P., Kreiner, J. G., US Patent 6,133,373 (2000) to Lauren International.
65. Lauren Manufacturing, Fluorolast™, www.laureninternational.com (2017).
66. solutions.3m.com (2021).
67. Pelseal Technology, LLC www.pelseal.com (2021).
68. Scheirs, J., *Fluoropolymers: Technology: Markets and Trends*, RAPRA Technology, Ltd., Shawbury, UK, p. 107, (2001).
69. Ross, Jr., E. E., Hoover, G. S., *Modern Fluoropolymers* (Scheirs, J., Ed.), Chapter 23, John Wiley & Sons, Ltd, Chichester, UK, (1997).
70. Blatz, P. S., US Patent 3,125,547 (1964), to E.I. DuPont de Nemours and Co.

71. Moore, A. L., Tang, W. T. Y., US Patent 3,929,934, to E.I. DuPont de Nemours and Co.

72. Larsen, H. R., US Patent 3,334,157 (1967), to Union Carbide, Canada.

73. Chisholm, P. S., US Patent 5,854,352 (1998) to Nova Chemical Ltd.

74. Duchesne, D., US Patent 6,242,548 (2001), to Dyneon, LLC.

75. Chandrasekaran, V. C., *Essential Rubber Formulary*, p. 42, William Andrew Inc., Norwich, NY, (2007)

76. Stivers, D. A., *The Vanderbilt Rubber Handbook*, 1st ed (Winspear, G. G., Ed.), R. T. Vanderbilt Co., Norwalk, CT, p. 201, (1968).

77. *Dyneon Fluoroelastomer Recipe Book*, 3M™ Dyneon™ Document 16258 (2020).

78. Lynn, M. L., Worm, A. T., *Encyclopedia of Polymer Science and Technology*, Vol. 7, (Mark, H. F., and Kroschwitz, J. I., Eds.), John Wiley & Sons, New York, p. 265, (1987).

79. *Technoflon® Guide to Fluoroelastomers*, Solvay Solexis, (2005).

80. Ciullo, P. A., Hewitt, N., *The Rubber Formulary*, Chapter IX, Silicones and Fluoroelastomers, William Andrew Publishing, Norwich, NY, (1999).

81. *Viton® Fluoroelastomer Selection Guide*, DuPont Dow Elastomers Technical Information (1998).

82. *Handling Precautions for Viton® and Related Chemicals*, DuPont Dow bulletin VT-100.1 (originally issued November 1980).

83. Pelosi, L. F., Moran, A. L., Burroughs, A. E., Pugh, T. L., The Volatile Products Evolved from Fluoroelastomer Compounds During Curing, *Rubber Chemistry and Technology*, *49(2): 367–374* (May–June, 1976).

84. *Viton® GFLT-S Fluoroelastomer (VIT128) Material Safety Data Sheet*, DuPont Dow Elastomers LLC (October 2002).

85. *Viton® A-401C Fluoroelastomer (VIT007A) Material Safety Data Sheet*, DuPont Dow Elastomers LLC (June 1999).

86. Baumgardner, M. M., Biobased Polymers, *Chemical & Engineering News*, *92*(43), 10, (2014).

10

Fluorinated Thermoplastic Elastomers

Jiri G. Drobny

10.1 Introduction

Thermoplastic fluoroelastomers (TPEs) are polymeric materials exhibiting elastic behavior similar to cross-linked rubber but can be processed by conventional thermoplastic methods without curing (cross-linking). This allows flash from molding and other scrap as well as post-consumer waste to be recovered and reused. They are essentially phase-separated systems [1]. Usually one phase is hard and solid at ambient temperature and the other one is soft and elastic. The hard phase forms the *physical cross-links*, which are thermoreversible [1] (Figure 10.1).

Often the phases are bonded chemically by block or graft polymerization. Such materials are most frequently referred to as A-B-A block copolymers often made by *living radical copolymerization* [1]. Another major group of thermoplastic elastomers are thermoplastic vulcanizates (TPVs) prepared by *dynamic vulcanization* [1, p. 148]. The products display a disperse morphology, where particles of cross-linked elastomer are dispersed in a thermoplastic matrix [1, p. 148] (Figure 10.2).

Considering the exceptional commercial success of hydrocarbon thermoplastic elastomers as a frequent replacement of conventional cross-linked (vulcanized) elastomers, it is logical that a similar concept is viable in the field of fluorinated elastomers. This is a particularly attractive concept, considering the rather involved chemistry of cross-linking fluoroelastomers discussed in Chapter 9.

10.2 Types of Fluorinated Thermoplastic Elastomers

The first two fluorinated thermoplastic elastomers (FTPEs) were developed and commercialized in Japan. One is a block-copolymer type, composed of a central soft fluoroelastomer block and multiple fluoroplastic hard segments (of ETFE or PVDF, depending on the grade). This type has been available commercially since 1982 and is produced by Daikin under the trade name DAI-EL™ Thermoplastic [2]. A schematic representation of the fluorinated thermoplastic elastomer is shown in Figure 10.3.

The second type is a graft copolymer type, comprising main-chain fluoroelastomers and side-chain fluoroplastics. This type was introduced commercially in 1987 by Central Glass Co. under the trade name Cefral Soft® [3].

Since then, there have been numerous programs to develop and commercialize fluorinated thermoplastic elastomers. The polymers involved are telechelics (or α, ω-difunctional derivatives, block copolymers, and graft copolymers [4]).

10.3 Methods to Produce Fluorinated Thermoplastic Elastomers

There are several methods used for the production of commercial FTPEs. One of them is referred to as "iodine transfer polymerization", which is similar to the "living" anionic polymerization used to make block copolymers such as styrene-butadiene-styrene (e.g., Kraton®). The difference is that this "living" polymerization is based on a free radical mechanism. The products consist of soft segments based on copolymers of vinylidene fluoride (VDF) with hexafluoropropylene (HFP) and optionally

DOI: 10.1201/9781003204275-13

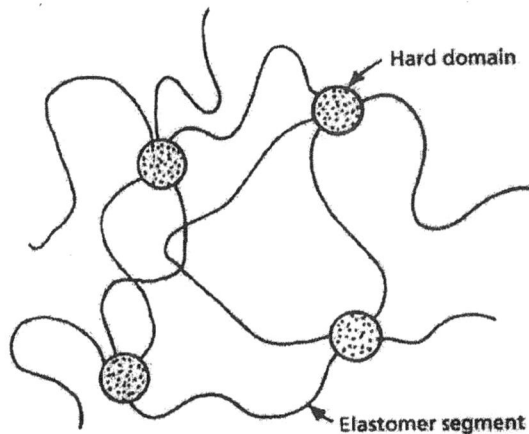

FIGURE 10.1   Phase structure of a block copolymer [2].

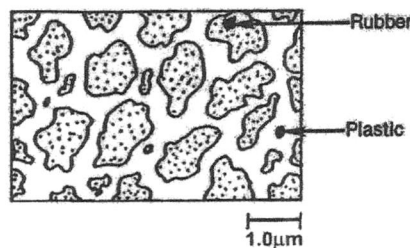

FIGURE 10.2   Morphology of a thermoplastic vulcanizate [3].

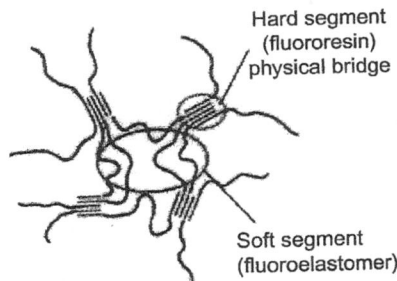

FIGURE 10.3   Schematic of a fluorinated thermoplastic elastomer. (Courtesy of Daikin)

with tetrafluoroethylene (TFE) and of hard segments that are formed by fluoroplastics such as ETFE or PVDF [5]. The other method is a two-step graft copolymerization using unsaturated peroxides, such as [$CH_2=CHCH_2OC$ (O)–O–O–tert-butyl] where the monomers involved are VDF and CTFE. In the second step, post-polymerizations mainly with VDF to form crystalline segments are repeatedly performed while successively raising the reaction temperature [5, p. 567]. More details on this are provided in [6–13].

   Another method to prepare thermoplastic fluoroelastomers is the extension of diiodo technology [14]. In the first stage iodine-terminated TFE/VDF/HFP terpolymers are synthesized by emulsion polymerization, using $I(CF2)_nI$ as the diiodo compound. The polymerization of the hard segment component takes place in the presence of the iodine-terminated terpolymer emulsion. One type contains E/TFE/HFP; the other contains polyvinylidene fluoride as the hard segment. Typically, the product contains an 85% soft segment of composition VDF/HFP/TFE and 15% hard segments of composition TFE/E/HFP. According to the basic patent the molecular weight of the hard segments has to be at least 10,000 and that of the central soft block at least 110,000 [14]. These FTPEs do not need curatives, metal oxides, fillers, or process aids and this makes them suitable for medical applications. Because of the presence of VDF or ethylene (E), they can be cross-linked by ionizing radiation if the required compression ratios used are larger than 10% [15].

Polyurethane-based FTPEs are produced by reacting fluorinated polyether diols with aromatic diisocyanates. The resulting block copolymers contain fluorinated polyether soft segments [16].

Another possible method of preparation of fluorinated TPE is dynamic vulcanization [1, p. 148]. Examples are blends of a fluoroplastic and fluorocarbon elastomer or a perfluoroplastic and a perfluoroelastomer containing curing sites or a combination of VDF-based fluoroelastomers and thermoplastics, such as polyamides, polybutylene terephtalate, and polyphenylene sulfide [17, 18].

Yet another method to prepare fluorinated block copolymers is atom transfer radical polymerization (ATRP) using pentafluorostyrene and styrene as monomers. The resulting copolymers can be linear or star-shaped. This technique generates products with unique properties (e.g., Li⁺ complexation) while preserving an excellent film-forming capability [19]. Because of that, these materials are interesting for solid state applications in batteries. Another product is hexaarm star fluoropolymers, which represent novel fluorinated nanoparticles.

10.4 Commercial Fluorinated Thermoplastic Elastomers and Their Properties

As mentioned earlier, the first block copolymer produced and commercialized by Daikin was DAI-EL™ Fluoroplastic. The high fluorine contact of the soft blocks gives the product excellent fluid resistance and a glass transition temperature of −8°C (18°F). The thermoplastic can be extruded and formed at temperatures above its melting range; after cooling, crystallization of the hard segments gives finished parts with good dimensional stability at temperatures up to about 120°C (230°F). Typical properties of the commercial product DAI-EL™ T-530 are summarized in Table 10.1. The first graft copolymer produced and commercialized by the Central Glass Company, Cefral Soft®, is no longer manufactured and marketed.

Daikin also offers another grade, namely DAI-EL™ Fluoro-TPV, that exhibits a very low permeability, high chemical resistance, as well as transparency and high flexibility. The polymer consists of a vulcanized fluoroelastomer dispersed in a fluorinated thermoplastic matrix as shown in Figure 10.2. The reported melting point is 220°C and the maximum service temperature is 150°C. Additional technical data are given in Tables 10.2–10.4 [20].

TABLE 10.1

Typical Properties of DAI-EL T-530

Property	Data	Test Method
Color	Light yellow transparent pellet	Visual observation
Specific gravity (23°C)	1.89	JIS K 6268
Melt flow rate	19.7	JIS K 7210
Melting point	Approx. 230°C	—
Physical properties (original)		
100% modulus	1.6 MPa	JIS K 6251
Tensile strength	11.8 MPa	JIS K 6251
Elongation at break	580	JIS K 6251
Hardness (Shore A)	61	Peak value
Physical properties after 15 Mrad (150 kGy) irradiation		
100% modulus	1.8 MPa	JIS K 6251
Tensile strength	17.7 MPa	JIS K 6251
Elongation at break	500%	JIS K 6251
Compression set	23%	150°C, 70 h, 25% compression

Source: Data Sheet DAI-EL T-530, Ver. 10 (May 2009), Daikin Industries Ltd.

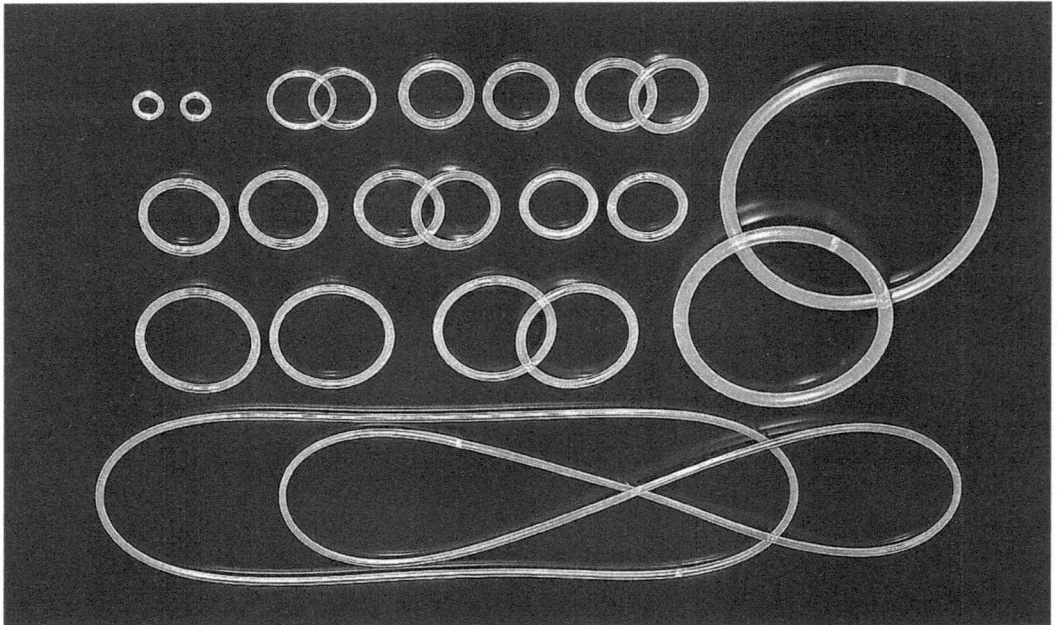

FIGURE 10.4   O-rings made from a thermoplastic fluoroelastomer.   (Courtesy of Daikin)

TABLE 10.2

Typical Properties of DAI-EL Fluoro-TPV, SV Series

Physical/Mechanical Properties				
Property	SV-1010	SV-1020	SV-1030	SV-1050
Hardness, Shore A	94	92	90	90
Tensile strength, MPa	32	24	17	11
Elongation at break, %	440	410	370	360
Electrical properties				
Dielectric constant @				
1 MHz	2.72	2.99	–	3.14
10 MHz	2.60	2.75	–	2.81
100 MHz	2.48	2.56	–	2.57
1 GHz	2.35	2.42	–	2.42
Loss factor @				
1 MHz	0.018	0.037	–	0.053
10 MHz	0.021	0.037	–	0.047
100 MHz	0.022	0.030	–	0.036
1 GHz	0.021	0.024	–	0.026
Fuel permeation and fuel resistance				
Permeation rate, CE10 60°C, g-mm/m^2-day	4.5	5.3	7.5	20
Volume change, CE85 60°C, %	1.0	1.4	1.5	–

Source: SM090131, Daikin America, Inc., 2012.

TABLE 10.3

Resistance to Automatic Transmission Fluid, DAI-EL TPV Versus FKM and FEPM

Property	DAI-EL TPV	FKM*	FEPM
Volume swelling, %	~0	2.8	8.1
Tensile strength retention, %	81	10	82

Source: Modified data from [20].
Note: Test conditions: 1000 @ 150°C.
* Copolymer, bisphenol curing system.

TABLE 10.4

Resistance to Biofuels, DAI-EL TPV Versus FKM

Property	DAI-EL TPV	FKM
Weight change, %		
250 h	~0	8
500 h	~0	38
1000 h	~0	48
Volume change, %		
250 h	~0	20
500 h	~0	80
1000 h	~0	98

Source: Data from [24].
Note: Test conditions: SME soybeans +H_2O (1 wt. %), test temperature 125°C.

A base-resistant thermoplastic fluoroelastomer has been developed by DuPont [21] using an iodine transfer polymerization technique. In this material, soft segments are of composition E/TFE/PMVE at about 19/45/36 mole % with a glass transition temperature of –15°C (5°F); the soft segments have composition E/TFE at about 50/50 mole % with a differential scanning calorimeter (DSC) melting endotherm maximum at about 250°C (482°F). The thermoplastic fluoroelastomer is readily molded at 270°C (518°F) to give good physical properties and excellent resistance to fluids including polar solvents, strong inorganic bases, and amines. This composition can be readily cross-linked with ionizing radiation after molding to obtain better properties, with no compounding required. The physical properties of the base-resistant thermoplastic fluoroelastomer are listed in Table 10.5; the enhanced fluid resistance is also shown. However, this material has not been offered commercially as yet.

Fluorinated thermoplastic vulcanizates (FTPVs) have been developed by Freudenberg NOK-GP and offered as FluoroXprene® [22]. They are essentially dynamically vulcanized blends of fluorocarbon elastomers and fluoroplastics, such as PVDF, ETFE, ECTFE, THV, FEP, and MFA, prepared in either a batch or continuous process. The continuous process, using a twin-screw extruder, is preferred. The morphology is that typical for a TPV, that is dispersed cross-linked FKM in the fluoroplastic matrix [23]. The fluid resistance is considerably better than that of FKM, mainly because the semi-crystalline fluoroplastic matrix protects the elastomeric particles. The fluorinated TPV exhibits superior fuel permeation resistance to that of FKM materials. Since these FTPVs contain VDF and ethylene monomeric units they can be cross-linked by ionizing radiation if desired. A comparison of the physical properties of FluoroXprene with a standard FKM elastomer is provided in Table 10.6 and a comparison of the fuel penetration of FluoroXprene with two types of FKM elastomers is shown in Table 10.7.

TABLE 10.5

Properties of Base-Resistant Fluorinated TPE

Polymer	Base-Resistant FTPE
Property	**Values**
Compression molded	
M_{100}, MPa	3.4
T_B, MPa	14.5
E_B, %	510
Irradiated, 15 Mrad (150 kGy)	
M_{100}, MPa	5.3
T_B, MPa	16.9
E_B, %	270
Compression set, % (pellets, 70 h/150°C)	37
Chemical resistance, % wt. gain after 3 days/25°C	
Acetone	3.6
Methanol	0.0
Dimethyl formamide	0.5
Toluene	1.1
Trichlorotrifluoroethane	100.0
Butylamine	1.9

Source: Data from [21].

TABLE 10.6

Comparison of the Physical Properties of FluoroXprene and an FKM Elastomer

Property	FKM Control	FluoroXprene
Hardness, Shore A	80 ~ 95	80 ~100
Tensile strength, MPa	6.0 ~ 12.0	2.0 ~ 25.0
Elongation at break, %	100 ~ 300	10 ~350
Compression set, % 70 h @ 150°C	15 ~50	27 ~55

Source: Freudenberg Sealing Technologies, FLUOROXPRENE® MATERIALS, Features and Benefits, www.fts.com, 2021.

TABLE 10.7

Comparison of the Fuel Permeation Resistance of FluoroXprene and FKM Elastomers

Elastomer Type	Permeation Rate g/m2/day	Permeation Constant g-mm/m2/day
FKM (terpolymer)	15	28
FKM (copolymer)	29	55
FluoroXprene	1 ~ 4	2 ~ 8

Source: Freudenberg Sealing Technologies, FLUOROXPRENE® MATERIALS, Features and Benefits, www.fts.com, 2021.
Note: Test conditions: ASTM D814, CE10 Fuel, 30 days @ 40°C.

10.5 Applications of Fluorinated Thermoplastic Elastomers

10.5.1 Chemical and Semiconductor Industries

The most common applications for thermoplastic fluorinated elastomers are seals in the chemical and semiconductor industries (O-rings, V-rings, gaskets, and diaphragms) because of their excellent chemical resistance and high purity [5, p. 573]. These parts are often cross-linked by ionizing (actinic) radiation without adding any other components to improve their mechanical properties [24]. An example of O-rings made from FTPE is shown in Figure 10.4. Other parts for these industries are tubing and the liners of multilayer hoses for corrosive gases or ultrapure water, and liners for vessels for inorganic acids (e.g., HF) [25].

10.5.2 Electrical, and Wire and Cable

Because of their flexibility, low flammability, and resistance to oil, fuel, and chemicals, FTPEs find use in the electrical and wire and cable industries as wire coating and as sheathing and the coating of cables [26, 27].

10.5.3 Other Applications

Other applications include tents and greenhouses, as laminates with polyester fiber-reinforced PVC, and as tubing, bottles, and packaging in food processing and in sanitary goods [5, p. 573].

REFERENCES

1. Drobny, J. G. *Handbook of Thermoplastic Elastomers, Second Edition*, Chapter 1, p. 3, Elsevier, Oxford, UK, (2014).
2. Tatemoto, M., *Int. Polym. Sci. Technol.*, *12*:4, (1985).
3. Kawashima, C., *Fusso Jushi Handbook* (Satokawa, T., Ed.), Nikkan Kogyu Shinbunsya, Tokyo, pp. 671–686, (1984).
4. Ameduri, B., Boutevin, B., *J. Fluoropolymer Chem.*, *126*:224, (2005).
5. Tatemoto, M., Shimizu, T. *Modern Fluoropolymers* (Scheirs, J., Ed.), John Wiley & Sons Ltd., Chichester, UK, p. 566, (1997).
6. Brinati et al., US Patent 6,207,758 (2001), to Ausimont, S.p.A.
7. Arcella, V., Brinati, G., Albano, M., Tortelli, V., EP 0,661,312 (1995) to Ausimont S.p.A.
8. Rees, R. W., US Patent 5,006,594 (Apr.9, 1991) to E.I. DuPont de Nemours and Company.
9. Malik, A. A. et al., US Patent 6,891,013 (2005), to Aero-Jet-General Corporation.
10. Cordis Project, funded by European Union, Grant Agreement ID: BRE20144 (February 1, 1993 to January 31, 1996).
11. Dossi et al., United States Patent Application Publication No. US2019/0211129 (2019) to Solvay Specialty Polymers, Italy, S.p.A.
12. Nakagawa, T, Tatemoto, M, U,S. Patent 4,158,678 (1979), to Daikin.
13. Kawachi, S, *Gummi Faser Kunststoffe*, *39*:162, (1986).
14. Moore, A. L. *Fluoroelastomers Handbook*, William Andrew Publishing, Norwich, NY, p. 119, (2006).
15. van Cleeff, A. *Modern Fluoropolymers* (Scheirs, J., editor), John Wiley & Sons, Ltd., Chichester, UK, p. 608, 1997.
16. Tonelli, C., Trombetta, T., Scicchitano, M. et al. *J. Appl. Polym. Sci.*, *59*, 311, (1996).
17. Logothetis, A. L., Stewart, C. W., U.S. Patent 4,713,418 (1987).
18. Novak, C. T. et al. US. Patent 5,371.143 (1984) to Minnesota Mining and Manufacturing Company.
19. Jankova, K., Hvilsted, S., *J. Fluorine Chem.*, *126*:248, (2005).
20. Drobny, J. G., *Fluoroelastomers Handbook*, *2nd* ed., Elsevier, Oxford, UK, p. 128 (2016).
21. Carlson, D. P., U.S. Patent 5,284,920, assigned to DuPont (1994).

22. Park, E. H., Paper presented at TPE TopCon 2010, Society of Plastics Engineers, Akron, OH, September 13–15, (2010).
23. Park, E. H., U.S. Patent 7,135,527, (2006), to Freudenberg-NOK General Partnership.
24. Tatemoto, M., Tomoda, M., Kawachi, M. et al., Kokai Tokkyo Koho Japanese Patent 62635 (1984).
25. Kawashima, C., Koga, S., *Jpn. Plast.*, *39*:98, (1988).
26. Cheng, T. C., Kaduk, B. A., Mehan, A. K. et al., U.S. Patent 4,935,467 (1988).
27. Kawamura, K., Kawashima, C., Koga, S., U.S. Patent 4,749,610 (1988).

11

Fluoro-Inorganic Elastomers

Jiri G. Drobny

Fluoro-inorganic (or semi-organic) elastomers belong to the group of elastomers which contain other elements in their backbones, unlike most other elastomers, having strictly carbon-carbon (organic) backbones. In this chapter, we focus on fluorosilicone elastomers (Section 11.1) containing the Si-O bonds and polyphosphazene elastomers (Section 11.2) containing nitrogen and phosphorus.

11.1 Fluorosilicone Elastomers

11.1.1 Introduction

In this context, the term "fluorosilicone" applies to polymers containing C–F bonds and Si–O bonds with hydrocarbon entities between them. Thus, the repeating structure may be generally written as $[R_fX (CH_2)_n]_x$ $(CH_3)_y SiO_z$, where R_f is the fluorocarbon group and the X group is a consequence of the chemistry chosen to link the R_f fluorocarbon group to the hydrocarbon spacer $(CH_2)_n$. It can be oxygen or sulfur, for example, but is not present in current commercial materials [1]. The length of the hydrocarbon spacer n is optimally 2 [1, p. 360]. An example of the structure of a common fluorosilicone polymer is shown below:

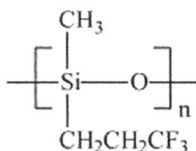

Current commercially available fluorosilicones are based on poly(methyltrifluo-ropropylsiloxane) (PMTFPS), or more accurately poly[methyl (3,3,3-trifluoropropyl) siloxane], as shown above. Certain polymers contain small amounts of a methylvinylsiloxane monomer which serves mainly as a cure site (see Section 11.1.2). In some cases PMTFPS is copolymerized with polydimethylsiloxane (PDMS) for a cost–benefit balance [1, p. 362]. The manufacture of monomers for fluorosilicones is discussed in some detail in [1, p. 362]. The use of the chemically stable, trifluoropropyl pendant substitution results in the polymer being very polar, which in turn gives the siloxane backbone resistance to nonpolar liquids such as fuel, solvents, and many oils. The inorganic silicon-oxygen backbone contributes greatly to the elastomer's low temperature flexibility as well as its high thermal stability even at very high temperatures. Fluorosilicone elastomers are referred to in ASTM D1418 and ISO1629 as FMQ or FVMQ. The ASTM name is fluoro-vinyl polysiloxane. Major manufacturers of fluorosilicone products are Dow, Momentive, Wacker, and Shin-Etsu Silicone.

Fluorosilicone polymers are optically clear and are available in a broad range of viscosities, from very low-viscosity fluids to very high-viscosity gums. The physical properties of the raw polymer – such as viscosity; resistance to nonpolar fuels, oils and solvents; specific gravity; refractive index; lubricity; solubility in polar solvents; the degree of crystallinity; and glass transition temperature (T_g) – depend on the structure, more specifically on the number of trifluoropropyl groups in the molecule. The mechanical properties of the polymer depend on the molecular weight, dispersity, and mole % of the present vinyl groups [2].

DOI: 10.1201/9781003204275-14

11.1.2 Polymerization Process for Fluorosilicone Elastomers

The most common method of preparation of PMTFPS is through the base-catalyzed ring-opening polymerization of the corresponding cyclic trimer [3, 4]. A specific cure site for peroxide curing is developed by incorporating 0.2 mole % of methyl vinyl siloxane [5]. Typically, fluorosilicone elastomers are copolymers of 90 mole % of trifluoropropyl siloxy- and 10 mole % of dimethyl siloxy- monomers, but the fluorosilicone content in commercial products ranges from 40 to 90 mole % [2].

The presence of fluorine increases the polarity to the level above the standard for methyl vinyl silicone (MVQ) rubber. Consequently, the fluorosilicone elastomers have considerably greater resistance to oils and many liquids (with the exception of some ketones and esters) with only slightly impaired resistance to low temperatures when compared with MVQ. Still, fluorosilicone elastomers have better low-temperature resistance than FKM elastomers. Moreover, when compared with FKM they have lower hardness, higher resilience, and considerably better bonding to other polymers and to metals.

11.1.3 Processing of Fluorosilicone Elastomers and Compounds

High-molecular-weight PMTFPS gums (molecular weight is typically 0.8 to 2.0 million) can be compounded by the addition of mineral fillers and pigments in a similar fashion to most common industrial elastomers. Fillers for such compounds are most commonly silicas (silicon dioxide) because they are compatible with the elastomeric silicon–oxygen backbone and are thermally very stable. They range in surface area from 0.54 to 400 m²/g and in average particle size from 100 to 6 nm. Because of these properties, they offer a great deal of flexibility in reinforcement. Thus, cured compounds can have Durometer A hardness values from 40 to 80. For example, the addition of high-surface-area fumed silica can increase the strength of the peroxide-cured elastomer by a factor of 10. Other fillers commonly used in fluorosilicones are calcium carbonate, titanium dioxide, and zinc oxide. A small amount of low-molecular-weight fluorosilicone diol processing fluid and a peroxide catalyst are added. Other additives, such as extending fillers, pigments, and thermal stability enhancers, are often added to meet final product requirements [6]. Suppliers of fluorosilicone gums also offer already fully formulated compounds.

Frequently, fluorosilicone elastomers are blended with PDMS silicones either to lower compound cost or to enhance properties of the silicone compound. Fluorosilicone elastomers can also be blended with fluoroelastomers to improve their low-temperature flexibility. The properties of cured fluorosilicone elastomers depend on the base polymer and compounding ingredients used. Essentially, the processing of fluorosilicone elastomers and the fabrication technology of parts are basically the same as that of silicone elastomers, though quite different from that of fluorocarbon elastomers [7, 8]. For details, see Section 11.1.6.

Mill and mold release is improved by the addition of a small percentage of dimethyl silicone oils or gums. Processing aids are mostly proprietary. Plasticizers are generally fluorosilicone oils of various viscosities. The lower the molecular weight, the more effective is the plasticizing action. On the other hand, the higher the molecular weight, the lower the volatility. This is critical when the service temperature is very high.

Fluorosilicone compounds can be processed by the same methods used for silicone elastomers based on PDMS. They can be milled, calendered, extruded, and molded. A large proportion of fluorosilicone compounds is used in compression molding. Molded parts produced in large series are made by injection molding, and parts with complex shapes are produced by transfer molding. Calendering is used to produce thin sheets and for the coating of textiles and other substrates [7, 8].

The cross-linking of fluorosilicones is done by essentially the same methods as conventional silicones. A comprehensive review of this subject is provided in [9]. Currently, there are three methods of cross-linking used in industrial practice:

* Peroxides (free radicals).
* Condensation reactions.
* Hydrosilylation addition.

For *peroxide cross-linking*, organic peroxides, such as dicumyl, di-*t*-butyl, and benzoyl peroxides, are used for gums and compounds in the amounts 1 to 3 phr (parts per 100 parts of rubber). Typical cure cycles are 5 to 10 min. at temperatures 115 to 170°C (239 to 338°F), depending on the type of peroxide used. Each peroxide has a specific use. A post-cure is recommended to complete the cross-linking reaction and to remove the residues from the decomposition of peroxide. This improves the long-term heat aging properties [1, p. 362]. Condensation and hydrosilylation addition methods are used for fluorosilicone liquid systems (see Section 11.1.6). Standard fluorosilicone polymers can also be cross-linked by ultraviolet radiation or by electron beam, but these methods are not commonly used [10].

11.1.4 Properties of Fluorosilicone Elastomer Compounds

11.1.4.1 Fluid and Chemical Resistance

In general, fluorosilicones exhibit a very good fuel and fluid resistance. The volume swelling in solvents decreases with increasing fluorine content. Cured fluorosilicone elastomers have good resistance to jet fuels, oils, hydrocarbons, and automotive fuels [10]. However, higher swelling in ketones and esters are observed [11]. Relatively low swelling is found in alcohol/fuel blends; once the solvents are removed the physical properties return to nearly the original unswollen state [12]. However, they have only fair resistance to polar fluids like alcohols and poor resistance to ketones, aldehydes, amines, and (non-petroleum based) brake fluids. Physical and mechanical properties of cured PMTFPS elastomer compounds are listed in Table 11.2.

11.1.4.2 Heat Resistance

Fluorosilicone elastomers have excellent heat resistance, although they have slightly lower high-temperature stability compared with PDMS [13]. The ultimate temperature stability depends on cure conditions and environment. A typical cured fluorosilicone elastomer (PMTFPS) aged for 1,350 hours at 200°C (392°F) will show a two-point reduction in Durometer hardness, a 40% reduction in tensile strength, and a 15% reduction in elongation. There are essentially two mechanisms of degradation: reversion (occurs in confinement) or oxidative cross-linking. The latter occurs by the radical abstraction of protons, which recombine to form additional cross-linking sites, and this ultimately leads to embrittlement of the vulcanizate [1, p. 364].

TABLE 11.1

Fluid and Chemical Resistance of Fluorosilicone Elastomer Compounds (ASTM D471)

Fluid	Immersion Conditions	Hardness Change (Points)	Volume Change (%)
IRM 903 Oil*	70 hrs./150°C	−4	0
Fuel A*	168 hrs./24°C	−5	+15
Fuel B*	168 hrs./24°C	−5	+20
Fuel C*	168 hrs./24°C	−14	+19
JP-8**	168 hrs./24°C	−9	+4.6
Toluene	168 hrs./24°C	−10	+20
Benzene	168 hrs./25°C	−5	+25
Heptane	168 hrs./24°C	−10	+25
Carbon tetrachloride	168 hrs./25°C	−5	+20
Methanol	336 hrs./25°C	−10	+4
Ethanol	168 hrs./25°C	0	+5
Hydrochloric acid (10%)	168 hrs./25°C	−5	0
Nitric acid (70%)	168 hrs/25°C	0	+5
Sodium hydroxide (50%)	168 hrs./25°C	−5	0
Steam	24 hrs. @ 100 psi	−5	nil

Source: Data from [10, 11].
* ASTM reference fluid;
** jet engine fuel.

TABLE 11.2

Properties of Typical Commercially Cured PMTFPS Elastomer Compounds

Property	Typical Range
Specific gravity (g/cm³)	1.35–1.65
Hardness (Shore A)	20–80
Tensile strength (MPa)	7.5–12 (at 22°C)
	2.4–4.1 (at 204°C)
Elongation (%)	100–600 (at 22°C)
	90–300 (at 204°C)
Modulus at 100% (MPa)	0.5–6.2
Compression set (%) (22 h/177°C)	10–40
Tear strength, die B[a] (kN/m)	17.5–46
Service temperature (°C)	−68 to +280
Bashore resilience[b] (%)	10–40
Volume swell, Fuel B, 24 h/23°C	15–23

Source: Data from [10, 14].
[a] Die B refers to a particular specimen shape in ASTM D624.
[b] Bashore resilience is measured by a falling metal plunger according to ASTM D2632.

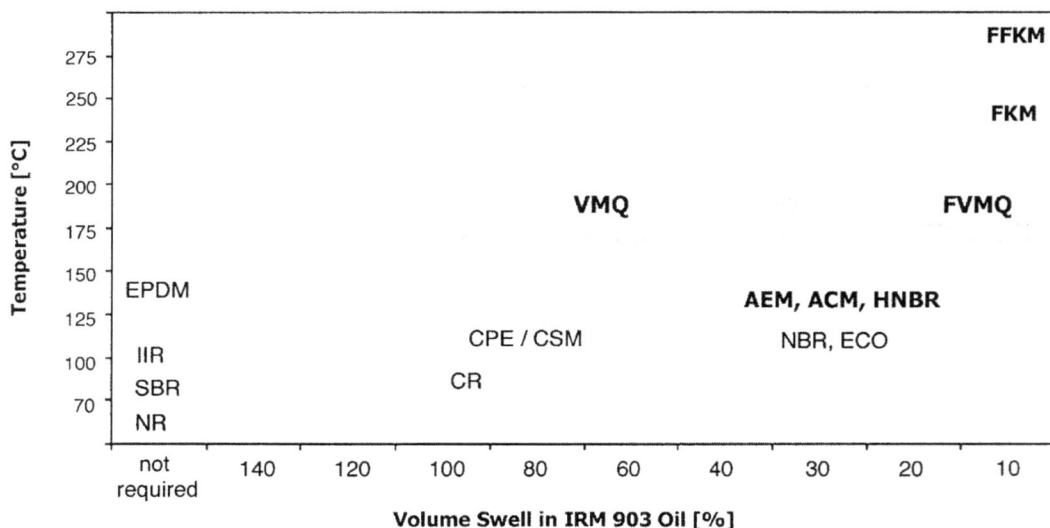

FIGURE 11.1 Positioning of high-performance elastomers according to ASTM D2000. (Courtesy of Momentive Performance Materials)

A comparison of combined swelling and heat resistance of FMQ elastomers with other high-performance elastomers and the positioning of FMQ among them according to ASTM D2000 is shown in Figure 11.1.

11.1.4.3 Low Temperature Behavior

The glass transition temperature T_g of PMTFPS is −75°C (−103°F). Moreover, it does not exhibit low-temperature crystallization at −40°C (−40°F) as PMDS does. Because of this and the low T_g of fluorosilicone elastomers they remain very flexible at very low temperatures. For example, the brittleness temperature by impact (ASTM D746B) of a commercial fluorosilicone vulcanizate was found to be −59°C (−74°F) [14]. This is considerably lower than the values typically measured on fluorocarbon

TABLE 11.3

Typical Electrical Properties of Selected Polymers

Polymer	Dielectric Strength E_B, V/mil	Dielectric Constant ε (100 Hz)	Dissipation Factor tan δ (100 Hz)	Resistivity ρ_V (Ω-cm)
Low density PE	742	2.2	0.0039	2.5×10^{15}
Natural rubber	665	2.4	0.0024	1.1×10^{15}
PDMS	552	2.9	0.00025	5.3×10^{14}
PMTFPS	350	7.0	0.20	1.0×10^{14}
FKM Elastomer	351	8.6	0.040	4.1×10^{11}

Source: Data from [19] and [1, p. 366].

elastomers. Fluorosilicones combine the superior fluid resistance of fluoropolymers with the very good low-temperature flexibility of silicones.

11.1.4.4 Electrical Properties

Electrical properties – dielectric constant (ε), representing polarization; dissipation factor (tan δ), representing relaxation phenomena; dielectric strength (E_B), representing breakdown phenomena; and resistivity (ρ_v), an inverse of conductivity – are compared with other polymers in Table 11.3 [15]. The low dielectric loss and high electrical resistivity coupled with low water absorption and retention of these properties in harsh environments are major advantages of fluorosilicone elastomers over other polymeric materials [1, p. 366].

11.1.4.5 Surface Properties

Surface energy is a fundamental property of polymers and can be expressed in terms of surface tension (force per unit length, mN/m) or surface free energy (free energy per unit area, mJ/m^2). These quantities are identical for liquids but not for solids [16]. Values of *polymer liquid surface tensions*, σ_{LV}, are directly measured and are dependent on the molecular weight of the polymer and the temperature of measurement; the values of *solid surface energy*, σ_S, shown in Table 11.4, were measured using water and methylene iodide [1, p. 370]. The solid surface energy of PMTFPS was measured directly at room temperature (20–25°C) and compared to the values of polydimethylsiloxane (PDMS) and polytetrafluoroethylene (PTFE) as shown in Table 11.4 [1, p. 364]. It can be seen that this type of fluorosilicone exhibits considerably lower surface energy than PTFE or PDMS.

11.1.5 Applications of Fluorosilicone Compounds

Fluorosilicones are used for sealing applications in the aerospace industry which require resistance to hot fuels, oils, and diester based lubricants. They are often a good choice for static sealing systems for a wide temperature range. The limitation to static applications is due to their poor abrasion resistance, relatively

TABLE 11.4

Surface Energy Values of Selected Polymers

Polymer	Solid Surface Energy, σ_s mJ/m2
PMTFPS	13.6
PDMS	22.8
PTFE	19.1

Source: Data from [1, p. 364] and [1, p. 370].
Note: PMTFPS = polymethyltrifluoropropyl siloxane; PDMS = polydimethyl siloxane; PTFE, polytetrafluoroethylene.

low tear strength, and only fair flex-cracking resistance, although some grades exist that have improved or even good flex-cracking resistance and compression set [1, p. 370].

Commercial fluorosilicone elastomer compounds are made from high-molecular-weight PMTFPS (molecular weight is typically 0.8 to 2.0 million) and are cross-linked mainly by organic peroxides. Such compounds contain some reinforcing filler (usually high-surface-area fuming silica), a small amount of low-molecular-weight fluorosilicone diol processing fluid, and a peroxide catalyst [17]. Other additives (e.g., extending fillers, pigments, thermal stability enhancers) are often added to meet final product requirements [6]. Frequently, fluorosilicone elastomers are blended with PDMS silicones either to lower compound cost or to enhance properties of the silicone compound. Fluorosilicone elastomers can also be blended with fluoroelastomers to improve their low-temperature flexibility. Properties of cured fluorosilicone elastomers depend on the base polymer and compounding ingredients used. The addition of heat stabilizers at service temperatures of more than 180°C is recommended. The current manufacturers of fluorosilicone elastomers offer not only raw (uncompounded) gums but also a large number of fully compounded materials for specific uses.

As mentioned earlier, cured fluorosilicone elastomer compounds are particularly suited for service where they come in contact with aircraft fuels, lubricants, hydraulic fluids, and solvents. Compared with other fuel-resistant elastomers, fluorosilicones offer the widest hardness range and the widest operating service temperature range of any material [12, 13]. The automotive and aerospace industries are the largest users of fluorinated elastomers [18]. Typical automotive applications are: fuel injector O-rings and turbocharger hose liners [1, p. 371]; fuel line pulsator seals and fuel line quick-connect seals; gas cap washers; vapor recovery system seals; electrical connector inserts; exhaust gas recirculating diaphragms [10]; fuel tank access gaskets; and engine cover and oil pan gaskets. In the aerospace industry fluorosilicone O-rings, gaskets, washers, diaphragms, and seals are used in fuel line connections, fuel control devices, electrical connectors, hydraulic line connectors, fuel system access panels [1, p. 371], and fuel delivery quick connection seals [1, p. 369].

An interesting application of fluorosilicone elastomers is their blending with vinylidene fluoride polymers to obtain materials having a unique combination of flexibility (improved flex life) at low temperatures (−30°C and below) and high mechanical strength. Vinylidene polymer is first exposed to ionizing radiation (e.g., a gamma ray source) and then mixed with the fluorosilicone. Radiation promoters (prorads) such as diallylphtalate and triallylisocyanurate are used to assist in grafting and compatibilizing the two polymers. Typical applications for such blends are the lining, coating, and jacketing of electrical and optical cables, films, and piezoelectric products [19].

11.1.6 Fluorosilicone Liquid Systems

Medium-molecular-weight PMTFPS with vinyl or hydroxyl end blocks are essentially liquids used for a variety of applications. Some of them are used for adhesives and sealants. They are cured either at ambient temperatures (room temperature vulcanization, or RTV) or at elevated temperature. One-part moisture-activated RTV sealants have been available commercially for many years. They generally consist of medium-molecular-weight PMTFPS vinyl or hydroxyl end blocks, a small amount of silica filler, and either a condensation or addition curing system. Because of their excellent resistance to jet engine fuels, they are used in military and civilian aerospace applications [20, 1, p. 369]. Two-part, heat-cured fluorosilicone sealants have been used in military aircraft applications and for sealing automotive fuel systems [1, p. 369].

Commonly, condensation reactions are used for cross-linking at ambient temperatures. The acetoxy-functional condensation system is widely used in fluorosilicone sealants. The cross-linking occurs after exposure to atmospheric moisture [12]. The limitation of this system is that it is effective for only thin layers. Moreover, it often requires up to 14 days to cure, and the acetic by-products may corrode certain substances.

Thicker sections can be cross-linked by hydrosilylation addition. This is the same chemistry used to produce fluorosilicone monomers with the vinyl functionality present in silicon. The catalyst reaction occurs between a vinyl group and silicone hydride [10, p. 4]. The advantage of this system is that it does not produce volatile by-products. On the other hand, the disadvantage is that it is available only as a two-part system [1, p. 363]. However, one-part, platinum-catalyzed products have been developed [10, p. 5]. The reaction is very rapid, and at room temperature it is completed in 10 to 30 min. It is accelerated with increasing temperature, and at 150°C (302°F) it is completed within a few seconds. This makes the compounds ideal for fast automated injection molding operations [10, p. 12]. One-part systems use the chemical complexing of the catalyst, which is activated at elevated temperatures, or its encapsulation in an impermeable shell, which is solid at room temperature and melts at elevated temperatures [10, p. 12].

FVMQ/VMQ liquid copolymers are available and designed for automated liquid injection molding systems ideal for the fabrication of small, intricate parts utilizing fast cycles, high yields, and automated molding processes [10].

The process called "liquid injection molding" (LIM) employs predominantly two-component, platinum (addition)-curing, liquid-pumpable compounds used usually in a 1:1 ratio. These are transferred by a hydraulic or pneumatic metered pumping system to a static mixer. The static mixer blends the two components into a homogeneous suspension; this mixing process activates the compound's platinum cure system. The liquid silicone material (LSR) from the static mixer flows to the injection unit. It is then injected into the mold cavity through a runner and gate system where it is held in the mold under high pressure and elevated temperature until the material is cured (that is, when the rubber is vulcanized). The mold is usually heated by water flowing inside the mold or by electrical heaters. The typical curing temperature is 140°C, but may be higher when required. Additionally, a cold runner system can be used to prevent the premature curing of the flowing silicone in the runners or its vicinity. The cycle time is established to reach an optimal level of cure. At the end of the cycle, the parts are removed or ejected from the cavities and the next cycle begins. The final step of the injection molding process of silicone rubber is the cooling of the product inside the mold, followed by a product demolding and final cooling in ambient conditions [21]. In general, post-curing is not required [21]. A schematic of a typical LIM equipment and the process is shown in Figure 11.2.

The advantages of liquid injection molding are [22]:

- Shorter cure times compared to traditional compression molding.
- High level of repeatability, good for tight tolerance/precision components.
- Superior clarity; material can be pigmented in-line with material flow to produce colors.
- Closed mold injection supports molding of complex geometries and over-molding.

The disadvantages of liquid injection molding vs. other rubber molding methods are [22]:

- Higher start-up/shutdown costs; better suited to high volume applications.
- Runner systems can lead to increased gross material weights when cold runner systems or other low waste options are not utilized.

Current LIM technology allows the molding of parts with inserts, flashless molded silicone components, silicone overmolded assemblies [22], and two-shot molding using special molds [21]. These techniques provide increased design flexibility for making multi-material, multi-color, and multifunctional components at a competitive cost. There are many applications where assembly injection molding of LSR and thermoplastics and/or metals plays a significant role in the industry.

A special class of fluorosilicone sealants are *channel sealants* or *groove injection sealants*, which are sticky, putty-like compounds that do not cure. They are used to seal the fuel tanks of military aircraft and missiles [1, p. 370].

The adhesion of fluorosilicone compounds requires surface treatment. For particularly difficult surfaces plasma treatment is necessary. However, for most common applications, satisfactory bonding is achieved by using a specialized primer [10, p. 12].

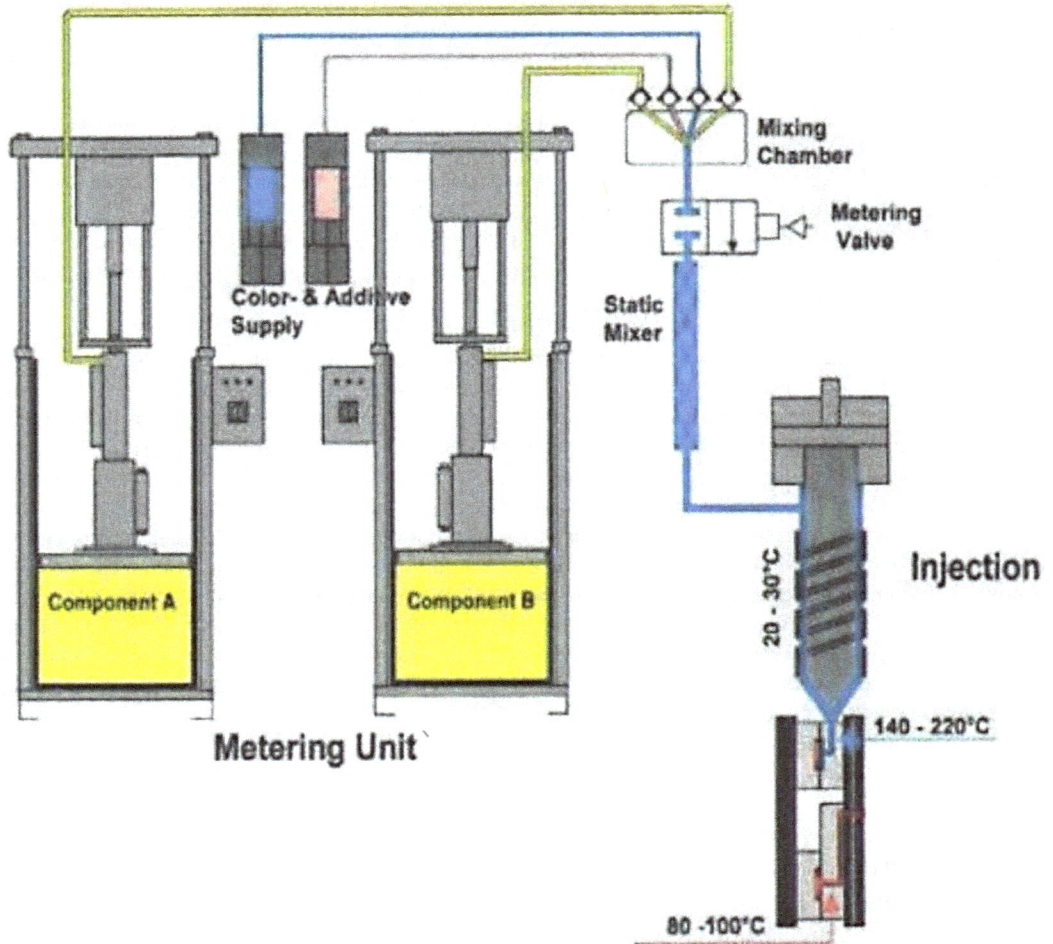

FIGURE 11.2 Schematic of liquid injection molding equipment and process. (Courtesy of SIMTEC Silicone Parts, LLC)

11.1.7 Toxicity and Safety

Under normal conditions, PMTFPS is relatively inert. Skin tests performed on albino rabbits have shown no dermal irritation or toxicity. In more than 40 years of industrial use of fluorosilicone compounds, no problems have been reported with respect to human dermal contact with these materials, uncured or cured [1, p. 366]. However, when exposed to elevated temperatures in air or are burned, toxic fluorinated compounds are formed and inhalation of such vapors is harmful. One toxic species is 3,3,3-trifluoropropionaldehyde (TFPA). The no-effect limit for heating in air is 150°C. At any rate, it is prudent to minimize direct contact with the materials and the exposure of personnel to fumes in the workplace and to follow recommendations from the supplier. More on this subject is discussed in Chapter 19.

11.2 Polyphosphazene Elastomers

11.2.1 Introduction

Polyphosphazene (or polyphosphonitrilic chloride) elastomers, like silicone elastomers, have a fully inorganic backbone, consisting of nitrogen and phosphorus. The basic building block is –N=P- and the pendant organic groups are attached to the phosphorus. The technology is over 100 years old [23], but the actual development work leading to commercial products was done only in the 1970s.

In general, the synthesis of polyphosphazene polymers is unique. In theory, an infinite number of polymers with a variety of properties can be derived from the common polymeric intermediate, poly(dichlorophosphazene) ($PNCl_2$), by replacing the chlorines with different nucleophiles.

11.2.2 Fluorinated Polyhosphazene Elastomers

Two commercial polyphosphazene elastomers were developed and marketed in the mid-1980s, namely poly(fluoroalkoxyphosphazene) elastomer (ASTM designation FZ) and poly(aryloxyphosphazene) elastomer (ASTM designation PZ) [24]. The fluorinated products were produced and marketed by Firestone Tire and Rubber under the trade name PN-F® and by the Ethyl Corporation under the trade name Eypel-F.

11.2.2.1 Preparation

Fluorinated polyphosphazenes are synthesized by a condensation reaction of fluorinated alcohols with a polydichlorophosphazene precursor [25]:

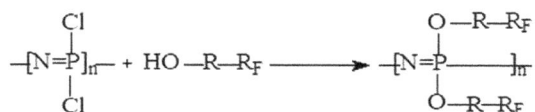

$$\begin{array}{c} Cl \\ | \\ -[N{=}P]_n{-} \\ | \\ Cl \end{array} + HO{-}R{-}R_F \longrightarrow \begin{array}{c} O{-}R{-}R_F \\ | \\ -[N{=}P{-}]_m \\ | \\ O{-}R{-}R_F \end{array}$$

where R is CH_2 and R_F is CF_3 or $(C_2F_4)_x$ with x = 1, 2, 3.

The structure of the commercial product (FZ Elastomer) is shown on the right side of the above equation [26].

11.2.2.2 Properties

The polymer is a soft gum, which can be compounded with carbon blacks and fillers and cured with sulfur, peroxides, or by ionizing radiation. FZ Elastomer offers: a broad service temperature range, namely from -55 to +175°C, or -67 to +347 °F [27], excellent flex fatigue resistance, damping properties, and resistance to chemicals and fluids [28]. Typical mechanical properties of this elastomer are listed in Table 11.5 and the swelling resistance data in Table 11.6.

FZ elastomers are different from most other elastomers in that they do not require the addition of plasticizers to have a very high low-temperature performance. Therefore their low-temperature performance will not deteriorate due to plasticizer extraction or migration as the parts age. The exceptionally high low-temperature performance is inherent in the unique nitrogen–phosphorus backbone. FZ elastomers have a glass transition temperature T_g value of -65°C, a low temperature retraction (TR-10) value of −58°C, and a low-temperature brittleness of -50°C [29].

TABLE 11.5

Typical Properties of FZ Elastomer Compounds

Property	Value
Specific gravity	1.70–1.85
Tensile strength, MPa (psi)	6.9–13.8 (1,000–2,000)
100% modulus, MPa (psi)	2.8–13.8 (400–2,000)
Elongation at break, %	75–250
Compression set, %[a]	15–55

Source: Data from [25] and [27].
[a] 70 h at 150°C.

TABLE 11.6

Swelling Resistance of FZ Elastomer Compounds

Medium	Conditions	Volume Swell, %	Hardness Change, Points
IRM 903*	166 h @ 325°F	1.8–2.5	−9
Water	1 week @ 23°C	3.5	−2
	1week @ 100°C	7.9	0
10% sulfuric acid	1 week @ 23°C	3.5	−2
	1 week @ 100°C	2.8	+1
10% ammonia	1 week @ 23°C	3.6	−1
	1 week @ 100°C	6.6	+12
10% sodium	1 week @ 23°C	2.0	−2
hydroxide	1 week @ 100°C	7.8	+2

Source: Data from [27].
* ASTM reference oil.

11.2.2.3 Applications

In summary, fluorinated polyphosphazene elastomers offer excellent flex fatigue resistance, damping properties, chemical resistance, and a broad service temperature range. Because of that they have been used in very demanding applications [26], including in the military, airspace, and automotive industry in specialty seals and diaphragms, and in oil field applications [28]. They were very successful throughout the 1980s. However, because of their high cost and relatively small volume market, they are currently not available commercially other than on special order. On the other hand research on these materials continues, leading to new or potentially new applications [30–35].

REFERENCES

1. Maxson, M. T., Norris, A. W., Owen, M. J. *Modern Fluoropolymers* (Scheirs, J., Ed.), John Wiley & Sons, Ltd, Chichester, UK, p. 359, (1997).
2. Waible, K., Maxson, T., *"Silikonkautschuk, Eigenschaften und Verarbeitung,"** Conference, Würzburg, Germany, September 20, 1995, Conference Proceedings, p. 2. (In German).
3. Saam, J. C., *Silicon-Based Polymer Science* (Zeigler, J. M. and Fearon, F. M. G., Eds.), *Advances in Chemistry Series 224*, American Chemical Society, Washington, DC, p. 71, (1990).
4. Chijnowski, J. *Siloxane Polymers* (Clarkson, S. J. and Semlyen, J. A., Eds.), Prentice-Hall, Englewood Cliffs, NJ, p. 1, (1993).
5. Hertz, Jr., D. L., *The Vanderbilt Rubber Handbook*, 13th ed. (Ohm, R. F., Ed.), R. T. Vanderbilt Co., New York, p. 239, (1990).
6. Maxson, M. T., *Gummi, Fasern Kunstst*. 12, p. 873, (1995).
7. Drobny, J. G., *Fluoroelastomers Handbook*, 2nd ed., Elsevier, Oxford, UK, p. 7, (2016).
8. DiPino, M. A., *The Vanderbilt Rubber Handbook, Fourteenth Edition* (Sheridan, M. F., Editor), R.T. Vanderbilt Company, Inc., Norwalk, CT, p. 355, (2010).
9. Thomas, D. R. *Siloxane Polymers* (Clarkson, S. J. and Semlyen, J. A., Eds.), Prentice-Hall, Englewood Cliffs, NJ, 1993, p. 567.
10. Waible, K., Maxson, T., *"Silikonkautschuk, Eigenschaften und Verarbeitung,"** Conference, Würzburg, Germany, September 20, 1995, Conference Proceedings, p. 3. (In German).
11. Gomez-Anton, M. R., Masegosa, R. M., Horta, A., *Polymer*, 28:2116, (1987).
12. Norris, A. M., Fiedler, L. D., Knapp, T. L., Virant, M. S., *Automotive Polym. Design*, *19*(April):12, (1990).
13. Knight, G. J., Wright, W. W., *Br. Polym. J.*, *21*, 199, (1989).
14. Mastromateo, R., Paper 900195, SAE Meeting, February 26 to March 3, 1990, Detroit, MI.
15. Ku, C. C., Liepins, R., *Electrical Properties of Polymers*, Hanser Publishers, Munich, p. 326, (1987).
16. Kobayashi, H., Owen, M. J., *Trends in Polym. Sci.*, *3*:330, (1995).

17. Fluorosilicone Rubber Product Selection Guide, Document S2094194/E2725, Dow Company (2018).
18. Maxson, M. T., *Aerospace Engineering*, December, 15 (1990).
19. Caporiccio, G., Mascia, L., US Patent 5,457,158 (Oct. 10, 1995) to Dow Corning Corporation.
20. *Silicone RTVs in Aerospace and Aviation*, Momentive, Waterford, NY, (2017), www.momentive.com
21. *LSR and LSR-2 Innovative Technology*, SIMTEC Silicone Parts, LLC, Miramar, FL (2021).
22. *Liquid Injection Molding Silicone (LIM), Liquid Silicone Rubber (LSR) Molding*, Datwyler Sealing Solutions USA, Inc., Vandala, OH, (2021), usa.datwyler.com.
23. Stokes N. H., *Am. Chem. J.*, *19*:782, (1897).
24. ASTM D1418, ASTM International, West Conshohocken, PA.
25. Kyker, G. C., Antkowiak, T. A., *Rubber Chem. Technol.*, *47*:32–39, (1974).
26. Lohr, D. F., Penton, H. R., *Handbook of Elastomers Second Edition*, (Bhowmick, A. K. and Stephens, H. L., Eds.) Chapter 21, CRC Press, Boca Raton, FL, (2001).
27. Polyphosphazene Rubber (PZ, FZ); FZ refers to the fluorinated polymer (55% fluorine), MatWeb, www.matweb.com (July 2021).
28. Jones, M. S., Polyphosphazene elastomers in the oil field. *Rubber World*, *264*(3):33–35, (1991).
29. Rose, S. H., US Patent 3,315,688 (June 2, 1970) to Horizon Research.
30. Cleria, M., Bertani, R., DeJaeger, R., Lora, S., *J. Fluorine Chem.*, *125*:329, (2004).
31. Matyjaszewski, K., Moore, M. K., White, M. L., *Macromolecules*, *26*:6741, (1994).
32. Alcock, H. R., Prange, R., *Macromolecules*, *34*:6858, (2001).
33. Alcock, H. R. *Polyphosphazenes: A Worldwide Insight* (Gleria, M., and DeJaegers, Eds.), Nova Science Publishers, Hauppage, NY, p. 1–22, (2004).
34. Amin, A. M. et al., *Polymer-Plastics Technology and Engineering*, *49*, 1399–1405, (2010).
35. Alcock, H. R., The background and scope of polyphosphazenes as biomedical materials. *Regen. Eng. Transl.Med.*, *7*, 66, (2021).

Part IV

Technology of Fluoropolymer Aqueous Systems

12

Characteristics and Properties of Fluoropolymer Aqueous Systems

Jiri G. Drobny

Aqueous polymeric systems have advantage over solutions in organic solvents in that they are safer to handle and do not require expensive solvent recovery systems to prevent environmental pollution. Moreover, preparing a solution of a polymer is an additional, often time-consuming operation, requiring specialized equipment and often an explosion-proof environment. Finally, most of the perfluoropolymers do not dissolve at all or only in exotic and very expensive solvents, so aqueous systems are the only liquid form in which they are available. The disadvantage of waterborne systems is that they very often dry much more slowly than the usual solvents (e.g., ketones, esters, chlorinated or fluorinated hydrocarbons) used for some of the fluoropolymers. Furthermore, some additives, such as surfactants, may not be removed completely from the dry film and may have adverse effects on the quality of the final product.

In general, waterborne fluoropolymer systems are handled and processed in a similar fashion to other organic coatings. They can be compounded by the addition of fillers, pigments and colorants, resins, and other additives; they can be, for example, viscosified or blended with other waterborne polymeric systems. Because of the nature of the base polymer, they differ in their processing behavior. For example, polytetrafluoroethylene (PTFE) dispersions are very shear sensitive, whereas the others are not. Coating formulations from dispersions of perfluoropolymers are usually very simple, containing only the necessary surfactant. If fillers and pigments are used, the amounts that can be added are limited; otherwise poor-quality films would be obtained. Another aspect to consider is the processing temperature of the polymer; some pigments and colorants cannot tolerate the high processing temperatures required for perfluoropolymers. On the other hand, some fluoropolymers can be compounded into highly filled and pigmented coatings and paints.

The following fluoropolymers are commercially available in aqueous systems: PTFE, perfluoroalkoxy (PFA), methylfluoroalkoxy (MFA), fluorinated ethylene propylene (FEP), polyvinylidene fluoride (PVDF), terpolymer of tetrafluoroethylene, hexafluoropropylene, vinylidene (THV Fluoroplastic), and fluorocarbon elastomers. This chapter covers these aqueous systems and their characteristics and properties. Their processing and applications are the subject of Chapter 13.

12.1 PTFE Dispersions and Coatings

12.1.1 PTFE Aqueous Dispersions

Aqueous dispersions of PTFE resin are hydrophobic, negatively charged colloidal systems containing PTFE particles with diameters ranging from 0.05 to 0.50 μm suspended in water. Commercial products have a resin content of approximately 60% by weight [1]. Most PTFE dispersions typically contain 6 to 10% of the weight of the resin of the nonionic wetting agent and stabilizer (essentially a surfactant). The specific gravity of such a dispersion is about 1.50.

PTFE resin dispersions are milky-white liquids, with a viscosity of approximately 20 cP and a pH of about 10. The resin contained therein has the characteristics of a fine powder, that is, a high sensitivity to shear.

DOI: 10.1201/9781003204275-16

Upon prolonged standing, the particles, which have a specific gravity of 2.2 to 2.3 [2], tend to settle with some classification of sizes. During storage, the particles settle gradually to the bottom of the container with gradually increasing clarification of the water phase on the top. Normally, the settled dispersion can be redispersed completely by gentle agitation. Essentially, stabilized dispersions have an indefinite shelf life as long as they are stirred occasionally [3] and kept from freezing.

Unstabilized PTFE dispersions are irreversibly coagulated by acids, electrolytes, and water-miscible solvents, such as alcohol and acetone, and by violent agitation, freezing, and boiling. A small addition of an ionic or nonionic surfactant stabilizes the dispersion so it tolerates some mechanical agitation and the addition of water-miscible solvents. It also slows sedimentation. The lower limit for an adequate stabilization is about 1% of the surfactant on the weight of the polymer; however, for most coating and impregnating operations the amount is more like 10% to assure good wetting and penetration. Higher amounts than that may increase the viscosity of the dispersion to an undesirable level. Moreover, for cases where the dried coating is sintered, an increased amount of surfactant (12% or more) increases the amount of time necessary to remove it in the baking stage of the process (see Chapter 13).

The pH of commercial PTFE dispersions as supplied is usually 10 and tends to drop during storage. Therefore, it is important to check the pH periodically and to maintain it by adding ammonium hydroxide to prevent souring. This is particularly important during warm and humid weather when conditions can promote the growth of bacteria. The bacteria feed on the surfactant present in the dispersion, causing a dark brown discoloration and a rancid odor. If souring occurs, the containers must be cleaned out and disinfected with a sodium hypochlorite solution. The ammonium hydroxide used for maintaining the pH is usually available as a reagent grade of 29%. At this concentration about 5 g/gal of dispersion should be sufficient [1, p. 3].

The relationship between the solids content and gravity is approximately linear and can be expressed by the equation [3, p. 1802]:

$$V = P\left(A - B/100AB\right) + 1/A, \tag{12.1}$$

where V is the specific volume of the dispersion, P is the percentage by weight of the polymer solids, A is the specific gravity of water (0.9985), and B is the specific gravity of the polymer solids (2.25). Examples of specific gravities for different solids are listed in Table 12.1. The determination of the specific gravity of a dispersion is made by a glass hydrometer.

The viscosity of PTFE dispersions increases proportionally with increasing solids content up to about 30 to 35% solids, and beyond that the increase is much more rapid. Dispersions with surfactants exhibit the same pattern, but the rate of increase is faster and depends on the type and amount of surfactant [3, p. 1802].

Chemours offers several PTFE dispersions with different characteristics with the brand name Teflon™ [4]:

- Teflon™ PTFE DISP 30 is a general purpose product.
- Teflon™ PTFE DISP 33 is a dispersion to create enhanced surface smoothness, adhesion, gloss, and weldability.

TABLE 12.1

Correlation Between the Solids Content of PTFE Dispersion and Its Specific Gravity

Solids Content, %	Specific Gravity	Amount of Solids, g/L
60	1.51	906
50	1.39	695
45	1.34	601
40	1.29	515
35	1.24	436

Source: Data from [3, p. 1802].

TABLE 12.2

Typical Property Data for Teflon™ PTFE DISP 30 Aqueous Dispersion*

Property	Unit	Typical Value
Solids content (% PTFE by weight)	%	60
Density of dispersion (at 60% solids)	g/cm³	1.51
Surfactant content on PTFE solids	%	6
Dispersion particle size, average diameter	μm	0.220
pH of dispersion	–	10
Brookfield viscosity (ASTM D2196), at 25°C (77°F)	mPa·s	25

Source: Teflon™ PTFE DISP 30, Fluoroplastic Dispersion, Document C-10146 (3/20), Chemours (2020).
* Not to be used for specification.

- Teflon™ PTFE DISP 35 is a dispersion used in a co-coagulation process.
- Teflon ™ PTFE DISP 40 is a dispersion with improved shear stability.

The typical properties of the general purpose dispersion are summarized in Table 12.2.

Daikin offers currently four grades of PTFE aqueous dispersions to be used for fabric impregnation and coating, film casting, release coatings, packing, and as a battery binder. Typical properties, features, and applications of these products are shown in Table 12.3.

Solvay currently offers five grades of PTFE aqueous dispersions used for fabric impregnation and coating, film casting, release coatings, packing, bearings, and seals. Typical properties and applications of these dispersions are shown in Table 12.4.

TABLE 12.3

Typical Properties and Applications of Polyflon™ PTFE Dispersions

Property/ Feature/ Application	Grade			
	D-210	D-210C	D-610	D-610C
Solids content, wt.%	59–61	59–61	59–61	59–61
Surfactant content, wt.% on solids	6.0–7.2	6.0–7.2	6.0–7.2	5.5–6.5
Specific gravity @ 25°C	1.50–1.53	1.50–1.53	1.50–1.53	1.50–1.53
Viscosity, cP @ 25°C, max.	35	35	35	35
Critical cracking thickness, μm	14	14	28	28
pH	8.5–10.5	8.5–10.5	8.5–10.5	9.5–10.5
Particle size, μm	0.22–0.25	0.22–0.25	0.26–0.30	0.26–0.30
Features	Low color and transparency; general purpose	High molecular weight; low color and transparency	High build; low color and transparency	High molecular weight; low color and transparency; high build
Typical applications	Glass cloth coating; impregnation; packing	Battery binder	Glass cloth coating; impregnation; packing; release coatings	Cast films

Source: Daikin ™ Polyflon PTFE Products, Document PB PTFE 001 Ro 11/23/2016, www.daikin-america.com.

TABLE 12.4

Typical Properties and Applications of Algoflon® PTFE Dispersions

Property, Feature Application	Grade				
	D-1610 F	D-1613 F	D-1614 F	D-2711 F	D-3511 F
PTFE content, wt.% on the mixture	60.0	60.0	60.0	27.5	59.0
Nonionic surfactant, wt.% on the mixture	3.5	2.8	3.5	1.7[a]	3.5
Density, g/cm^3	1.51	1.51	1.51	1.19	1.50
pH	>9.0	>9.0	>9.0	10.3	>9.0
Brookfield viscosity, mPas @ 20°C	—	24	25	—	20
Particle size, nm	250	240	240	250	240
Features	Good shear stability; Excellent anti-dripping effect	Excellent shear stability	Good shear stability	Can be easily coagulated; can be processed with a vast range of fillers	Superior film building behavior; very good surface finish and gloss; high shear stability
Typical applications	Addition to plastics as anti-drip in flame-retardant formulation	Impregnation; cast films	Impregnation of woven packings and yarns	Co-coagulation with fillers for production of bearings and seals	Coating; impregnation

Source: Algoflon® PTFE, Solvay Specialty Polymers; Revised 3/7/2022.
[a] Anionic dispersant.

12.1.2 PTFE Industrial Coatings

Because of the unique properties of PTFE, its aqueous dispersions can also be used for a variety of industrial coatings, such as protective and nonstick. Such coatings are compounds prepared from PTFE dispersions for specific applications. While Chemours is one of the major manufacturers and suppliers of dispersions, as mentioned in previous sections, the company is also an established manufacturer and supplier of industrial coatings. Chemours provides PTFE nonstick and protective coatings with operating temperatures up to 400°C (500°F), excellent chemical resistance, a low coefficient of friction, and good abrasion qualities. If the coating is applied to metal or any other substrate, a primer is required. They are essentially two-coat (primer/topcoat) systems. Common types are [5]:

- Liquid primer/one coat (two-part).
- Liquid primer/one coat (premixed).
- Liquid topcoat.
- High build liquid topcoat.

These products can be either clear or pigmented (colored).

12.2 Other Perfluoroplastic Dispersions and Coatings

12.2.1 FEP Dispersions

Commercial aqueous dispersions of FEP are supplied with 54 to 55% by weight of hydrophobic negatively charged particles with the addition of approximately 6% by weight of a mixture of nonionic and

TABLE 12.5

Typical Property Data for a Teflon™ FEPD 121 Aqueous Dispersion*

Property	Unit	Typical Value
Solids content (% FEP by weight)	%	55
Density of dispersion (at 55% solids)	g/cm³	1.41
Surfactant content on FEP solids	%	5.5
Dispersion particle size, average diameter	μm	0.18
pH of dispersion	—	10
Brookfield viscosity (ASTM D2196), at 25°C (77°F)	mPa·s	25
Melting temperature (ASTM D2116)	°C (°F)	260(500)
Melt flow rate (MFR 372/5.0), (ASTM D2116)	g/10 min.	8

Source: Teflon™ FEPD 121, FEP Aqueous Dispersion, Document C-10097 (3/20), Chemours (2020).
* Not to be used for specification.

anionic surfactants based on the polymer content. The particle size range is 0.1 to 0.26 pm. The nominal pH of the dispersion is 9.5, and the viscosity at room temperature is approximately 25 cP (25 mPa.s) [6]. The product offered by Chemours is Teflon™ FEPD 121 and its typical properties are shown in Table 12.5.

12.2.2 FEP Industrial Coatings

FEP copolymer industrial coatings melt and flow during baking to provide nonporous films. They provide excellent chemical resistance and, in addition to a low friction coefficient, they have excellent nonstick properties. The maximum use temperature is 204°C (400°F) [5]. They can be supplied as clear or pigmented (colored).

12.2.3 PFA/MFA Dispersions and Coatings

12.2.3.1 PFA/MFA Dispersions

Commercial aqueous dispersions of PFA contain 50% or more by weight of PFA particles and typically 5% of surfactants on the polymer content [6]. An example is Teflon™ PFAD 335D whose typical properties are summarized in Table 12.6. MFA dispersion, offered by Solvay, contains 55% by weight of MFA, 4% of nonionic surfactant on the polymer content, and a specific gravity of 1.41 [7].

TABLE 12.6

Typical Property Data for Teflon™ PFAD 335D Aqueous Dispersion*

Property	Unit	Typical Value
Solids content (% PFA by weight)	%	60
Density of dispersion (at 60% solids)	g/cm³	1.50
Surfactant content on PFA solids	%	5.5
Dispersion particle size, average diameter	μm	0.20
pH of dispersion	—	10
Brookfield viscosity (ASTM D2196), at 25°C (77°F)	mPa·s	25
Melting temperature (ASTM D2116)	°C (°F)	305 (581)
Melt flow rate (MFR 372/5.0), (ASTM D2116)	g/10 min.	2

Source: Teflon™ PFAD 335D, PFA Aqueous Dispersion, Document C-10109 (3/20), Chemours (2020).
* Not to be used for specification.

TABLE 12.7

Typical Properties of a Polyflon™ PTFE D-310 Modified PTFE Dispersion

Property	Value
Solids content (by weight), %	59–61
Surfactant content, wt.% on solids	6.0–7.2
Specific gravity @ 25°C	1.50–1.53
Viscosity @ 25°C, cP (mPa•s)	35 max.
Critical cracking thickness, μm	12
pH	8.5–9.5
Particle size, μm	0.22–0.25

Source: Daikin Polyflon™ PTFE Technical Data Sheet, TDS-PTFE D005 REV 02/9/16 Daikin - America.

12.2.3.2 PFA Industrial Coatings

Like FEP industrial coatings, PFA coatings melt and flow during baking to provide a nonporous film. They provide a high continuous-use temperature of 400°C (500°F), which is the same as a PTFE polymer. Some PFA coatings can have film thickness up to 1,000 μm (40 mils). They can be supplied as clear or pigmented (colored).

12.2.4 Modified PTFE Dispersions

Modified PTFE aqueous dispersions are manufactured without the use of a fluorosurfactant and have been specifically designed for use in release coatings, cookware, and glass cloth impregnation in applications where improved gloss, adhesion, surface smoothness, and wear properties are desired.

Modified PTFE dispersions exhibit good wetting properties, high shear stability, and a good film forming behavior [8]. An example is Polyflon™ D-310 (a product of Daikin) whose properties are shown in Table 12.7

12.2.5 Dispersions of PTFE Micropowders

PTFE micropowders (also known as fluoroadditives) are available in aqueous dispersions stabilized with a nonionic surfactant. Unlike other PTFE dispersions, they are based on a low molecular weight PTFE and are designed as an additive in host systems, in order to impart some of the unique properties of PTFE. When properly processed, they retain the properties of PTFE after service at 260°C (500°F) and its useful properties at −240°C (− 400°F) [9]. In addition they provide: inertness to nearly all industrial chemicals and solvents, excellent dielectric properties, enhanced nonstick properties, and the lowest coefficient of friction of any solid material. An example of this kind of product is Zonyl™ MPD 1700 (a product of Chemours). A summary of the properties of this aqueous dispersion is shown in Table 12.8.

TABLE 12.8

Typical Properties of a Zonyl™ MPD 1700 Fluoroadditive Dispersion

Property	Value
Solids content (% PTFE by weight), %	60
Density of dispersion (at 60% solids), g/cm^3	1.50
Surfactant based on PTFE solids, %	6
Dispersion particle size (average diameter), μm	0.210
pH of dispersion	10.5

Source: Zonyl™ MPD 1700, PTFE Additive Dispersion, Document C-10004(3/18), The Chemours Company, teflon.com/zonyl (2018).

TABLE 12.9

Grades of PVDF Aqueous Dispersions

Grade	Properties		
	Crystallinity	**Processing Temperature, °C (°F)**	**Flexibility**
Solef® XPH – 838	High	>170 (338)	Standard
Solef® XPH – 882	Low	140 (284)	Good
Solef® XPH – 859	None	60 (140)	Excellent
Solef® XPH – 884	Low	150 (302)	Good

Source: *Solef® PVDF Aqueous Dispersions for Lithium Batteries*, Document D05/2015, Version 12, Solvay Specialty Polymers www.solvay.com (2015)

12.3 Other Fluoroplastic Dispersions

12.3.1 Dispersions of PVDF

There are several commercially available aqueous dispersions based on PVDF. Solvay offers a new generation of water-based dispersions with proprietary chemical modification (not blending materials), manufactured via emulsion polymerization. These products have the special feature of nano-sized primary particles of PVDF in the shape of spheres, contain 25–35% solids, and are stable to shear stress [10]. A summary of the properties of the individual grades is shown in Table 12.9.

Arkema offers PVDF-based products that are hybrid dispersions containing different ratios of PVDF and a proprietary acrylic resin [11]. A summary of the properties of these dispersions is shown in Table 12.10.

The processing and applications of PVDF-based aqueous dispersions are discussed in Chapter 13.

12.3.2 Dispersions of THV Fluoroplastics

There is currently only one commercial grade of aqueous dispersion of THV Fluoroplastics, namely THV 340Z offered by 3M Dyneon. This is a dispersion of a hydrophobic polymer with almost spherical particles, having a size of around 120 nm with a solids content of 50%. THV 340Z contains 5.0% of ionic surfactant and has a pH value of above 9 [12]. It can be processed in a similar fashion as dispersions of

TABLE 12.10

Properties of Kynar Aquatec® Dispersions (Not for Specification)

Property	Grade		
	FMA-12	**ARC**	**CRX**
Solids content by weight, %	46	44	44
Solids content by volume, %	38	34	—
MFFT,* °C	12	27	15
Wet density (g/mL)	1.15	1.178	1.18
Dry density (g/mL)	1.37	1.50	—
PVDF/acrylic resin ratio	50:50	70:30	70:30
% VOC, as supplied	<1	<1	<2
Viscosity Brookfield 30 s⁻¹	100	100	100
Dispersion pH	8.0	8.0	8.0
Shelf life,** months	18	18	18

Source: Kynar Aquatec® Technical Data Sheets, KynarAquatec.com, ARKEMA (2022)
* Minimum film formation temperature;
** protect from freezing.

TABLE 12.11

Typical Properties of 3M™ Dyneon™ Fluoroplastic 340Z Dispersion

Property	Value
Solids content, % by weight	50
Surfactant content (based on solids), %	5.0
Mean particle size (diameter), nm	120
Dispersion pH	>9
Viscosity Brookfield (D = 30, s^{-1}), mPa•s	7
Melting temperature, °C (°F)	145 (293)

Source: Data from *3M™ Dyneon™ Fluoroplastic THV 340Z*, Product Data Sheet, Dyneon, April 2016 www.dyneon.eu.

other melt-processable fluoropolymers, namely, for coating fabrics or casting thin films (see Chapter 13). Typical properties of the THV Fluoroplastic aqueous dispersion are shown in Table 12.11.

12.4 Fluorocarbon Elastomers in Latex Form

A certain number of fluorocarbon elastomers are used in latex form. Typical latex products are based on VDF/HFP/THF terpolymers (about 68% fluorine), which are readily polymerized to relatively stable dispersions containing 20–30% solids. The dispersions are further stabilized by pH adjustment and addition of anionic or nonionic hydrocarbon soaps. They can be further concentrated up to about 70% solids by creaming [13]. Such highly concentrated latex can be used for a variety of applications, using standard latex technology, mainly for coating where high chemical resistance and high thermal stability are required (see Chapter 13). Moreover FKM latex can be blended with other compatible fluoropolymers and specialty polymers to attain specific properties. Coatings can be cross-linked chemically or by electron beam. An example of a commercial FKM latex is Tecnoflon® TN Latex, offered by Solvay [14]. Typical properties of this product are shown in Table 12.12.

Additional fluorocarbon elastomer products are chemically cross-linked coatings, which can be two or one-component systems, curing at ambient or elevated temperatures. Examples of the currently available products of this kind are two-component coatings offered by Daikin and a one-component water-based coating developed by Lauren International Co., which uses hydrolyzed and stabilized aminosilanes [15]. The products offered by Daikin are DAI-EL LATEX GL-252EA (black) and DAI-EL LATEX GLS-213RA (red). Both aqueous dispersions provide a coating layer displaying elasticity and heat resistance after curing at elevated temperatures [16]. For bonding to silicone rubber, a solvent based primer DAI-EL LATEX GLP-103SR is used. More details are shown in Tables 12.13–12.15.

Other commercial fluoroelastomer latex products are AFLAS® 150 CS Latex and AFLAS® 300S Latex, both based on an alternating tetrafluoroethylene-propylene copolymer with a cure site monomer in

TABLE 12.12

Typical Properties of Tecnoflon® TN Latex

Property	Unit	Value
Appearance	—	White liquid
Solids content, by weight	%	~70
Fluorine content of the polymer	%	68
Specific gravity		
Latex		1.45
Polymer		1.86

Source: Tecnoflon® TN Latex Specialty Grade, Technical Data Sheet, Solvay Specialty Polymers Revised 02/23/2015 www.solvay.com.

TABLE 12.13

Typical Properties of DAI-EL LATEX GLS-213 RA

Property	Value
Solids content by weight, %	50
Viscosity, mPa•s	80
pH	7.5
Sintering temperature range, °C	120 to 350
Tensile strength of cured film, MPa	13.5
Elongation at break of cured film, %	200
Contact angle, degrees	
Water	114
n-hexane	59

Source: DAI-EL LATEX GLS-213RA, Technical Datasheet, tds-gls-213ra-E_ver01_Mar_2018, Daikin Industries, Ltd., 2018.

TABLE 12.14

Typical Properties of DAI-EL LATEX GL-252EA

Property	Value
Solids content by weight, %	51
Viscosity, mPa•s	100
pH	7.6
Sintering temperature range, °C	120 to 250
100% modulus, MPa	5.6
Tensile strength of cured film, MPa	7.6
Elongation at break of cured film, %	170
Hardness, IRHD*	73
Contact angle, degrees	
Water	91
n-hexane	49

Source: DAI-EL LATEX GL-252EA, Technical Datasheet, tds-gl-252ea-E_ver01_Mar_2018, Daikin Industries, Ltd., 2018.
* International rubber hardness degree, Wallace hardness meter value after 30 seconds.

TABLE 12.15

DAI-EL LATEX GLP-103SR Primer for Fluoroelastomer Coatings

Characteristics	Values
Active ingredient,* %	15
Drying temperature, °C	100–150
Adhesion to silicone rubber, kN/m	0.5
Adhesion without primer, kN/m	0.1

Source: DAI-EL LATEX GLP-103SR Technical Datasheet, tds-glp-103sr-E_ver01_Mar_2018, Daikin Industries, Ltd., 2018.
* Solvent-based coating.

an aqueous suspension [17, 18]. Typical properties of these two products are shown in Tables 12.16 and 12.17 respectively.

At the time of writing, there is no known record of any other commercial product that is a latex form of a fluorocarbon elastomer that would include brand and typical properties.

TABLE 12.16

Typical Properties of Aflas® 150CS Latex

Property	Unit	Value
Appearance	—	White, milky
Solids content (by weight)	%	30–35
Specific gravity	—	1.05–1.20
Particle size	Nm	50–200
pH	—	7–9

Source: AFLAS 150CS Latex Product Information Sheet 09/2015.

TABLE 12.17

Typical Properties of Aflas® 300S Latex

Property	Unit	Value
Appearance	—	White, milky liquid
Solids content (by weight)	%	30–40
Specific gravity	—	1.05–1.20
Particle size	μm	20–50
pH	—	3.0–10.0

Source: AGC AFLAS 300S TFE/P Latex Properties, Mat/Web matweb.com (2022).

REFERENCES

1. *Teflon® PTFE, Dispersion Properties and Processing Techniques*, Bulletin No. X50G, E. I. du Pont de Nemours & Co., Wilmington, DE, Publication E-55541-2, p. 2.
2. Renfrew, M. M., Lewis, E. E., Polytetrafluoroethylene, *Ind. Eng. Chem.*, *38*:870, (1946).
3. Lontz, J. F., Happoldt, Jr., W. B., Heat resistant, chemically resistant plastic, *Ind. Eng. Chem.*, *44*:1800, (1952).
4. *Teflon™ Aqueous Dispersions*, teflon.com/en/products, Chemours (2022).
5. McKeen, L. W., *Fluorinated Coatings and Finishing Handbook, Second Edition*, Chapter 9, Elsevier, Oxford, UK, (2016).
6. Gangal, S. V. *Encyclopedia of Polymer Science and Technology*, Vol. *16* (Mark, H. F. and Kroschwitz, J. I., Eds.), John Wiley & Sons, New York, 1989, p. 624, (1989).
7. *Hyflon® D 5010X, Technical Information, MFA Dispersion for Impregnation*, Solvay Solexis, 2006.
8. *Daikin Polyflon™ PTFE Products*, TDS – PTFE – D005 REV 02/9/16, Daikin America. (2022).
9. *Zonyl™ MPD 1700, Additive Dispersion*, Document C– 10004 (3/18), The Chemours Company (2018)
10. *Solef® PVDF Aqueous Dispersions for Lithium Batteries*, Document D05/2015 Version 12, Solvay Specialty Polymers, www.solvay.com, (2018).
11. *Kynar Aquatec® Data Sheets*, Arkema, Inc. KynarAquatec.com (2022).
12. *3M™ Dyneon™ THV 340Y Product Data Sheet*, 3M Dyneon 3m.com (2022).
13. Drobny, J. G., *Fluoroelastomers Handbook, Second Edition*, Elsevier, Oxford, UK, p, 125, (2016).
14. *Tecnoflon® TN Latex, Specialty Product*, Technical Data Sheet, Solvay Specialty Polymers, R 02/2015, Revision 2.0 www.solvay.com
15. Kirochko, P., Kreiner, J. G., *A New Waterborne Fluoroelastomer Coating*, Paper presented at the Conference Fluorine in Coatings III, Grenelefe, Orlando, FL, 25–27 (1999).
16. *DAI-EL Latex, Fluoroelastomer Coating*, Daikin Industries Ltd. daikinchemical.com, 2022.
17. *Aflas® 150 CS Latex*, Product Information Sheet 09/2015, AGC Chemicals America, www.agcchem.com
18. *AGC Aflas® 300S TFE/P Latex, Property Data*, MatWeb, matweb.com, 2022.

13

Processing and Applications of Fluoropolymer Aqueous Systems

Jiri G. Drobny

13.1 Introduction

In general, waterborne fluoropolymer systems are handled and processed in a similar fashion to other organic coatings. They can be compounded by the addition of fillers, pigments and colorants, resins, and other additives; they can be, for example, viscosified or blended with other waterborne polymeric systems. Because of the nature of the base polymer, they differ in their processing behavior. For example, common polytetrafluoroethylene (PTFE) dispersions are very shear sensitive, whereas others are not. Coating formulations from dispersions of perfluoropolymers are usually very simple, containing only the necessary surfactant. If fillers and pigments are used, the amounts that can be added are limited; otherwise poor-quality films would be obtained. Another aspect to consider is the processing temperature of the polymer: some pigments and colorants cannot tolerate the high processing temperatures required for perfluoropolymers. On the other hand, some fluoropolymers can be compounded into highly filled and pigmented coatings and paints.

13.2 Processing and Applications of PTFE Dispersions

The major utility of PTFE dispersions is that they allow the processing of PTFE resin, which cannot be processed as an ordinary polymeric melt, because of its extraordinarily high melt viscosity, or as a solution, because it is insoluble. Thus, PTFE dispersions can be used: to coat fabrics and yarns; to impregnate fibers, nonwoven fabrics, and other porous structures; to produce antistick and low-friction coatings on metals and other substrates; and to produce cast films.

Surfactants are an essential ingredient for the sufficient wetting of various substrates including sintered PTFE and for the formation of continuous films by uniform spreading on substrates such as metals, glass, ceramics, and PTFE. Generally, 6 to 10% of a nonionic surfactant (e.g., alkylaryl polyether alcohol), based on the polymer content, is sufficient to impart wetting and film-forming properties. Sometimes a small amount of a fluorosurfactant (typically below 0.1% on the polymer content) can be added to increase the efficiency of the nonionic surfactant.

To convert a dispersion into a sintered PTFE film, four distinct steps are required: (1) casting (dipping or flowing out) onto a supporting surface; (2) drying to remove water; (3) baking to remove the surfactant; and (4) sintering to obtain a clear coherent film.

For the successful production of a cast film, the dispersion has to wet the supporting surface and spread uniformly. In drying, the thickness of the deposited layer is very important. If the deposit is too thick, it develops fissures and cracks, referred to as "mudcracking." These flaws cannot be eliminated in the sintering step. Thus, for each formulation there is a critical cracking thickness, which represents a limit below which cracking will not occur in a single application. This depends mainly on the particle size range, the amount of surfactant used, and the solids content. A typical value under optimum conditions is 0.0015 in.

DOI: 10.1201/9781003204275-17

(0.038 mm) [1]. For thicker films multiple coats have to be applied. For a properly formulated dispersion, recoating over an unsintered or sintered coating is not a problem.

13.2.1 Impregnation

Properly compounded PTFE dispersions are suitable for impregnation because of their low viscosity, very small particles, and ability to wet surfaces. The surfactant aids the capillary action and wetting interstices in a porous material. After the substrate is dipped and dried, it may or may not be sintered. This depends on the intended application. In fact, the unsintered coating exhibits sufficiently high chemical resistance and an antistick property. If required, the coated substrate may be heated to about 290°C (555°F) for several minutes to remove the surfactant. Lower temperatures and longer times are used if the substrate cannot tolerate such a high temperature. In some cases, the impregnated material is calendered or compressed in a mold to compact the PTFE resin and to hold it in place.

13.2.2 Fabric Coating

13.2.2.1 Equipment

The largest proportion of PTFE dispersions is used for coating glass fabric. Equipment used for that purpose is a vertical coating tower consisting of three heated zones, namely, the drying zone, baking zone, and sintering zone (Figure 13.1). There are several systems for heating these zones, and the choice of the heating system depends on the type of product to be made. The most common heating is by circulating hot air heated by gas burners. Air in drying and baking zones can also be heated by circulating hot oil. Infrared heating is another choice, and its use has been growing over the past decade. Each method has its advantages and disadvantages. *Gas heating* is very effective but requires a relatively long startup time and is rather inflexible and difficult to control. *Hot oil heating* is very precise and effective but requires a very long start-up time and represents a very high investment cost and high operation costs. *Infrared heating systems* are very flexible and relatively inexpensive. They can be switched on and off very quickly, and if they burn out replacement is very simple. Their disadvantage is nonuniform temperature across the width

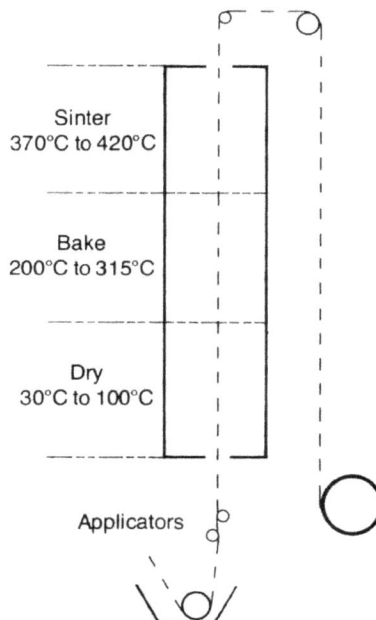

FIGURE 13.1   Schematic of a modern PTFE coating tower. (Courtesy of DuPont)

of the web and overall temperature control. In modern production coating towers, the heating zones are heated independently.

A typical coating line consists of fabric pay-off, the tower, and a take-up for the coated goods. Some lines also include a festoon (accumulator) to avoid the need for stopping the line if the web is being spliced or if the take-up has to be stopped for roll change or a problem. The take-up may be arranged as a turret for a faster roll change.

A stainless-steel dip tank is at the bottom of the coating tower. Inside the tank is a submerged roll, sometimes called a "dip bar", which may be made from stainless steel or PTFE [2]. The roll may be locked or rotated. A rotating bar most frequently has sleeve bearings, which are lubricated by the liquid in the tank. Some designs use multiple rolls (typically three; Figure 13.2); this arrangement reduces differences in pickup between the two sides of the web.

If the process requires the wiping of excess liquid from the web, applicators of different design are used. The most wiping is achieved by sharp knives, the least with horizontally opposed, spring-loaded, fixed-gap, metering rolls [2, p. 6]. In actual industrial practice the most common wiping devices are round-edged knives, wire-wound rods (Meyer rods) with varied wire diameter, and smooth bars (Figure 13.3).

The larger the wire diameter, the greater thickness of the coating applied. Free dipping (i.e., without wiping) is another possibility. The amount of coating can also be controlled by the amount of solids and by viscosity and to a degree by the web speed.

Coating towers can be of straight-up design or can be built in an up-and-down configuration. In the latter design, the drying and baking zones are in the first ("up") part of the tower, and the sintering zone, including a section for web cooling, is in the "down" part. This design saves space but has the disadvantages that sometimes the baked, unsintered coating is picked up by the rolls on the top of the tower, that the residual surfactant decomposes vigorously at the entry to the sintering zone, and that the products of decomposition condense on the top of the tower, mainly on the rolls. An example of this design is shown in Figure 13.4. Regardless of the design, each tower has an exhaust system on the

FIGURE 13.2 Dip tank with three dip bars.

FIGURE 13.3   Detail of coating of glass fabric by PTFE dispersion.

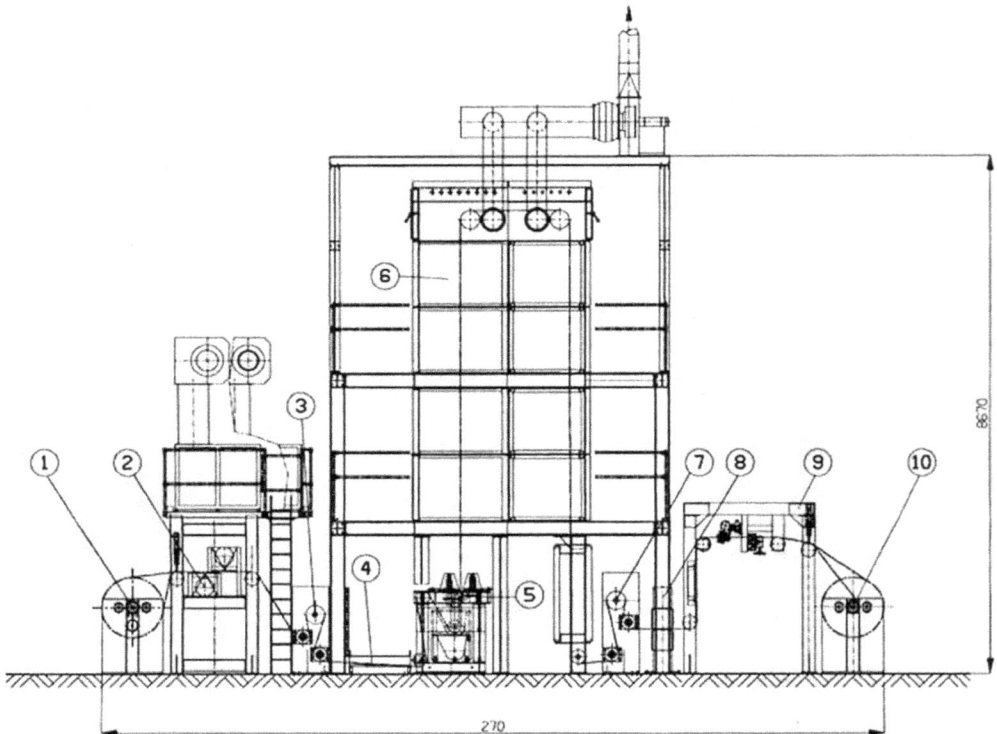

FIGURE 13.4   Up-and-down design of a coating tower: (1) Unwinding unit; (2) Fabric cleaning; (3) Tensioning unit; (4) Operator platform; (5) Dip tank; (6) Oven (drying and baking zones); (7) Tensioning unit; (8) Beta gauge; (9) Inspection table and trimming device; (10) Wind-up unit. *Note*: the sintering zone is on the right side (descending) part of the oven. (Courtesy Gebrüder Menzel Maschinenfabrik, GmbH & Co.)

FIGURE 13.5   Production fabric coating tower in operation.

top for the removal of the volatile decomposition products, which is often coupled with a combustion system to eliminate air pollution. Coating towers are built to process webs up to about 5 m (15 ft.) wide, although the majority of them are in the 2 to 3 m (6 to 9 ft.) range. A typical production coating tower is shown in Figure 13.5.

The coating speeds depend on the type of fabric and can vary between 0.3 and 14.0 m/min (1 to 45 ft./ min). The speed is also limited by the process, which can be drying, baking, or sintering. Too high a speed may cause blisters due to insufficient removal of water in the drying zone or insufficient removal of surfactant in the baking zone, which may cause a fire in the sintering zone or discoloration of the coating and impaired rewetting of the coating in the subsequent pass. Insufficient sintering (off-white, dull coating) is another possible consequence of too high a coating speed.

The temperatures in the individual zones may vary according to the type of fabric being coated. However, the goal is to remove water and other volatile components from the coating without boiling and before the web reaches the baking zone and then to remove surfactant in the baking zone. The sintering is almost instantaneous once the temperature is above the crystalline melting point of the PTFE resin. Typical air temperature ranges in the individual zones are:

Drying zone: 80 to 95°C (176 to 203°F).
Baking zone: 250 to 315°C (480 to 600°F).
Sintering zone: 360 to 400°C (680 to 750°F).

If any of the zones is heated by infrared systems, the surface temperature of the heating elements is set and controlled to attain the required temperature.

13.2.2.2 Formulations

As supplied, PTFE dispersions contain a nonionic wetting agent (surfactant), which provides them with good wetting properties and a minimum tendency to foam [3]. Generally, nonionic surfactants are preferred because they are less likely to induce abnormal viscosities due to thixotropic effects. Other acceptable surfactants are anionic types. Cationic wetting agents are not used because they tend to flocculate the dispersion [4]. The surfactants used in the formulation can be decomposed at the temperatures required for baking, minimizing residual contamination. Other additives, such as mineral fillers or colorants, may be added to achieve the desired properties or appearance. Fillers and pigments must be added in the form of a paste, either purchased or prepared in house using ball mills or dispersion mills. Often the original PTFE dispersion is diluted by water to attain the required coat thickness.

The viscosity of a dispersion may be increased by adding a water-soluble viscosifier. There are several types, for example, hydroxyethyl cellulose (HEC) or acrylic viscosifiers, which are added in the form of an aqueous stock solution. To utilize their viscosifying effects fully, the pH of the formulation must be increased to a certain optimum value, typical for a given viscosifier.

When preparing formulations by blending or adding mineral fillers to PTFE dispersions, only mild agitation must be used. Propeller-type stirrers are best suited for that. High-speed, high-shear mixing is likely to result in coagulation.

13.2.2.3 Coating Process

Glass and aramid fabrics (e.g., Kevlar or Nomex, DuPont) are currently the only fabrics that can withstand the high temperatures required for the sintering of PTFE resin. Thus, they can be coated by it without being degraded greatly in the process. It should be noted that the degradation of these synthetic fabrics is faster than that of glass fabrics. Glass fabrics come with a starch-based treatment (sizing) that is necessary in the weaving operation. However, this treatment interferes with the coating and has to be removed by *heat cleaning*, which means that the fabric is heated to high temperatures to burn off the starch and other organic compounds used in the sizing. This operation is frequently the first pass of the coating process, in which the glass fabric passes through the tower without being coated. In some contemporary designs an infrared heater is placed between the let-off and the dipping tank. This way the treatment is removed in this heater, and the fabric can be coated immediately. This heater also often contains a vacuum cleaner, which removes any loose contaminants from the surface of the fabric.

Depending on the fabric construction and required thickness, the number of passes can be as high as 12 or even more. In some cases undiluted dispersions (with typically 60% solids) are used for at least some of the passes. Typical amounts of PTFE in the coated fabrics are anywhere from 15% for porous fabric to 85% for heavily coated fabrics. Each coat must be below the critical cracking thickness to avoid *mud cracking*. As an alternative, several unsintered coats are applied, and the coated fabric is then calendered prior to sintering to seal the mud cracks. The calander rolls are heated to temperatures ranging from 148 to 177°C (300 to 350°F) and operate at a pressure of about 1 ton/linear in. to be effective. Wetting agents must be removed completely in the baking zone to prevent the coating from being picked off by the rolls. Calendering is also used to flatten the fabric and to bury filaments, which could be broken during coating [4, p. 6]. Calenders used for this purpose consist either of one chrome-plated steel roll and a compressed paper backup roll, or of two chrome-plated steel rolls. In the former design, only the steel roll can be heated. The result is that there is a difference in appearance between the two sides of the web, which, depending on the application, may or may not be a disadvantage. The advantage is that the roll combination exerts a gentler pressure. In the design with two steel rolls, both rolls can be heated, and both sides of the calendered web are smooth. The disadvantage is that the nip may sometimes be too harsh on the glass fibers. However, in general, this design is more effective for the compaction of unsintered PTFE and consequently in sealing the mud cracks.

To obtain PTFE-coated fabrics that can be heat-sealed or laminated at lower temperatures, a thin coat of diluted aqueous dispersions of FEP or PFA is applied on top of the PTFE coating.

13.2.2.4 Lamination

PTFE-coated fabrics can be laminated in electrically heated presses at temperatures in the range of 360 to 400°C (680 to 750°F) and pressures of around 3.4 MPa (500 psi). If the fabric has a coat of FEP or PFA, the lamination temperatures are reduced to be above the melting points of the respective resins. Such fabrics can also be laminated on equipment operating continuously. Another lamination process is based on laminating two substrates having an unsintered baked layer of PTFE on the surface. The two adjacent unsintered layers act as a pressure-sensitive adhesive and as a result of fibrillation under pressure [5] form a bond sufficiently strong mechanically to survive a free sintering (without pressure). After sintering, the bond will have the heat resistance of PTFE. This process is suitable for the lamination of coated fabrics as well as for lamination of PTFE cast films with PTFE-coated fabrics or a combination of PTFE with other materials in a continuous fashion [5]. A schematic of this technology is shown in Figure 13.6.

13.2.2.5 Applications of PTFE Coated Fabrics

The largest volume of PTFE-coated fabric is used in construction. The typical application is as roofing material for large structures, often with minimum support. Such fabrics are produced from heavy medium or high strength glass fabrics with six to seven different coats, such as a prime coat, buffer coats, bulk coats (viscosified PTFE compound with glass beads), and often with a thin top coat of diluted FEP

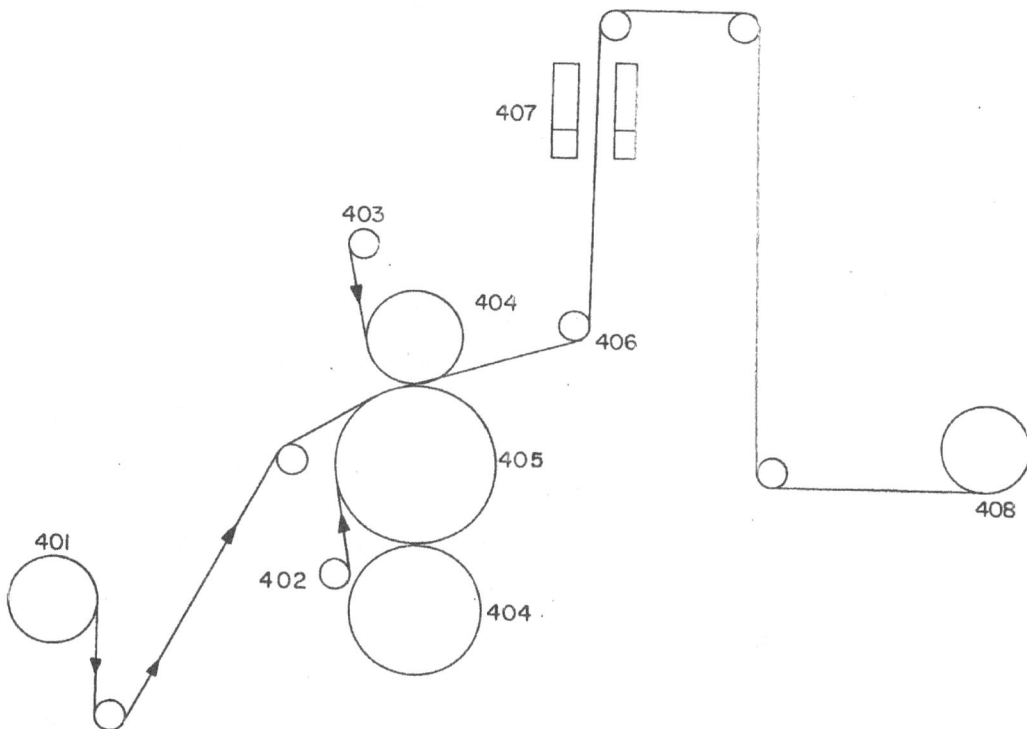

FIGURE 13.6 Schematic of the laminating machine and process using unsintered coating; 401, 402, 403 are pay-off rolls; 405 is a heated steel roll; 406 is a packed roll; 407 is the sintering oven; 408 is a take-up roll.
Source: US Patent 5,141,800 [5].

FIGURE 13.7   Sample of PTFE-coated roofing fabric. (Courtesy of DPA)

dispersion as a sealing surface. Coated and laminated architectural fabrics generally used for air supported or tension supported roofing systems must meet demanding specifications based on tests specified by ASTM Standard D-4851. An example of a roofing fabric is shown in Figure 13.7.

Typical applications are sports buildings (Figure 13.8), shopping malls, airports (Figure 13.9), industrial warehousing, and cooling towers [6, 7]. In the past, large tents for huge tent cities in the Middle

FIGURE 13.8   PTFE roof of Shellcom Stadium Sendai in Miyagi, Japan. (Courtesy of Chukoh Chemical Industries, Ltd.)

FIGURE 13.9 PTFE roof of the International Terminal, Denver Airport. (Courtesy of DPA)

East were made from PTFE-coated glass fabric. A unique architecture using PTFE-coated glass fabric is shown in Figure 13.10. Other architectural applications are acoustic fabrics, façades, and shading applications [8].

Another use of PTFE-coated glass fabrics is in radomes. Radomes are structures or enclosures designed to protect antennae and associated electronics from the surrounding environment and elements

FIGURE 13.10 Arabian Tower Hotel in Dubai. (Courtesy of Skyspan Europe, GmbH)

FIGURE 13.11   Spherical radome from a PTFE-coated fabric. (Courtesy of St-Gobain Performance Plastics)

such as rain, snow, UV light, and strong wind. The name "radome" is derived from the words radar and dome. While there are many designs, the most common ones are spherical and planar types. As for materials, there are several used for radomes. PTFE is one of the most effective because it has low values of dielectric constant and dissipation factor and therefore has virtually no effect on the function of the protected equipment. An example of a spherical type is shown in Figure 13.11 and that of a planar type in Figure 13.12.

Other applications include belts in the ceramic and food industries, release sheets for laminates from composites, cooking sheets in fast-food establishments, release sheets for baked goods, industrial drying

FIGURE 13.12   Planar radome covered by PTFE-coated composite. (Courtesy of Textiles Coated International)

belts, in the electrical and electronic industries, and as reinforcement for high-temperature pressure-sensitive tapes.

13.2.3 Cast Films

13.2.3.1 Process and Equipment

As pointed out earlier, a PTFE homopolymer cannot be processed by melt extrusion because of its extremely high melt viscosity. Thus, other methods, such as skiving from compression molded and sintered billets (see Chapter 6) and by casting from dispersions, were found to prepare films. The original method for casting films from PTFE dispersions employs polished stainless-steel belts, which are dipped into a properly compounded dispersion. The thin coating of the liquid is then dried, and the dry powdery layer is subjected to baking and sintering. To obtain a good-quality film, the thickness of the film has to be below the critical value to prevent mudcracking.

The equipment used is again a vertical coater with heated zones, very similar to the coating tower for fabric [9]. The speed of the belt is slow, about 0.3 to 1 m/min (1 to 3 ft./min), and there are no applicators used to remove excess dispersion. The amount of coating picked up by the belt is controlled mainly by the solids content of the dispersion and by the belt speed. A production machine is built with multiple stages. Thus, after a film is sintered, it is recoated in the next stage. At the end of the machine, there is a device designed to strip the finished films from both sides of the belt and to wind them up into rolls. A schematic of this process is shown in Figure 13.13.

The advantage of this method is that each layer can be made from a different type of dispersion. For example, clear and pigmented layers can be made or the top layer can be prepared from an FEP or PFA

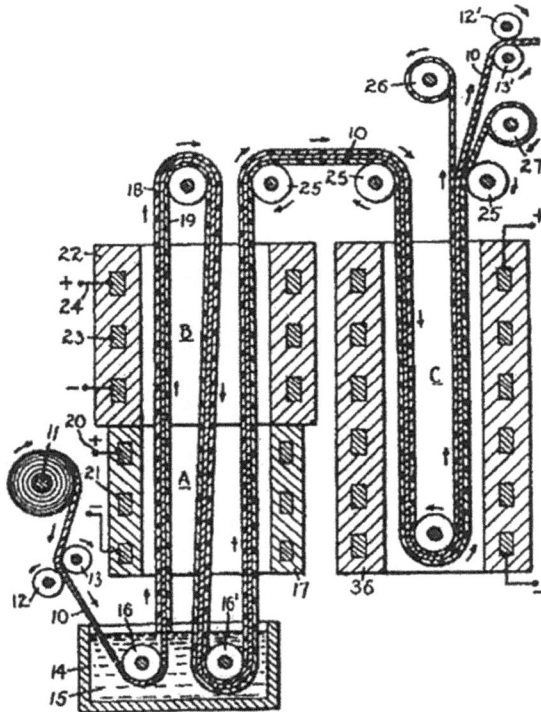

FIGURE 13.13 Schematic of the steel belt coater for the production of fluoropolymer cast films.

Source: (US Patent 2,852,874 [8]. 10 Stainless steel belt; 15 Fluoropolymer dispersion; 16, 16' Dip rolls; 26, 27 Film stripping system; A Drying zone; B Sintering zone; C Baking zone.

dispersion to obtain films that can be heat-sealed or laminated. In fact, films with both surfaces heat-seal-able can be produced by this method. In such an instance, PFA is applied as the first coat onto the belt and FEP as the last coat, because PFA can be stripped from the steel belt, whereas FEP would adhere to it and be impossible to strip. Another possibility is to make films with an unsintered last coat, which can be used for lamination with substrates coated with unsintered PTFE using the lamination method described in Section 13.2.2.4.

An improved process and equipment for cast PTFE films have been developed that have considerably higher productivity than the method and equipment described earlier [10]. The process essentially uses a vertical coater with multiple stages. The carrier belt has to be made from a material of low thermal mass, which can tolerate repeated exposure to the sintering temperature and has surface properties such that it can be wetted by the dispersion; the film can then still be stripped without being stretched. There are several possible belt materials, but Kapton-H (DuPont) was found to be particularly suitable because of its heat resistance, dimensional stability, and surface characteristics [10]. The production speeds used in this process are 3 to 10 m/min (9 to 30 ft./min). A schematic of this coater is shown in Figure 13.14.

Unlike in the coater with steel belts, in this process equipment applicators such as wire-wound bars are used, which may be designed to rotate to assure a better, more uniform coating. The dip tanks are similar to those used in fabric coating, also using an immersed dip bar. To prevent coating defects due to shear, the dip tanks have double walls and are chilled by circulated water to temperatures below 19°C (66°F), the first-order transition temperature.

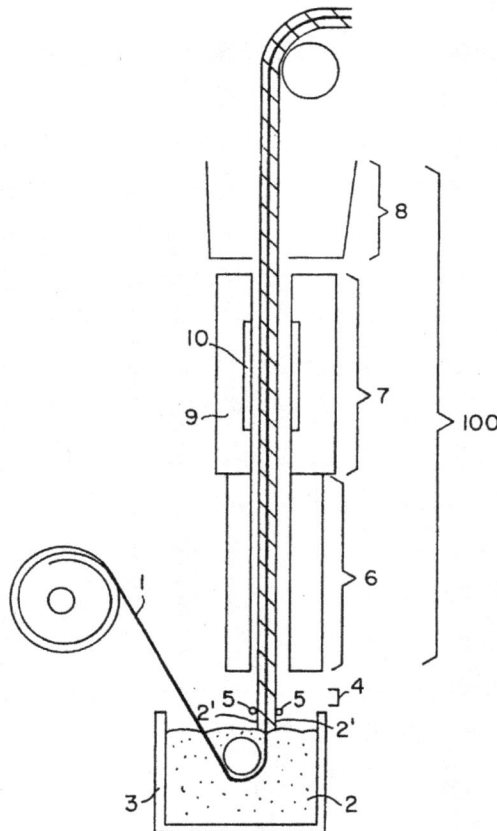

FIGURE 13.14 Schematic of the high-speed coater for fluoropolymer films. The coating tower (100) consists of (4) a metering zone, (5) metering bars, (6) a drying zone, (7) a baking zone, and (8) a cooling plenum.

Source: US Patent 5,075,065 [9].

The dispersions used in this process are formulated in such a way that they wet the carrier sufficiently and tolerate higher shear at the wire-wound bars due to a relatively high coating belt speed. This is accomplished by a combination of nonionic surfactants (e.g., octyl phenoxy polyethoxy ethanol) and fluorosurfactants. This subject is discussed in detail in [10].

13.2.3.2 Applications of PTFE Cast Films

Because of the nature of the manufacturing process, there is no melt flow, and consequently the cast films do not exhibit anisotropy typical of extruded films. PTFE cast films have higher tensile strength, elongation, and dielectric strength [10]. Another advantage is that they can be produced in layered form from different dispersions (e.g., two colors, with one layer clear, others pigmented, or with one layer having static-dissipative properties, or with one or both layers consisting of melt-processible perfluoropolymers, such as FEP or PFA). If suitable tie-layers are used, it is possible to produce a combination of PTFE film with a bonding layer based on a fluoroelastomer or other fluoropolymer, such as lower-melting THV fluoroplastic or PVDF. Films with lower-melting bonding layers can be laminated with substrates that normally would not tolerate the high PTFE sintering temperatures.

Cast PTFE films can be laminated with different substrates, most frequently with PTFE-coated glass and aramid fabrics. Specialty laminates of films and coated fabrics are used for protective clothing. They provide protection against liquid and vapor, chemicals, chemical and biological warfare agents, and are used in the chemical industry, firefighting, and the military. An example of a protective suit is shown in Figure 13.15.

FIGURE 13.15   Protective clothing made from fluoropolymer laminates. (Courtesy of DPA)

Cast films also can be metallized, in particular with aluminum, for use in electronics. Other applications include use as release films for the manufacture of composite materials for aerospace vehicles, in the electronics and electrical industries, as well as selective membranes in the chemical industry.

13.2.4 Processing and Applications of Modified PTFE Dispersions

Dispersions of modified PTFE are designed for coating processes and are particularly recommended for the production of coated fabrics and laminates. They exhibit good wetting properties, high shear stability, and good film forming behavior. They are especially suited for topcoat passes to obtain products with improved surface finish and high gloss [11].

13.2.5 Processing and Applications of PTFE Micropowders

Dispersions of micropowders are used as additives to paints and coatings, for demolding, and for improved lubrication [12].

13.2.5.1 Addition to Paints

PTFE micropowders are added to decorative, masonry, aircraft, and marine paints. The major benefits of micropowder dispersions for paints include [13]:

- Matte-based paints that are traditionally difficult to clean become easier with PTFE micropowders.
- Marine coatings exhibit excellent anti-fouling properties.
- Gloss coatings have a smoother surface and fewer imperfections due to the small particle size.

13.2.5.2 Addition to Elastomers

Lubricants are added to elastomer formulations to improve the coefficient of friction and wear properties versus solid lubricants, waxes, stearates, soaps, plasticizers, and oils.

Adding lubricant powders to natural rubbers and synthetic elastomers during processing gives finished moldings many of the surface slip characteristics of PTFE. These characteristics include: improved mold release, lower static and dynamic coefficients of surface friction, abrasion resistance, elimination of stick slip, and improved tear strength.

13.2.5.3 Addition to Oils and Greases

PTFE micropowders are ideal for improving lubrication in applications that experience extreme pressures, temperatures, and environments. They are also used in applications where conventional additives such as graphite and molybdenum are unsuitable [13].

In addition, PTFE additives offer cleanliness, an important characteristic for greases used in food, pharmaceutical, and dairy equipment. Because PTFE is not flammable, it is ideal for applications where the lubricants are exposed to gases and other potential fire hazards [13].

13.2.6 Other Processing and Applications of PTFE Aqueous Dispersions

PTFE aqueous dispersions are applied onto *metal substrates* by spraying, dipping, flow coating, electrodeposition, or coagulation to provide chemical resistance, nonstick, and low-friction surfaces. Nonstick cookware and bakeware are made from dispersions specifically formulated for that purpose with the use of a primer for the metal. In some cases an additional layer (protective layer) between the primer and the top coat is applied. After coating, the parts are dried and sintered. Since PTFE coating tends to be porous, it can be combined with PFA [14]. An example of cookware coating is Silverstone, which is a specialty line of highly abrasion-resistant nonstick finishes produced by DuPont. Silverstone coatings are three-coat (primer/midcoat/topcoat) systems formulated with PTFE and PFA. The characteristics of

Silverstone coatings are similar to other PTFE coatings; however, durability is greatly increased because of the proprietary formulation. A ceramic reinforced version with higher scratch and abrasion resistance is also available. The maximum continuous service temperature of these coatings is 290°C (550°F). An addition of polyphenylene sulfide (PPS) or polyamide imide improves the adhesion of the primer to the metal surface [15]. Besides cookware, PTFE or blends of PTFE and PFA or PTFE and FEP are used to coat, for example, industrial rollers, pipes, storage tanks, pumps, probes, catheters and other medical devices, paper cutters, and drill bits.

PTFE fibers are spun out of aqueous dispersions, which are mixed with a matrix-forming medium and forced through a spinneret into a coagulating bath. Then the matrix material is removed by heating, and the fibers are sintered and oriented (drawn) in the molten state to develop their full strength [16].

Another use of PTFE dispersions is the preparation of a variety of compositions with other materials, such as mineral fillers, or with other polymers in powdered form by coagulation. The dispersion of the other component is blended with the PTFE dispersion, and the blend is then coagulated. The resulting composition can be processed by extrusion with lubricants or by compression molding [17].

13.3 Processing and Applications of Other Fluoropolymer Aqueous Systems

13.3.1 Aqueous Dispersions of FEP and PFA/MFA

Aqueous dispersions of these two melt-processible perfluoropolymers are processed in a way similar to PTFE dispersion. FEP dispersions can be used for coating fabrics, metals, and polyimide films. They are very well suited for bonding seals and bearings from PTFE to metallic and nonmetallic components and as nonstick and low-friction coatings for metals [18]. FEP can be fused completely into a continuous film in approximately 1 minute at 400°C (752°F) or 40 minutes at 290°C (554°F) [18, p. 624]. PFA and MFA dispersions are used to coat various surfaces, including glass fabric, glass, and metals.

Fabrics coated with FEP and PFA/MFA can be laminated and heat sealed into, for example, protective garments or canopies. FEP-coated polyimide films are used in electronics and as a wire tape. FEP-based anticorrosion coatings are used in the chemical industry and as chemical barriers [18, p. 624]. A thin coating of FEP or PFA/MFA can be used as a hot melt adhesive for a variety of substrates, including PTFE-coated fabrics and laminates.

13.3.2 Aqueous Dispersions of PVDF

As pointed out in Chapter 12, there are several commercially available aqueous dispersions containing PVDF alone or in combination with acrylic resins. They can be processed as coatings of fabrics or into thin cast films using equipment described in Section 13.2 on the processing of PTFE dispersions. The fusing temperature for films made from a PVDF homopolymer is 230 to 250°C (446 to 482°F) [19], which is low enough for coating polyester fabrics and casting films on carriers that tolerate this temperature. The thin films cast from PVDF aqueous dispersions can be pigmented and used for decorative surfaces [19, p. 543]. PVDF aqueous dispersions alone or their blends with acrylic aqueous systems are used for coating fabrics or as decorative or protective coatings [19, 20, and 21]. One known application is as a coating for the Stockholm Globe Arena (now named Avicii Arena) shown in Figure 13.16.

PVDF aqueous dispersions are well-suited for use as the binder in lithium ion batteries, offering many advantages in the formulation of the electrodes [20]. PVDF is very stable and delivers a reliable performance having:

- Better cohesion between binder and active material.
- Improved adhesion to metal collector.
- Highly stable functional groups vs. styrene butadiene rubber (SBR).
- Lower binder content for improved energy density.
- Higher capacity at high C-rate for improved power performance (the C-rate represents the rate at which the battery provides energy).

FIGURE 13.16   PVDF protective coating on the building of the Stockholm Globe Arena (now Avicii Arena). (Courtesy of Arkema)

13.3.3 Aqueous Dispersions of THV Fluoroplastics

Currently, the only available commercial grade of aqueous dispersion of THV Fluoroplastics is THV 340Z. This is a dispersion of a hydrophobic polymer with almost spherical particles with a size of around 120 nm with a solids content of 50%. THV 340Z contains 5% of ionic surfactant and a pH value of above 9. It can be processed in a similar fashion to dispersions of other melt-processible fluoropolymers, namely, for coating fabrics or casting thin films. Because of the relatively low melting temperature of the base polymer (145°C or 293°F) [22], it can be used to coat polyester fabrics and can be cast on carriers, tolerating the processing temperature (e.g., polyester film). A THV Fluoroplastic aqueous dispersion, when properly compounded, can be foamed by whipping in a fashion similar to elastomeric latexes [23]. Coated and laminated fabrics (the original application of THV resins) are used in many fabric applications where flexibility, weatherability, or low permeability is required.

Typical applications of aqueous dispersion are protective covers, tarpaulins, awnings, and chemical-protective garments [24]. Thin THV Fluoroplastic films can be laminated onto temperature-sensitive substrates, such as plasticized polyvinyl fluoride (PVC) and polyester. In certain situations, where the barrier film needs to be optically clear, the optical clarity of THV Fluoroplastic is an additional feature, particularly in signage applications [24]. Another application for THV Fluoroplastic films is in the protective film for solar module constructions [24, p. 269].

13.3.4 Fluorocarbon Elastomers in Latex Form

The compounding techniques used for fluorocarbon latexes are similar to those used for standard latexes; that is, solid ingredients are first dispersed in water with the use of surface active agents; liquid ingredients are then prepared as emulsions prior to their addition to the latex. The dispersions of solids are prepared in ball mills or high-speed mills (e.g., a Kady Mill).

Fluorocarbon elastomers in the form of highly concentrated latex (typically 70% solids by weight) can be used in the coating of fabrics to produce, for example, protective garments and expansion joints. They can be blended with other compatible fluoropolymers and specialty polymers to attain specific properties. Coatings can be cross-linked chemically or by electron beam [25].

An example of a commercial highly concentrated terpolymer FKM latex is Tecnoflon® TN Latex offered by Solvay [26]. This product can be used to make rubber coated fabrics and protective gloves, for fiber impregnation, protection films over substrates with lower chemical resistance or to reduce friction, and in general for any coating or impregnation where chemical resistance and/or thermal stability are critical factors. Although Tecnoflon® TN latex is resistant to most of the commonly used solvents and chemicals, it is not resistant to ketones, esters, and low-molecular-weight polar solvents. When compounded, the latex is mixed with some ingredients to meet specific requirements. These include:

- Curing agents.
- Acid acceptors.
- Fillers.
- Pigments.
- Emulsifiers.

The compounds are usually cured with polyamines (di-, tri-, and tetraamines) such as:

- N.N'- dicinnamilydene -1,6 hexanediamine (2.5 to 5 phr).
- Hexamethylene diamine carbamate (1 to 3 phr).
- Triethylene tetramine (TETA) (1 to 3 phr).

Zinc oxide is suggested as an acid acceptor (8 to 15 phr) because other metal oxides and hydroxides may cause coagulation due to their high pH [26]. Reinforcing fillers are used for improving physical properties and they should be dispersed in an adequate filler using an emulsifier. The same process is used for pigments and the acid acceptor. Not all fillers are suitable; there are some constraints: the pH should not be lower than 5 or higher than 8; a high specific gravity tends to settle too fast; carbon black should be added as an aqueous dispersion, because it is difficult to disperse when added to the compound directly. An example of a typical FKM latex compound is shown in Table 13.1.

The mixed compound contains typically about 50% solids and is applied to the substrate by spreading or dipping. Typically, one pass 0.5 mm (20 mils) is applied. After that the coat is dried. To achieve a good quality dried coat, a drying temperature not higher than 60°C (140°F) is recommended.

The curing time and temperature are strongly related to the amount of the amine used and its amount. As an example, a compound containing 1.5 phr TETA can be cured for 1 hour at 90°C (194°F). The physical properties of a compound cured for 1 and 2 hours at 90°C is shown in Table 13.2.

A post-cure is recommended if high temperature resistance is required. An example of physical properties from the compound shown in Table 13.1 cured for 1 hour at 50°C (122°F) and post-cured for 1 hour at 250°C (482°F) is shown in Table 13.3.

Chemically cross-linked coatings can be two or one-component systems curing at ambient or elevated temperatures [27]. An example of a one-component water-based coating is that developed by Lauren International, Inc., covered by US Patent 5,854,342. This is a water-borne fluoroelastomer coating

TABLE 13.1

Example of a Typical FKM Latex Compound

Ingredient	Amount, phr
Tecnoflon® TN Latex	145
Zinc oxide (acid acceptor)	10
TETA (curing agent)	1.5
Nyad® 400 (calcium metasilicate (filler))	20
Sodium lauryl sulfate (emulsifier)	1.0
Chromium oxide (pigment)	5.0
Total	**182.5**

Source: Data from [24].
Note: phr = Parts per hundred parts of rubber.

TABLE 13.2

Typical Cured Physical Properties of a Compound

Property	Unit	Value	
		A	B
100% modulus	MPa	1.4	1.4
Tensile strength	MPa	3.1	3.3
Elongation at break	%	700	550

Source: Data from [24].
Note: A = cured for 1 hour at 90°C (194 °F); B = cured for 2 hours at 90°C (194°F).

TABLE 13.3

Typical Physical Properties of the Compound After Post-Cure

Property	Unit	Value
100% modulus	MPa	3.3
Tensile strength	MPa	6.5
Elongation at break	%	3.30

Source: Data from [24].
Note: Cured for 1 hour at 50°C (122°F) and then post-cured for 1 hour at 250°C (482°F).

composition, comprising an aqueous dispersion of a fluoroelastomer polymer, an amino/polyamino-siloxane curative and, from 0 to about 40 parts by weight of an additive filler per 100 parts by weight of the polymer. The patent covers the method of coating a substrate with the fluoroelastomer coating composition and fluoroelastomer films cured with amino/polyamino-siloxane curatives.

The fluoroelastomer polymers used in this coating can include any copolymerizable monomers, but preferably include copolymers of vinylidene fluoride and hexafluoropropylene, or terpolymers of vinylidene fluoride, hexafluoropropylene, and tetrafluoroethylene. Other examples of fluoroelastomers include those modified with monomers that provide enhanced properties, for example, copolymerization with perfluoro(methylvinylether) to improve low temperature performance. Mixtures of the above fluoroelastomers may also be employed. Such a system cures to optimum properties in about 1 hour at 100°C (212°F) [27, 28].

Other similar products are DAI-EL LATEX GLS-213RA and DAI-EL LATEX GL-252-EA offered by Daikin and Aflas® 150CS Latex and Aflas® 300 S Latex based on the TFE/P fluoroelastomer, a product of AGC Chemicals.

The Daikin products are cured by the addition of a proprietary curative at elevated temperatures and are used to provide an elastic coating with heat resistance, chemical resistance, and low friction on an elastic substrate [29]. For the application of the water-based coatings mentioned above to silicone elastomer surfaces, a special solvent-based primer DAI-EL LATEX GLP-103SR is used to enhance their adhesion [29].

The AGC products are used as elastomeric surface treatments of various substrates such as metals, ceramics, plastics, other elastomers, textiles, and paper. The incorporated cure site monomer allows this material to be cured by peroxide systems or by electron beam irradiation [30, 31].

13.4 Health and Safety

Aqueous dispersions of fluoropolymers are in general neutral to moderately alkaline, with the exception of certain coatings for metals that are strongly acidic. Some additives in the aqueous phase of the dispersion may irritate the eyes or skin. Therefore, it is advisable to use protective garments, goggles, or a facial shield. If the liquid comes in contact with skin, the affected spot must be flushed with water immediately. If the liquid comes in contact with the eyes, they must be flushed immediately and medical help provided as soon as possible. In general, it is important to obtain and review the appropriate Material Safety Data Sheet (MSDS), which are documents that contain information on the potential hazards (health, fire, reactivity, and environmental) and how to work safely with the given chemical product supplied.

When aqueous dispersions are processed at elevated temperatures, particularly above the melting point of the dispersed polymer, the same health and safety precautions must be taken as when corresponding resins in solid form are processed (see Chapter 19).

REFERENCES

1. *Teflon® PTFE, Dispersion Properties and Processing Techniques*, Bulletin No. X50G, E. I. du Pont de Nemours & Co., Wilmington, DE, Publication E-55541-2, p. 2.
2. Renfrew, M. M., Lewis, E. E., Pyrolysis of polytetrafluoroethylene, *Ind. Eng. Chem.*, *38*:870, (1946).
3. Lontz, J. F., Happoldt, Jr., W. B., Teflon tetrafluoroethylene resin dispersion: A new aqueous colloidal dispersion of polytetrafluoroethylene, *Ind. Eng. Chem.*, *44*:1800, (1952).
4. *Teflon® PTFE, Dispersion Properties and Processing Techniques*, Bulletin No. X50G, E. I. du Pont de Nemours & Co., Wilmington, DE, Publication E-55541-2, p. 3.
5. Effenberger, J. A., Enzien, F. M., Keese, F. M., Koerber, K. G., U.S. Patent 5,141,800 (August 25, 1992) to Chemical Fabrics Corporation.
6. Forster, B., *J. Coated Fabrics 15*:25, (July, 1985).
7. Fitz, H., *Tech. Rdsch. (Bern)*, *75*(51/52), 20th December 10 (1983).
8. SHEERFILL® *Architectural Membranes*, Saint-Gobain Performance Plastics www.sheerfill.com, (2021).
9. Petriello, J. V., U.S. Patent 2,852,811 (September 1958).
10. Effenberger, J. A., Koerber, K. G., Latorra, M. N., Petriello, J. V., U.S. Patent 5,075,065 (December 24, 1991) to Chemical Fabrics Corporation.
11. *Daikin Polyflon® PTFE Ptoducts*, TDS-PTFE- D005 REV 02/9/16, Daikin America (2022).
12. *Zonyl™ MPD 1700 Additive Dispersion*, C-10004 (3/18), The Chemours Company (2018).
13. 5 Key Products That Are Enhanced with PTFE Micropowders, AGC Chemicals Americas agcchem.com (2022).
14. Tsai, T.-H., U.S. Patent 5,462,769 (October 31, 1995).
15. Ulrich, H. *Polymer Powder Technology* (Narkis, M., and Rosenzweig, N., Eds.), John Wiley & Sons, Ltd., Chichester, UK, 1995, p. 14.
16. Steuber, W., U.S. Patent 3,051,545 (August 28, 1962) to E.I. du Pont de Nemours & Co.
17. Sperati, C. A. *Handbook of Plastic Materials and Technology* (Rubin, I. I., Ed.), John Wiley & Sons, New York, p. 125, (1990).
18. Gangal, S. V. *Encyclopedia of Polymer Science and Technology*, Vol. *16* (Mark, H. F. and Kroschwitz, J. I., Eds.), John Wiley & Sons, New York, p. 611, (1989).
19. Dohany, J. E., Humphrey, J. S. *Encyclopedia of Polymer Science and Technology*, Vol. *17* (Mark, H. F. and Kroschwitz, J. I., Eds.), John Wiley & Sons, New York, p. 542, (1989).
20. *Solef® PVDF Aqueous Dispersions for Lithium Batteries*, D 05/2014/ R 02/2015/ Version 1.2, Solvay Specialty Polymers www.solway.com (2015).
21. *Kynar Aquatec® Data Sheets*, Arkema Inc KynarAquatec.com (2022).
22. *3M™ Dyneon™ THV 340Z Product Data Sheet*, 3M Dyneon (2016).
23. *THV Coating, Publication IPR/8-4-94*, American Hoechst Corporation, Leominster, MA (1994).
24. Hull, D. E., Johnson, B. V., Rodricks, I. P., Staley, J. B., *Modern Fluoropolymers* (Scheirs, J., Ed.), John Wiley & Sons, Ltd, Chichester, UK, p. 268, (1997).
25. Arcella, V., Ferro, R. *Modern Fluoropolymers* (Scheirs, J., Ed.), John Wiley & Sons, Ltd, Chichester, UK, p. 86, (1997).
26. *Tecnoflon® TN Latex, Tcchnical Data Sheet*, Document R02/2015/ Version 2.0, Solvay Specialty Polymers, www.solvay.com (2015).
27. Kirochko, P., Kreiner, J. G., *A New Waterborne Fluoroelastomer Coating*, Paper presented at the Conference Fluorine in Coatings III, Grenelefe, Orlando, FL, pp. 25–27, (1999).

28. Kirochko, P, Kreiner, J. G., U.S. Patent 5,854,342 to Lauren International Inc. (1998).
29. *DAI-EL Latex, Fluoroelastomer Coating*, Daikin Industries Ltd. daikinchemical.com, (2022).
30. *Aflas® 150 CS Latex*, Product Information Sheet 09/2015, AGC Chemicals America, www.agcchem.com (2022).
31. *AGC Aflas® 300S TFE/P Latex, Property Data*, MatWeb, matweb.com (2022).

Part V

Other Fluoropolymers

14

Specialty Fluorinated Polymers

Sina Ebnesajjad

This chapter describes a number of specialty fluorinated polymers which possess unique properties. It is doubtless that fluorine (F) is a special element beyond all others. F substitution for hydrogen (and many other elements) in organic compounds is relatively easy, simply because of its voracious appetite for grabbing electrons. The underlying reason is F is the most electronegative among all elements, thus forming C–F, the strongest carbon bond.

F substitution for hydrogen in a chemical compound imparts unique and desirable effects to organic compounds and polymers. For example it increases or decreases polarity, and alters chemical activity in both directions. Other properties include elevation of biological activity, for example as use in pharmaceuticals and agrochemicals. The presence of F in a compound elevates its thermal and oxidative stability and chemical resistance in combination with excellent physical and mechanical properties.

Several partially and fully fluorinated polymers have been developed to take advantage of the unique impact of F on their properties. Examples of F substitution in common polymer chemistries include polymethyl siloxane, polyolefins, fluorinated elastomers, acrylic and methacrylic polymers, and perfluorinated ether polymers. Another innovation has been the development of unique polymers by the alteration of crystallinity of the fluorinate polymers including fully amorphous ones.

14.1 Amorphous Fluoropolymers

Amorphous fluoropolymers are prepared by the introduction of specialty monomers. Those monomers are larger than typical fluoropolymer monomers and usually contain sizable groups that disrupt crystallization of the polymer. The reaction proceeds via free radical copolymerization, which can be carried out in either aqueous or nonaqueous media. Amorphous fluoropolymers (AFs) are thermoplastics and dissolve in halogenated solvents and can be applied as coatings. AFs are resistant to nearly all chemicals, frequently compared with PTFE, except for their solvents.

Commercial examples of AF include Teflon® AF (Chemours Co., www.Teflon.com), Cytop™ (AGC Chemicals Americas, www.agcchem.com), and CyclAFlor®. CyclAFlor is based on Cytop chemistry and is manufactured by Chromis Technologies. Development of customized amorphous fluoropolymers has been reported by the polymerization of monomers such as perfluoro-2,2-dimethyl-1,3-dioxole (PDD) and a perfluorinated vinyl ether with sulfonyl fluoride side chains [2].

The first commercial amorphous fluoropolymer products were developed by the Chemours Co. based on the copolymerization of TFE and perfluro-2,2 dimethyl-1,3 dioxide (PDD) and trade-named Teflon® AF. In addition to the outstanding chemical, thermal, and surface properties of perfluorinated polymers they also exhibit unique electrical, optical, and solubility characteristics [3, 4]. The structure of Teflon® AF is displayed in Figure 14.1. A list of the properties of Teflon® AF are:

- Low refractive index.
- Optical clarity.
- Excellent UV stability and transmission capability.
- Low dielectric constant (1.89–1.93), even at gigahertz frequencies.

DOI: 10.1201/9781003204275-19

$$(---CF-CF--)_x--(--CF_2--CF_2)_y$$

$$\begin{array}{ccc} & / \quad \backslash & \\ & O \quad\quad O & \\ & \backslash \quad / & \\ & C & \\ & / \quad \backslash & \\ & CF_3 \quad CF_3 & \end{array}$$

PDD   **TFE**

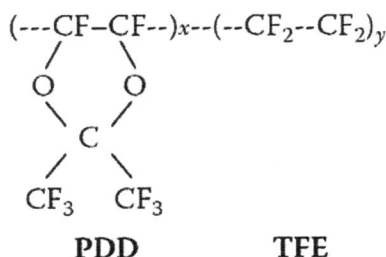

FIGURE 14.1   Molecular structure of Teflon® AF.

- Low dissipation factor and moisture sensitivity/moisture absorption.
- Solubility in perfluorinated solvents.
- Dimensional stability.
- Reduced mold shrinkage.
- High compressibility and gas permeability.

PDD is prepared in four steps from hexafluoroacetone and ethylene oxide. Hexafluoroacetone condenses with ethylene oxide to form a highly stable dioxolane ring in quantitative yields. Exhaustive chlorination followed by chlorine–fluorine exchange results in 2,2-bistrifluoromethyl-4,5-dichloro-4,5-difluoro-1,3-dioxolane in a yield exceeding 90%. Dechlorination of this dioxolane with magnesium, zinc, or a mixture of titanium tetrachloride and lithium aluminum hydride yields a PDD monomer, a clear, colorless liquid boiling at 33°C. It is highly reactive, and therefore it must be stored with trace amounts of a radical inhibitor [3].

PDD copolymerizes with tetrafluoroethylene and other fluorinated monomers by a free radical mechanism, aqueously or non-aqueously, quite easily. Examples include vinylidene fluoride (VDF), chlorotrifluoroethylene (CTFE), vinyl fluoride (VF), and propylvinyl ether (PVE). PDD forms an amorphous homopolymer with a glass transition temperature (T_g) of 335°C [3, 4].

The available grades are Teflon® AF 1600 and Teflon® AF 2400, produced by copolymers of PDD and tetrafluoroethylene with respective glass transition temperatures of 160 and 240°C. They have a perfluorinated structure and exhibit similar high-temperature stability, chemical resistance, low surface energy, and low water absorption. They are soluble in several perfluorinated solvents at room temperature and have high optical transmission across a broad wavelength region from the ultraviolet (UV) to the near infrared [5, 6].

Other properties include lower refractive indexes and dielectric constants, and a gas permeability which is higher than that of the semicrystalline perfluoropolymers. The refractive index of Teflon® AF is the lowest known for any solid organic polymer (the respective values for Teflon® AF 1600 and AF 2400 are 1.31 and 1.29 at 20°C at the 589.6 line) as seen in Table 14.1. The presence of the dioxole structure in the chain imparts high stiffness and a high tensile modulus [5, 6]. Table 14.2 illustrates some of the similarities of Teflon AF to commercial perfluoropolymers and some differences between them.

Teflon® AF can be subjected to a variety of melt processes including extrusion, compression, and injection molding. Compression molding is usually done at temperatures 100°C above those of glass transition. Extrusion and injection molding conditions depend on the type of part to be produced. Solution-based methods are spin coating and dipping from solutions in perfluorocarbon solvents. Spin coating is used to obtain an ultrathin, uniform-thickness coating on flat surfaces. Nonplanar surfaces can be coated

TABLE 14.1

Dependence of Refractive Index of Teflon® AF on Glass Transition Temperature [7].

Temperature, °C	100	150	200	250	300
Refractive index	1.316	1.307	1.298	1.293	1.290

TABLE 14.2

Teflon® AF Compared with Other Teflon® Fluoropolymers [5, 6]

Similarities	Differences
High-temperature stability	Noncrystalline, amorphous
Excellent chemical resistance	Soluble at ambient temperature in fluorinated solvents
Low surface energy	Transparent
Low water absorption	Lower refractive index
Limiting oxygen index (LOI) > 95	Stiffer
	High gas permeability

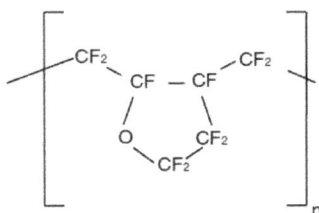

CF$_2$=CF-O-CF$_2$-CF$_2$-CF=CF$_2$
Perfluoro(3-butenyl vinyl ether)

Poly perfluoro(3-butenyl vinyl ether)

FIGURE 14.2   Structure of the Cytop™ monomer and polymer [3, 8].

by dipping. Other solution-based methods are spraying as a paint for more thickly coated layers. Ultrathin layers without the use of solvent are applied by laser ablation methods [3].

Another amorphous perfluoropolymer of this type has been developed by Asahi Glass under the trade name Cytop®. Unlike semicrystalline fluoropolymers, Cytop® is exceptionally transparent, with visible light transmission levels exceeding 95%. This polymer is prepared by the cyclopolymerization of perfluoro(3-butenyl vinyl ether) (PBVE). The monomer and polymer structures can be seen in Figure 14.2 [8]. Table 14.3 illustrates some of the similarities of Cytop™ to commercial perfluoropolymers and some differences between them.

TABLE 14.3

Physical and Mechanical Properties of Cytop™ Compared with Competitive Polymers

Property	CYTOP	PTFE	PFA	PMMA	Remarks
Glass transition temperature (°C)	108	130	75	105~120	By DSC
Melting point (°C)	Not observed	327	310	160 (isotactic)	By DSC
Density (g cm^{-3})	2.03	2.14 ~ 2.20	2.12 ~ 2.17	1.20	At 25°C
Contact angle of water (degrees)	110	114	115	80	At 25°C
Critical surface tension (dyne cm^{-1})	19	18	18	39	At 25°C
Water absorption (%)	<0.01	<0.01	<0.01	0.3	60°C, H$_2$0
Tensile strength (kg cm^{-2})	390	140~350	280~320	650~730	
Elongation at break (%)	150	200~400	280~300	3~5	
Yield strength (kg cm^{-2})	400	110~160	110~150	(650)	
Tensile modulus (kg cm^{-2})	12,000	4,000	5,800	30,000	

Note: PTFE = polytetrafluoroethylene; PFA = perfluoroalkoxy; PMMA = polymethyl methacrylate.

14.2 Fluorinated Acrylates

The homopolymers and copolymers of fluoroalkyl acrylates and methacrylates are most useful for practical applications. These polymers are used in the manufacture of: plastic lightguides (optical fibers); resists; water, oil, and dirt-repellent coatings; other advanced applications. There are several, rather complex, methods to prepare α-fluoroalkyl monomers (e.g., α-phenyl fluoroacrylates, α-(trifluoromethyl) acrylic and its esters, and the esters of perfluoromethacrylic acid). Generally, α-fluoroacrylates polymerize more readily than corresponding nonfluorinated acrylates and methacrylates, mostly by a free radical mechanism [9]. The copolymerization of fluoroacrylates has been carried out in bulk, solution, or emulsion initiated with peroxides, azobisisobutyronitrile, or γ-irradiation. Fluoroalkyl methacrylates and acrylates also polymerize by an anionic mechanism, though the polymerization rates are considerably slower than those of radical polymerization [10].

The homopolymers of poly(phenyl) α-fluoroacrylate (PPhFA) have a considerably higher glass transition temperature than the usual acrylates. Its T_g is 180°C, and it resists to temperatures above 270°C. Its shortcoming is a rather low resistance to UV radiation. Other polymers, such as poly(fluoroalkyl methacrylates), exhibit exceptional optical properties. Poly(fluoroalkyl α-fluoroacrylates) (PFAFAs) combine optical properties with increased resistance to elevated temperatures. Homopolymers and copolymers of perfluoroalkyl acrylates and methacrylates exhibit the lowest critical surface tension (y_c) of all polymers, including PTFE. Values of y_c for these polymers, depending mainly on the length of the molecules, composition, branching, and terminal groups of fluoroalkyl side chains, may be as low as 10.6 dyne/cm as compared with 18.5 to 19.0 dyne/cm for PTFE [9, 11]. Additional, extensive discussions of fluorinated acrylic esters can be found in the literature [12].

14.3 Fluorinated Polyurethanes

Introducing F into polyurethane resins results in changes in properties similar to those seen with other polymers. Chemical, thermal, hydrolytic, and oxidative stability is improved. On the other hand, the polymer becomes more permeable to oxygen. Surfaces become more biocompatible, and the capability to bond to other substances in contact with them is diminished [13, 14].

Fluorourethanes are used in products ranging from hard, heat-resistant electrical components to biocompatible surgical adhesives. The properties of a specific fluorourethane resin are determined by the raw materials and the manufacturing process used.

Raw materials used for the production of fluorinated polyurethanes are:

- Fluorinated alcohols, typically straight-chain alcohols with all but α and β carbons fluorinated, which are most often used.
- Fluorinated diols.
- Fluorinated polyols with molecular weights between 500 and 10,000, which are preferred for most applications.
- Fluorinated acrylic polyols.
- Isocyanates and polyisocyanates, mostly nonfluorinated types, which are used because of their considerably lower cost in comparison with fluorinated ones.
- Miscellaneous fluorinated precursors, such as amines, anhydrides, oxiranes, alkenes, and carboxylic acids, which are used for special properties.

The most frequently used method to prepare fluorourethanes commercially is the well-known addition reaction of polyisocyanates with polyols. F is most frequently introduced through the polyol component, since fluorinated polyisocyanates are relatively difficult to obtain and considerably more expensive than the non-fluorinated kind. Examples of fluorinated alcohols for polyurethane resins are listed in Table 14.4 [14].

TABLE 14.4

Fluorinated Alcohols for Polyurethane Resins [14]

Alcohol	Application
CF_3CH_2OH	Hard contact lenses
$CF_3CH_2CH_2OH$	Antifogging coating for glass
$(CF_3)CHOH$	Solvents
$CF_3CHFCF_2CH_2OH$	Water-based coatings
$F(CF_2)nOH$	Oil, water, and soil-resistant textile finishes
[n = 6–12]	
[n = 7]	Leather substitutes
[n = 3–5]	Oil and water-resistant textile finishes
$F(CF_2CF_2)_nCH_2CH_2OH$	
[n = 3–6]	Oil and water-resistant textile finishes
[n = 3–7]	Emulsion polymers
$F(CF_2CF_2)_nCH_2CH_2SH$	
[n unspecified]	Oil and water-resistant textile finishes
$C_7F_{15}CH_2OH$	Rigid insulating foams
$HOCH_2CF_2CF_2OCF(CF_3)CF_2OCF=CF_2$	Elastomers
$C_6F_{13}(CH_2)_2S(CH_2)_3OH$	Cladding for optical fibers
$H(CF_2CF_2)_5CH_2OH$	Coating for magnetic recording tape
$C_8F_{17}CH_2CH_2OH$	Coatings for textiles and leather
$C_8F_{17}CH_2CH_2OCH_2CH_2OH$	Housings for office machines

Another manufacturing method involves irradiation by UV light, in which acrylic-modified polyurethane resins are used. Reactive fluorinated oligomers are reviewed in Head et al. [15].

Fluorinated polyurethanes may also be prepared by treating the surface of an unfluorinated material with the cold plasma of elemental F [16] or carbon tetrafluoride [17].

14.4 Fluorinated Thermoplastic Elastomers

Considering the exceptional commercial success of hydrocarbon thermoplastic elastomers as a frequent replacement of conventional cross-linked (vulcanized) elastomers, it is logical that a similar concept is viable in the field of fluorinated elastomers. This is a particularly attractive concept, considering the rather involved chemistry of cross-linking fluoroelastomers (see Chapter 5). More detailed discussion of fluorinated thermoplastic elastomers (FTEPs) is provided in Chapter 10.

14.5 Copolymers of Chlorotrifluoroethylene and Vinyl Ether

Fluoropolymers offer several advantages, such as chemical resistance, resistance to UV and weather in general, and excellent dielectric properties. However, because of their poor solubility in solvents and high processing temperatures, their use as protective and dielectric coatings is limited. The method used for some applications and for substrates that can tolerate the high temperature for film formation is powder. PTFE, FEP, PFA, ETFE, and PVDF are applied as antistick and anticorrosion coatings. PTFE, FEP, PFA, and PCTFE are commonly applied as aqueous dispersions, and the coating after drying requires high temperatures. From this group, only PVDF is used widely in water dispersion or as a solution in an organic solvent for weather-resistant coatings [18]. Table 14.5 shows some of the conventional fluoropolymers used for coatings.

The subject of CTFE/vinyl ether copolymers is covered in detail by Takakura [18].

TABLE 14.5

Commercial Fluoropolymers Used in Coatings [18]

Polymer	Dispersion Type	Baking Temperature, °C	Use
PTFE	Aqueous	350	Nonstick cookware, wire coating, release coatings
FEP	Aqueous	280	Nonstick coating, wire coating, hot melt adhesive
PFA, MFA	Aqueous	330	Release coating, hot melt adhesive
PVDF	Aqueous, solvent	180	Architectural coatings, protective coatings, paints, outdoor signs
THV fluoroplastics	Aqueous, solvent	> 150	Protective coatings, optical coatings
PVF	Latent solvent	> 200	Weather resistant paints and coatings, outdoor signs
CTFE/VE	Aqueous, solvent	RT	Weather resistant paints and coatings

Notes: PTFE = polytetrafluoroethylene; FEP = fluorinated ethylene propylene; PFA = perfluoroalkoxy; MFA = methylfluoroalkoxy; PVDF = polyvinylidene fluoride; THV = tetrafluoroethylene, hexafluoropropylene, and vinylidene; PVF = polyvinyl fluoride; CTFE = chlorotrifluoroethylene; VE = vinyl ether.

In an effort to find a fluoropolymer resin suitable for coatings that are soluble in organic solvents and capable of forming a film at ambient temperature, the Asahi Glass Co. Ltd. developed copolymers of CTFE and vinyl ether – poly(fluoroethylene vinyl ether) (PFEVE) – and made them available under the trade name LUMIFLON™. PFEVE is an amorphous fluoropolymer that is soluble in organic solvents, can form transparent films at ambient temperature, and is compatible with hardeners and pigments. All these characteristics make it suitable as a base resin for paints [19].

Compared with dispersion paints based on PVDF, PFEVE solvent-based paints have the distinct advantage of forming films at ambient temperature, as mentioned previously. PVDF requires temperatures of 250°C (482°F) or higher to form a continuous film. Moreover, PFEVE can be cured chemically even at room temperature, while PVDF does not contain curable sites. The lower pigment compatibility of PVDF and the higher baking temperature restrict the choice and amount of pigments that can be used [18].

The most recent developments are water-based dispersions of PFEVE [20], some of them cross-linkable with waterborne isocyanates. Another FEVE-based polymer, developed and commercialized recently by the Asahi Glass Co. Ltd., has the trade name LUMISEAL™ and is used for high-performance sealants, similar to silicone sealants, with the advantage of eliminating the staining problem associated with the latter [21].

14.6 Perfluorinated Ionomers

This group of resins is based on copolymers of tetrafluoroethylene and perfluorinated vinyl ether containing a terminal sulfonyl fluoride group. After this precursor, which is melt processable, it is fabricated into the desired physical shape, and the pendent sulfonyl fluoride groups are converted into sulfonate groups by reaction with a solution of sodium or potassium hydroxide. A conversion to other ionic forms is possible by ion exchange. Products developed over the last decades contain –COONa and –CF$_2$COONa group as an alternative to the –SO$_3$Na group. The physical and electrochemical properties of perfluorinated ionomers are determined by the ratio of the comonomers used for their synthesis [22].

An excellent source of information for specialty fluoropolymers has been published by Banerjee [23]. This covers processing properties and commercial aspects – as well as a practical assessment of the advantages and disadvantages of specialty fluoropolymers compared to other materials.

REFERENCES

1. Chromis Technologies, https://chromistechnologies.com, (2022).
2. Feiring, A. E., Lousenberg, R. D., Majumdar, S., Murnen, H., Nemser, S., Shangguan, N, Development and Applications of Custom Amorphous Fluoropolymers (CAF), Vol 2, Materials for Energy, Efficiency and Sustainability: TechConnect Briefs (2017).
3. Hung, M.-H., Resnick, P. R., Smart, B. E., Buck, W. H., *Polymer Materials Encyclopedia* Vol. *4* (Salamone, J. C., Ed.), CRC Press, Boca Raton, (1995).
4. Teflon AF® Resins, Chemours Co, www.teflon.com/en/products/resins/amorphous-fluoropolymer, (2021).
5. Resnick, P. R., Buck, W. H., Teflon® AF: A Family of Amorphous Fluoropolymers with Extraordinary Properties, in Hougham, G., Cassidy P. E., Johns, K., Davidson, T., eds., *Fluoropolymer 2 - properties*, Springer Link, (2002).
6. Resnick, P., Buck, W. H. *Modern Fluoropolymers* (Scheirs, J., Ed.), John Wiley & Sons, Ltd, Chichester, UK, (1997).
7. DuPont Co., http://www2.dupont.com/Teflon_industrial/en_US/], (2008).
8. Sugiyama, N. US Patent 6,221,987 Patent, assigned to AGC Inc, 24, (2001).
9. Chuvatkin, A. A., Panteleeva, I. Yu. *Modern Fluoropolymers* (Scheirs, J., Ed.), John Wiley & Sons, Ltd, Chichester, UK, (1997).
10. Narita, T. et al., *Macromol. Chem.*, *187*:731, (1986).
11. Pittman, A. G., *Fluoropolymers* (Wall, L. A., Ed.), Wiley-Interscience, (1972).
12. Shimizu, T. *Modern Fluoropolymers* (Scheirs, J. Ed.), John Wiley & Sons, Ltd, Chichester, UK, (1997).
13. Brady, R. F., Jr., Clean Hulls Without Poisons: Devising and Testing Nontoxic Marine Coatings, *1999 Mattiello Memorial Lecture. Journal of Coatings Technology*, *72*(900), (2000).
14. Brady, R. F. Jr. *Modern Fluoropolymers* (Scheirs, J., Ed.), John Wiley & Sons, Ltd, Chichester, UK, (1997).
15. Head, R. A., Powell, R. L., Fitchett, M., *Polym. Mater. Sci. Eng.*, *60*, (1989).
16. Ozerin, A. N., Rebrov, A. V., Feldman, V. I., Krykin, M. A., Storojuk, A. P., Kotenko, A. A., Tul'skii, M. N., *React. Funct. Polym.*, *26*:167, (1995).
17. Benoist, P., Legeay, G., *Eur. Polym. J.*, *30*, (1994).
18. Takakura, T., *Modern Fluoropolymers* (Scheirs, J., Ed.), John Wiley & Sons, Ltd, Chichester, UK, (1997).
19. Munekata, S., *Progr. Org. Coatings, 16*, (1988).
20. Yamauchi, M. et al., *Europ. Coatings J.*, (1996).
21. Takakura, T. *Modern Fluoropolymers* (Scheirs, J., Ed.), John Wiley & Sons, Ltd, Chichester, UK, (1997).
22. W. Grot, *Fluorinated Ionomers*, 2nd ed, Elsevier, Oxford, UK, (2011).
23. Banerjee, S., *Handbook of Specialty Fluorinated Polymers*, Elsevier, Oxford, UK, (2015).

15

Applications of Specialty Fluorinated Polymers

Sina Ebnesajjad

This chapter presents select applications of a few specialty fluorinated polymers which have unique properties. Substitution of fluorine for hydrogen and other elements in organic monomers imparts unique, unusual, and extreme properties to their polymers. The properties of specialty fluoropolymers continue to meet the new demands of advancing technologies in the twenty-first century [1]. For example, amorphous fluoropolymers are ideal coatings for high-bandwidth, low-attenuation, optical fibers and optical waveguides.

15.1 Applications of Amorphous Perfluoropolymers

Commercial examples of amorphous fluoropolymers include Teflon® AF (Chemours Co., www. Teflon.com), Cytop™ (AGC Chemicals Americas, www.agcchem.com), and CyclAFlor® (Chromis Technologies, https://chromistechnologies.com). CyclAFlor is based on Cytop chemistry and is manufactured by Chromis Technologies. The development of customized amorphous fluoropolymers has been reported by polymerization of monomers such as perfluoro-2,2-dimethyl-1,3-dioxole (PDD) and a perfluorinated vinyl ether with sulfonyl fluoride side chains [2]. A summary of the beneficial properties of amorphous fluoropolymers is:

- A completely amorphous thermoplastic fluoropolymer.
- Possibility to dial in the desired values of glass transition temperatures (T_g) to obtain desired end-use properties and characteristics.
- Near complete light transmission in the ultraviolet light range.
- Significant transmission in the infrared light range.
- An intermediate value of coefficient of friction between semicrystalline perfluorinated and partially fluorinated polymers.
- Commercially useful mechanical and physical properties at temperatures approaching 300°C.
- Limited solubility in selected perfluorinated solvents.
- One of the lowest refractive index values among polymers.
- The lowest dielectric constant value among plastics even in the 10^9 Hz range.

The main application of amorphous perfluoropolymers is as a coating for optical fibers, as an antireflective coating, as a low dielectric coating, in the electronics industry (e.g., photoresists) [3–4], and as a low-dielectric-constant insulator for high-performance interconnects [5].

The subject of amorphous perfluoropolymers has been covered extensively in a number of publications [2, 3, 6].

DOI: 10.1201/9781003204275-20

15.2 Applications of Fluorinated Acrylates

15.2.1 Textile Finishes

The low surface energy of fluoroacrylate polymers makes them suitable for use as water and oil-repellent coatings for fibers and textiles. At present, the largest volume of fluoroacrylates and methacrylates produced in the world is used in this application. A large proportion of these are aqueous dispersions of copolymers of perfluoroalkyl acrylates and perfluoroalkyl methacrylates. These materials successfully compete with other fluorine-containing compounds for the same application [7]. To achieve the required water and oil-repellent effect, it is necessary to use copolymers having a perfluoroalkyl pendant group with at least seven atoms. An acrylic polymer with such a structure has an ultimate surface tension value γ_c = 10.6 dyne/cm, almost half that of PTFE [8, 9].

Commercial products, used widely for the treatment of textiles, mainly apparel, soft furnishings, and carpets, are known under the trade names Scotchgard (3M), Teflon (DuPont), Asahigard (Asahi Glass), and Unidyne (Daikin). This subject is reviewed in detail in [10].

15.2.2 Optical Fibers

At present, fluorinated acrylic ester polymers are commercially used as cladding materials for polymeric optical fibers (POFs), which have cores made of poly(methyl methacrylate), and in some cases for silica optical fibers (polymer-coated fibers, or PCFs). These applications, which are quite sizable, take advantage of the low refractive index that is unique to fluoropolymers [11]. The cladding of POFs is most frequently made from polymethacrylates or poly(2-fluoroacrylates) that have rather short fluoroalkyl side groups, such as CF_3CH_2-, HCF_2CH_2-, $CF_3CF_2CH_2$-, and $(CF_3)_2CH$– [12]. In addition, they are often copolymerized with other acrylic ester monomers to adjust the required properties. Although VDF-based resins are also used in claddings, fluorinated acrylic ester polymers exhibit better transparency and lower attenuation loss [11]. PFAFAs exhibit significantly lower refractive indexes and higher thermal stability than similar poly(fluoroalkyl methacrylates) (PFAMAs), but they are considerably higher in cost, which prohibits their wider use [8].

15.2.3 Other Applications

In electronics, fluoroacrylates are used as resistors in high-density electronic integrated circuits and as protective coatings in printed circuit boards [11]. They are also often used in xerographic processes for negative charge control [13–15].

In optical applications, in addition to optical fiber claddings, special optical adhesives matching the refractive indexes of optical glass components based on fluorinated epoxyacrylates and epoxymethacrylates are used [11]. Fluoroalkyl methacrylates are frequently incorporated as comonomers with siloxanyl methacrylates into contact lenses for the improvement of oxygen permeability [16].

Fluorinated acrylic ester polymers are also used as surface modifiers to promote blending instead of coating, imparting functionality of the fluoroalkyl groups to the surfaces of other resins or paints. Since they tend to accumulate on the surface of the substrates facing the air, the surface is easily modified. The modifiers are usually added to paints to enhance leveling or dispersing pigments and sometimes to improve moisture-proof qualities and are blended to resins to give oil and water repellency, or low-friction properties. In addition, they are used as antiblocking agents [11].

15.3 Applications of Fluorinated Polyurethanes

15.3.1 Surface Coatings

Because of their low surface energy, resistance to chemicals and corrosive agents, and resistance to weathering, fluorinated polyurethanes are very well suited as protective coatings. They can be deposited in a desired location and thickness with the added advantage of curing mostly at ambient temperatures.

15.3.2 Solvent-Based Coatings

The majority of surface coatings are solvent-borne and based on fluorinated ethylene vinyl ether (FEVE) polyol resins. These are readily dissolved in conventional solvents, such as toluene, xylene, or butyl acetate [17]. They are valuable for their resistance to abrasion, corrosion, staining, impact, thermal shock, water, ice, and weather, and demonstrate a high gloss, gloss and color retention, durability, hardness, and adhesion to metals, glass, concrete, and many plastics. Their applications range from floorings to luxury automobiles [18]. FEVE polyols may be modified by acrylic resins to improve optical properties and hardness, and to lower costs [19, 20]. They can be modified by other methods to achieve specific properties [18].

Coatings based on hexafluoroacetone (HFA) are often modified by the addition of powdered polytetrafluoroethylene. They exhibit toughness and hardness typical for conventional polyurethane coatings combined with low surface energy and ease of cleaning of PTFE. The optimum amount of PTFE added appears to be about 24% by volume. Such coatings have been used successfully as anticorrosion and antifouling coatings on ships and small boats and as protective coatings on tanks and large structures [18].

Other coatings prepared from a variety of polyol resins, such as fluorinated acrylic resins, copolymers containing tetrafluoroethylene, vinylidene fluoride, and hexafluoropropylene, exhibit generally high gloss, good gloss retention, high resistance to weathering, and water repellency [18].

15.3.3 Water-Based Coatings

The addition of small amounts of fluorinated polyols to conventional aqueous polyurethane coatings can improve their water resistance and mechanical properties considerably [21]. FEVE resins can be applied as water-based coatings and cured at 160°C (320°F) for 25 min. to produce coatings with high gloss and good water and weather resistance [22]. They may also be cured with water-based hardeners [23].

15.3.4 Powder Coatings

Fluorourethanes can also be applied using powder coating technology. Resins suitable for this should have T_g values between 35 and 120°C (95 and 248°F) to optimize flow out and cure at the annealing temperature. Blocked isocyanates, which form free isocyanates after being heated above certain temperatures, are frequently used [24]. Certain FEVE copolymers with hydroxyl and carboxyl functionalities combined with blocked isophorone diisocyanate are suitable for powder coating technology [25].

15.3.5 Treatments of Textile, Leather, and Other Substrates

Surface treatment of textiles, leather, glass, wood, and paper is the second largest application for fluorinated polyurethanes. The coatings are applied in a one-step treatment and impart resistance to soil, water, oil, and stains as well as a smoothness to fabrics and leather that resists removal by many cycles of laundering or dry cleaning [18].

15.3.6 Medical and Dental Applications

Soft and hard contact lenses with good oxygen permeability, optical clarity, flexibility, and biocompatibility; dental composites; denture linings; surgical adhesives; catheters; and hydrophobic microporous membranes are examples of applications of fluorinated polyurethanes in the medical and dental fields [18].

15.3.7 Cladding for Optical Fibers

Because of their low refractive index (less than 1.43), low permeability of water and weather vapor, very low water absorption, and good adhesion to glass and optical polymers, fluoropolyurethanes are suitable for the cladding of optical fibers. A variety of specialty resins are used for this purpose, which are most frequently photocurable.

15.3.8 Elastomers

Elastomers based on fluorinated polyurethanes exhibit good mechanical properties and resistance to solvents, chemicals, cold, and heat [26]. Formulations for fluorourethane elastomers are frequently modified with siloxanes to optimize certain properties [27]. FEVE-based polyols are used to manufacture elastomeric automobile bumpers and interior trim components [28].

15.3.9 Other Applications

The versatility of fluorinated polyurethanes is further demonstrated by these applications:

- Electrical (printed circuit boards, recording media, insulations).
- Printing (printing heads, thermal recording media).
- Heat exchangers (coatings inside to prevent the formation of deposits).
- Binders of explosives [29] .
- Sealants (e.g., for use in liquid crystal display panels).
- Protective coatings for concrete, stone, and fibrous materials.
- Antifogging coatings for mirrors and optical components [30].

The subject of fluorinated polyurethanes is covered extensively by R. F. Brady [18].

15.4 Applications of Fluorinated Thermoplastic Elastomers

The most common applications for thermoplastic fluorinated elastomers are seals in the chemical and semiconductor industries (O-rings, V-rings, gaskets, and diaphragms) because of their excellent chemical resistance and high purity [31, 32]. Because of their flexibility, low flammability, and resistance to oil, fuel, and chemicals, FTPEs find use in the electrical and wire and cable industries as wire coating and as the sheathing and coating of cables [33, 34]. Other applications include tents and greenhouses, as laminates with polyester fiber-reinforced PVC, and as tubing, bottles, and packaging in food processing and in sanitary goods [35]. More details on this subject are discussed in Chapter 10.

15.5 Applications of Copolymers of CTFE and Vinyl Ether

PFEVE-based coatings are used as protective coatings for large architectural structures, such as office buildings and bridges, where on-site coating and curing are required. Other applications are in transportation (automobiles, trains, and ships) and as protective coatings on signs and solar panels.

15.6 Applications of Perfluorinated Ionomers

Commercial products are available mainly in the membrane form from Chemours Co. as Nafion® membranes, from Asahi Glass as Flemion™ membranes, and as Aciplex® from Asahi Chemical [36].

The major areas of application are in the field of aqueous electrochemistry. The most important application for perfluorinated ionomers is as a membrane separator in chloralkali cells [37]. They are also used in the reclamation of heavy metals from plant effluents and in the regeneration of streams in the plating and metals industry [38]. The resins containing sulfonic acid groups have been used as powerful acid catalysts [39]. Perfluorinated ionomers are widely used in worldwide development efforts in the field of fuel cells, mainly for automotive applications as polymer electrolyte fuel cells (PEFCs) [88–93]. The subject of fluorinated ionomers is discussed in much more detail in [38].

REFERENCES

1. Banerjee, S., *Handbook of Specialty Fluorinated Polymers*, Elsevier, Oxford, UK, 2015.
2. Feiring, A. E., Lousenberg, R. D., Majumdar, S., Murnen, H., Nemser, S., Shangguan, N, Development and Applications of Custom Amorphous Fluoropolymers (CAF), Vol 2, Materials for Energy, Efficiency and Sustainability: TechConnect Briefs 2017.
3. Resnick, P., Buck, W. H. *Modern Fluoropolymers* (Scheirs, J., Ed.), John Wiley & Sons, Ltd, Chichester, UK, 1997.
4. Sugiyama, N. *Modern Fluoropolymers* (Scheirs, J., Ed.), John Wiley & Sons, Ltd, Chichester, UK, 1997.
5. Cho, C.-C. Wallace, R. M., Files-Sesler, L. A., *J. Electron. Mater.*, *23*, 1994.
6. Hung, M. H., Resnick, P. R., Smart, B. E., Buck, W. H., *Polymer Materials Encyclopedia* Vol. *4* (Salamone, J. C., Ed.), CRC Press, Boca Raton, FL, 1995.
7. Ishikawa, N., *Fluorine Compounds, Modern Technology and Application*, Mir, Moscow (1984), translated from Japanese.
8. Chuvatkin, A. A., Panteleeva, I. Yu *Modern Fluoropolymers* (Scheirs, J., Ed.), John Wiley & Sons, Ltd, Chichester, UK, 1997.
9. Pittman, A. G., *Fluoropolymers* (Wall, L. A., Ed.), Wiley-Interscience, New York, 1972.
10. Kissa, E., *Handbook of Fiber Science and Technology, Vol. II, Chemical Processing of Fibers and Fabrics, Functional Finishes, Part B* (Lewin, M. and Sello, S. B., Eds.), Marcel Dekker, New York, Chapters 2 and 3, 1984.
11. Shimizu, T. Modern Fluoropolymers (Scheirs, J. Ed.), John Wiley & Sons, Ltd, Chichester, UK, 1997.
12. Ohomori, A., Tomihashi, N., Kitahara, T., U.S. Patent 4,729,166 (January 19, 1988) to Daikin Kogyo Co., Ltd.
13. Nomura, Y., Aoki, M., Nemoto, S., Kokai Tokkyo Koho Japanese Patent 53-97435, August 25, 1978.
14. Shigeta, K., Takahashi, J., Ohmori, A. et al., Kokai Tokkyo Koho Japanese Patent 61-12069, June 7, 1986.
15. Yabuuchi, N., Aoki, T, Kokai Tokkyo Koho Japanese Patent 62-39878, February 20, 1987.
16. KoBmehl, G., Fluthwedel, A., Schafer, H., *Makromol. Chem.*, *193*, 1992.
17. Shimizu, T. *Modern Fluoropolymers* (Scheirs, J. Ed.), John Wiley & Sons, Ltd, Chichester, UK, 1997.
18. Brady, R. F., Jr. *Modern Fluoropolymers* (Scheirs, J., Ed.), John Wiley & Sons, Ltd, Chichester, UK, 1997.
19. Nakao, I., Japanese Patent 06 281 231, September 6, 1994.
20. Hirashima, Y., Maeda, K., Tutsumi, K., Japanese Patent 07 76 667, March 20, 1995.
21. Yang, S., Xiao, H. X., Chen, W. P., Kresta, J., Frisch, K. C., Higley, D. P., *Progr. Rubber Plast. Technol.*, *7*, 163, 1991.
22. Okazaki, H., Fujii, S., Tonomura, S., Japanese Patent 04 131 165, May 1, 1992.
23. Kodama, S., Yamauchi, M., Hirino, T., Kitahata, H., Japanese Patent 07 179 809, July 18, 1995.
24. Sugimoto, K., Saka, J., Japanese Patent 05 186 565, July 27, 1993.
25. Yasumura, T., Kobayashi, S., Komoriya, H., European Patent Application EP 556 729, August 25, 1993.
26. Menough, J., Rubber World, p. 9, January 1989.
27. Koike, N., Sato, S., Japanese Patent 06 234 923, August, 23, 1994.
28. Maruyama, T., Nakamoto, M., Japanese Patent 03 167 276, July 19, 1991.
29. Hoeller, R., Rudolf, K., European Patent Application EP 316 891, May 24, 1992.
30. Honda, T., Kaetsu, I., Japanese Patent 06 172 675, June 21, 1994.
31. Tatemoto, M., Tomoda, M., Kawachi, M. et al., Kokai Tokkyo Koho Japanese Patent 62635, April 10, 1984.
32. Kawashima, C., Koga, S., *Jpn. Plast.*, *39*, 1988.
33. Cheng, T. C., Kaduk, B. A., Mehan, A. K., Taft, D. D., Weber, C. J., Zingheim, S. C., U.S. Patent 4,935,467, June 19, 1988 to Raychem Corp.
34. Katsuragawa, S., Kawashima, C., Masaki, T., U.S. Patent 4,749,610, assigned to Central Glass Co. Ltd, June 7, 1988.
35. Tatemoto, M., Shimizu, T., *Modern Fluoropolymers* (Scheirs, J., Ed.), John Wiley & Sons, Ltd, Chichester, UK, p. 575, 1997.
36. Grot, W., Fluorinated Ionomers, 2nd ed., Elsevier, Oxford, UK, 2011.

37. Dotson, R. L., Woodard, K. E., *Perfluorinated Ionomer Membranes, ACS Symposium Series* 180 (Eisenberg, A. and Yeager, H. L., Eds.), ACS, Washington DC, 1982.
38. Grootaert, W. M., *Encyclopedia of Polymer Science and Engineering*, Vol. *16* (Mark, H. F. and Kroschwitz, J. I., Eds.), John Wiley & Sons, New York, 1989.
39. Olah, G. A., Iyer, P. S., Prakash, G. K. S., Perfluorinated Resinsulfonic Acid (Nafion-H®) Catalysis in Synthesis, *Synthesis*, 1986.

Part VI

Effects of Temperature and Other Variables on Fluoropolymers

16

Effect of Temperature on Fluoropolymers

Sina Ebnesajjad

16.1 Introduction

Two properties came to the attention of Roy Plunkett after the serendipitous discovery of a white powder which later proved to be polytetrafluoroethylene (PTFE). The properties standing out were resistance chemicals and heat [1, 2]. This material was not soluble in organic solvents or strong inorganic acid bases. Unexpectedly, it became incandescent in flame, but when the flame was removed it did not burn. In contrast, polyethylene burnt and degraded upon exposure to flame or extensive heat. To be melted, PTFE had to be heated to >300°C. These two properties were needed in the pre-World War II era and in post-War industries. PTFE thus took off, first to meet military needs and then industrial requirements [3].

The thermal stability of PTFE has been the subject of extensive studies as indicated by numerous published studies beginning shortly after its inception [4] and continuing on into the twenty-first century [5, 6]. PTFE has the highest melting point, processing, and continuous use temperatures among all commercial fluoropolymers. Table 16.1 shows a comparison of these temperatures for a number of common fluoropolymers. In spite of their stability, when fluoropolymers degrade at elevated temperatures they produce toxic compounds.

In this chapter, the thermal stability of some of the commercial fluoropolymers are reviewed. PTFE is the first polymer discussed because of its importance throughout its over three-quarters of a century existence as well as having the largest consumption among all fluoropolymers.

16.2 Thermal Stability of PTFE

PTFE resins are stable when used at or below their maximum continuous use temperature 260°C. When the temperature exceeds the melting point of PTFE (342°C) degradation begins to accelerate slowly. A major point of acceleration in degradation is when polymer temperatures reach 380°C. This is the reason why the maximum PTFE process temperature is held below 380°C [7].

The fundamental properties of fluoropolymers stem from the atomic structure of fluorine and carbon and their covalent bonding in specific chemical structures. The backbone is formed of carbon–carbon bonds and the pendant groups are carbon–fluorine bonds. Both C–C and C–F are extremely strong bonds. The basic properties of PTFE are derived from these two very strong chemical bonds. The size of the fluorine atom allows the formation of a uniform and continuous shield around the carbon–carbon bonds which protects them from chemical attacks including by oxygen.

The rate of PTFE decomposition and degradation products depends on temperature, time at temperature, pressure, and the type of environment. The course and products of the thermolysis of PTFE differ under air (O_2) versus under a vacuum or inert gas. Degradation onset takes place at nearly the same temperature under air and nitrogen atmospheres. Degradation ends at a somewhat higher temperature, about 620°C, under nitrogen. The biggest difference between air and nitrogen or vacuum is in the products of degradation. Degradation under a vacuum or under an atmosphere favors the formation of a tetrafluoroethylene monomer. Under a vacuum, PTFE decomposes into a nearly pure monomer [8–10].

DOI: 10.1201/9781003204275-22

TABLE 16.1

Melting Point, Processing, and Continuous Use Temperatures of Commercial Fluoropolymers

Fluoropolymer*	Melting Point, °C	Processing Temperature, °C	Continuous Use Temperature, °C
PTFE (1st and 2nd)	342, 342	<380	−260 to 260
PFA	305	380	−268 to 260
MFA	280	360	240
FEP	260	360	−240 to 205
ETFE	218–280**	360	−200 to 165
ECTFE	230–240	280–310**	140–150**
PCTFE	215	265	−240 to 180
PVDF	170	232	−20 to 140
PVDF Copolymer	115–170**	232–249**	100–150**
PVF	195	165***	−70 to 105

* PFA = perfluoroalkoxy polymer; MFA = copolymer of TFE and perfluoromethyl vinyl ether; FEP = copolymer of TFE and hexafluoropropylene; ETFE = ethylene tetrafluoroethylene copolymer; ECTFE = ethylene chlorotrifluoroethylene copolymer; PCTFE = polychlorotrifluoroethylene; PVDF = polyvinylidene fluoride.
** Grade-dependent.
*** PVF dispersion in a solvent like dimethyl acetamide is extruded.

The mechanism of PTFE degradation in the presence of oxygen (e.g., in air) is thermo-oxidative, resulting in the generation of oxygenated products. The products of PTFE degradation include carbonyl fluoride (COF_2), fluorine monoxide (COF_2), carbon monoxide, and carbon dioxide. Other degradation compounds include TFE, hexafluoropropylene (HFP), and likely small amounts of highly toxic perfluoroisobutene (PFIB) detected at temperatures higher than 475°C. At temperatures exceeding 700°C PTFE degrades into TFE and forms larger fluorinated compounds such as octafluorobutane.

Care must be taken to avoid exposure to PFIB during the processing and fabrication of parts. In fact PFIB is classified as a Schedule 2 substance of the Chemical Weapons Convention. That means it can either be used as a chemical weapon or used in the manufacture of chemical weapons [11].

Any moisture that comes in contact with the degradation products will react with some compounds, such as carbonyl fluoride, and generate hydrofluoric acid. Carbonyl fluoride is also highly toxic. Another example is fluorine monoxide (F_2O) which degrades into fluorine gas and oxygen when heated. It is a strong oxidizer and reacts with water rather explosively, producing hydrofluoric acid and gaseous oxygen.

The rate and extent of degradation is measured by the loss of weight using a thermogravimetric analysis (TGA) technique. A thermogravimetric analyzer comprises a precision balance with a sample pan placed inside a furnace equipped with a programmable temperature controller. The most common mode of TGA operation is to increase the temperature at a constant rate while measuring the sample mass continuously. Typically, the temperature is increased until the sample is completely degraded. McKeen has published extensive TGA data and other thermal properties of commercial fluoropolymers in his 2021 handbook [12].

In manufacturing processing, degradation is determined by *indirect* measurements of molecular weight. Exposure to temperatures exceeding 370–380°C leads to a reduction in molecular weight which results in an increase in specific gravity and the heat of fusion of PTFE, as measured by differential scanning calorimetry (DSC) [7].

In short PTFE has become the gold standard for the stability of all fluoropolymers and indeed all polymers. In the ensuing sections, the thermal stability of perfluorinated and partially fluorinated polymers are discussed and at times compared to the stability of PTFE.

16.3 Thermal Stability of Perfluorinated Copolymers of Tetrafluoroethylene

Polytetrafluoroethylene is the most thermally stable among commercial fluoropolymers. Structurally, it is perfectly linear as it does not contain any branches, unlike a polymer such as polyethylene (Figure 16.1). The polymerization regime and the C–F bonds prevent branching. This is why it has to be polymerized

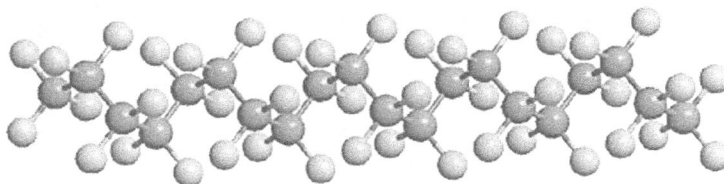

FIGURE 16.1   Ball and stick representation of the molecular structure of polytetrafluoroethylene.

with other monomers to reduce its molecular weight and thus its melt viscosity. Comonomers impart pendent groups to the TFE chains which prevent excessive crystallization of the smaller molecules of the copolymer.

16.3.1 Fluorinated Ethylene Propylene Copolymer (FEP)

The molecular structure of FEP is displayed in Figure 16.2. The CF_3 pendent group, when present in sufficient quantity, reduces the crystallinity of FEP while reducing its melting point. The carbon atoms to which the pendent CF_3 are bonded are tertiary while the other backbone carbon atoms are secondary.

The presence of the CF_{3-} branch leads to some changes in FEP properties including its thermal stability. The tertiary carbon atoms are points of stress because of the spatial size of the CF_3 group. The CF_3 group is tetrahedral as opposed to the CH_3 group which is planar or near-planar. As such CF_3 groups are susceptible to oxidation at elevated temperatures, thus resulting in lower thermal stability for FEP compared to PTFE. Figure 16.3 shows the rate of weight loss of several copolymers of TFE and PTFE, indicating more extensive weight loss for FEP than PTFE at temperatures in the range 300–400°C. Based on the data in Table 16.1 and Figure 16.3 one can conclude the order of thermal stability is: ETFE < FEP < MFA < PFA < PTFE. The thermal stabilities of FEP and PFA are fairly close.

During the early decades of FEP use there was an understanding that PFIB was not one of the degradation products of FEP in air at temperatures below 400°C. Recent studies by manufacturers, especially Chemours Corp (formerly DuPont), have shown PFIB indeed forms at temperatures as low as 360°C as opposed to PTFE. The main reason is availability and the use of a more sensitive analytical technique such as electron-capture gas chromatography.

FEP degrades thermally with the evolution of HFP as a principal degradation product, together with a smaller amount of PFIB. Both gases have been observed in laboratory experiments in nitrogen at temperatures of 360 to 400°C. In similar tests conducted in air, carbonyl fluoride (COF_2) is also formed. When held at 400°C for one hour FEP generates 30 parts per million PFIB which is significant considering its

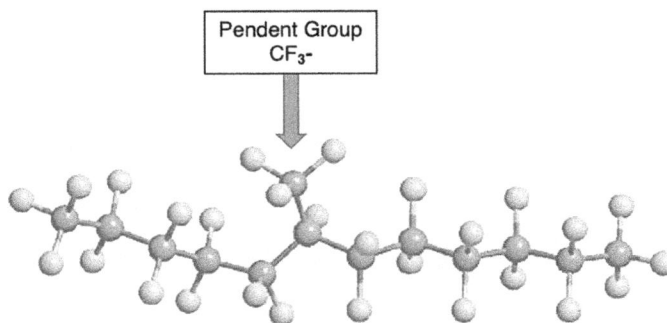

Pendent Group
CF_3-

FIGURE 16.2   Ball and stick representation of the molecular structure of fluorinated ethylene propylene copolymer.

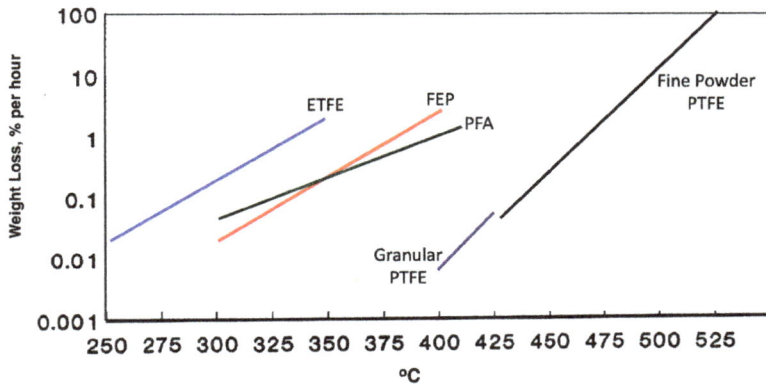

FIGURE 16.3 Comparison of the weight loss of fluoropolymers, post-thermal equilibrium, as a function of temperature by thermogravimetric analysis [13].

threshold limit value (TLV) of PFIB of 0.01 ppm. This is an important finding because of the extreme toxicity of PFIB [13].

16.3.2 Thermal Stability of PFA

There are very few published studies of the thermal stability of PFA resins. They are quite stable and can be heated up to 425°C for short periods of time during processing [14]. Higher temperature reduces melt viscosity, allowing more efficient melt processing. Figure 16.3 exhibits a comparison of the thermal stability of PFA and PTFE. Clearly PTFE is thermally more stable than PTFE, indicated by the 1% weight loss per hour at 400–425°C for PFA as compared to 0.01% weight loss per hour for PTFE.

The pendent group of PFA is an ether bond (–O–) with a perfluoro alkyl tail. An example of the pendent tail is displayed in Figure 16.4 in which the ether bond tail is perfluoropropyl (C_3F_7–). In spite of the thermal stability of the ether bond the molecular structure of PFA is less stable at temperatures above its melting point than PTFE is. When PFA degrades at 400°C in air it produces carbonyl fluoride, similarly to PTFE. Beyond 400°C the course of PFA degradation is likely to be similar to PTFE. This is due to the few branch points in PFA accompanied by long sections of TFE chains, assuming PFA contains 3.5% by wt. PPVE. That amounts to one PPVE molecule per 73 TFE monomer units in contrast to one HFP molecule per ten monomer units.

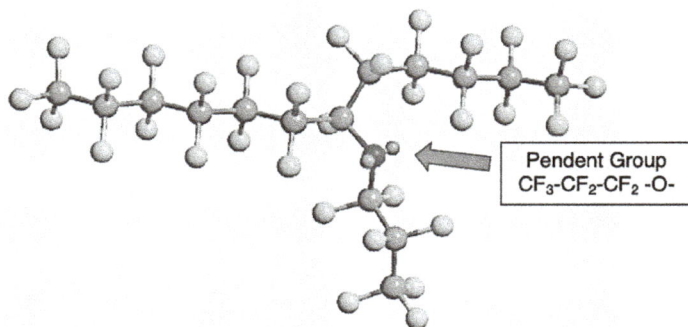

FIGURE 16.4 Ball and stick representation of the molecular structure of tetrafluoroethylene–perfluoropropyl vinyl ether copolymers.

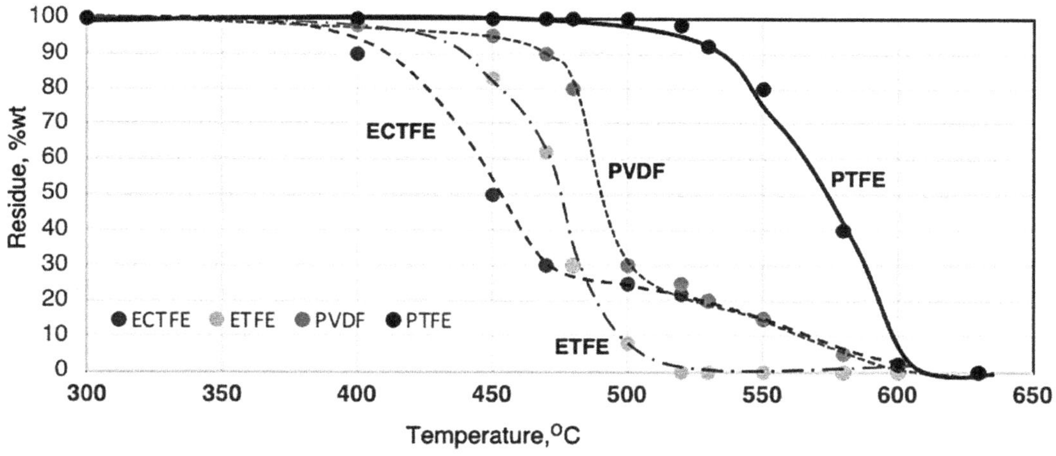

FIGURE 16.5 Thermogravimetric analysis of partially fluorinated fluoropolymers in air [16].

16.3.3 Thermal Stability of ETFE

ETFE copolymers [(-CH_2-CH_2-CF_2-CF_2-)$_n$] with an ethylene to tetrafluoroethylene molar ratio of around 1 melt in the range 218–280°C. ETFE contains defects in the form of diads, a sequence of two of the same monomer links. ETFE degrades to its oligomers at temperatures exceeding 340°C by the cleaving of the main chain at the diads or longer ethylene chains, particularly in the presence of oxygen [7].

Figure 16.3 shows significant degradation of ETFE at temperatures as low as 250°C. When heated to 350°C ETFE degrades and produces TFE, perfluoro butyl ethylene, carbonyl fluoride, and carbon monoxide [15].

Figure 16.5 shows an ETFE degradation graph which crosses over that of ECTFE at about 480°C. The degradation of ETFE is completed at about 510°C while ECTFE's decomposition ends at about 580°C. The last 25–30% of ETFE simply degrades more rapidly than ECTFE. From a practical standpoint ETFE is considered more stable than ECTFE.

16.3.4 Thermal Stability of ECTFE

ECTFE copolymers [(-CH_2-CH_2-$CFCl$-CF_2-)$_n$] with an ethylene to chlorotrifluoroethylene molar ratio of around 1 melt in the range 230–240°C. A carbon–chlorine bond (78.5 kcal/mol) is weaker than the C–F bond (105.4 kcal/mol). That means degradation of ethylene chlorotrifluoroethylene begins at a lower temperature than ethylene tetrafluoroethylene, as seen in Figure 16.5. The C–Cl bond is degraded by a thermo-oxidative mechanism leading to dehydrochlorination.

What is left behind is an unsaturated double bond (C=C) in the polymer backbone. This double bond destabilizes the backbone of the ECTFE polymer resulting in additional dehydrochlorination and leaving behind an unsaturated carbon chain (=C=C=C=C=C=C=). This polyene chain is quite resistant to combustion, forming a layer of char that retards additional degradation of the ECTFE sample. This is why, after reaching 480°C, ECTFE degradation slows down below that of ETFE.

16.3.5 Thermal Stability of PCTFE

PCTFE is a homopolymer of chlorotrifluoroethylene [(-$CFCl$-CF_2-)$_n$] with a melting point around 215°C. PCTFE degrades when heated to elevated temperatures because of the relative weakness of C–Cl. The slow degradation of PCTFE in air begins at 260°C and accelerates at 299°C (Figure 16.6). Figure 16.7 shows a comparison of the thermal stability of PCTFE with PTFE in a vacuum and in oxygen. Hydrochloric and hydrofluoric acids are produced by the reaction of the decomposition fragments and any moisture. Clearly PCTFE is thermally less stable than PTFE [17].

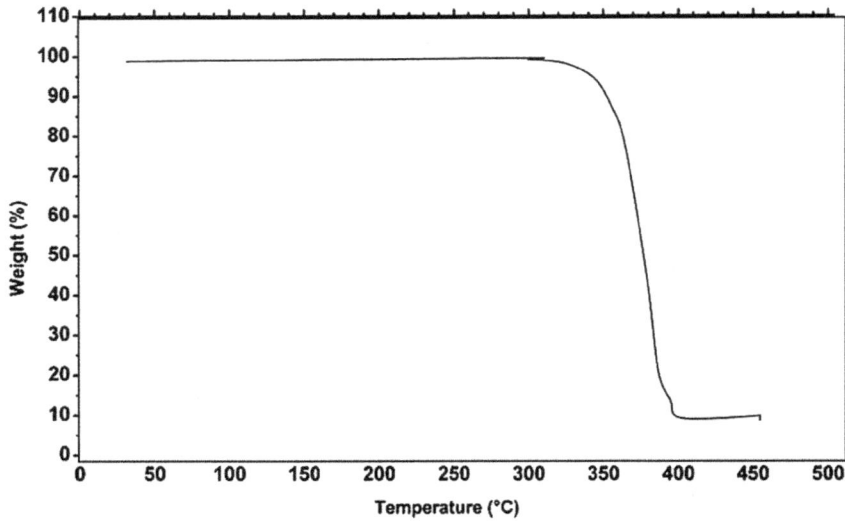

FIGURE 16.6   Thermogravimetric analysis of Arkema Voltalef® PCTFE [18].

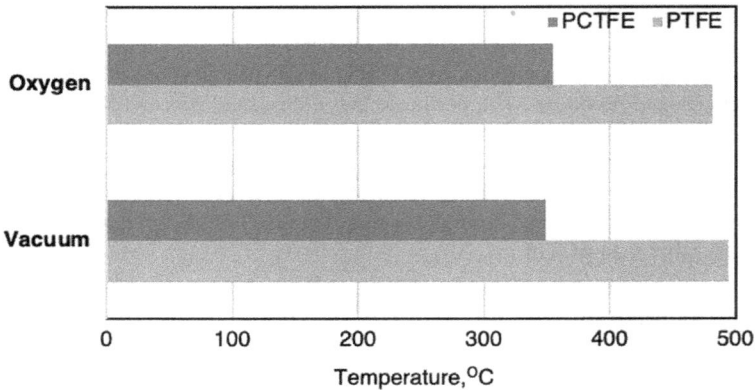

FIGURE 16.7   Degradation temperature of PTFE and PCTFE in a vacuum and oxygen for a 25% weight loss in two hours [19].

16.3.6 Thermal Stability of PVDF

PVDF has the second largest consumption after PTFE among the thermoplastic fluoropolymers. The thermal stability of PVDF has been studied extensively. According to Figure 16.5 it is more stable than ETFE and ECTFE overall. In applications, PVDF is only second to PTFE from the thermal stability standpoint.

One study by thermogravimetric analysis shows a small weight loss beginning at 200°C, probably associated with residual volatile impurities. The main onset of weight loss occurs from about 420°C; there is a first stage resulting in a 63% weight loss by 460°C. In the final stage of decomposition the remaining 37% of PVDF degrades, mainly at 600–700°C [20]. Figure 16.8 shows a typical TGA for a PVDF homopolymer.

The main product of the degradation of PVDF is hydrofluoric acid, both in a vacuum and in air because of the ready availability of hydrogen. In a vacuum at 372–500°C, PVDF produces hydrogen fluoride and a waxlike material consisting of chain fragments of low volatility [22].

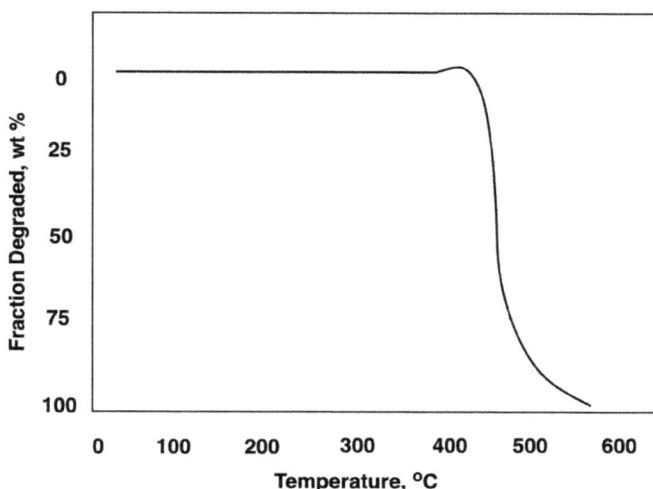

FIGURE 16.8 Thermogravimetric analysis under air of a Solef® PVDF homopolymer [21].

REFERENCES

1. US Patent 2,230,654, R. J. Plunkett, assigned to DuPont Feb 4, 1941.
2. Ebnesajjad S. Introduction to fluoropolymers - materials, technology and applications. 2nd ed. Elsevier, Oxford, UK, 2020.
3. The discovery of Teflon, Education in Chemistry, Royal Society of Chemistry, https://edu.rsc.org/news/the-discovery-of-teflon/2020492.article, October 31, 2008.
4. Lewis, E. E., Naylor, M. A., Pyrolysis of Polytetrafluoroethylene. *J. Am. Chem. Soc.*, *69*, 8 1947.
5. Ellis, D. A., Mabury, S. A., Martin, J. W., Muire, D. C. G., Thermolysis of Fluoropolymers as a potential source of halogenated organic acids in the environment, letters to Nature, *Nature*, *412*:19, July 2001.
6. Henri, V., Dantras, E., Lacabanne, C., Dieudonne, A. G., Koliatene, F., Thermal ageing of PTFE in the melted state, Polymer Degradation and Stability, Elsevier, 171, 1–8, 2020.
7. Ebnesajjad S., *Fluoroplastics Handbook: Melt processible fluoropolymers*, Vol. 2, 2nd ed., Elsevier, Oxford, UK, 2015.
8. U.S. Department of Health, Education, and Welfare, Public Health Service, Center for Disease Control, National Institute for Occupational Safety and Health. Criteria for a recommended standard occupational exposure to decomposition products of fluorocarbon polymers. DHEW (NIOSH) Publication # PB274727; September, 1977.
9. Kaplan, H. L., Grand, A. F., Switzer, W. C., Gad, S. C., Acute Inhalation Toxicity of the Smoke Produced by Five Halogenated Polymers. *J Fire Sci 2*: 153–172, 1984.
10. Williams, S. J., Baker, B. B., Lee, K. P. Formation of Acute Pulmonary Toxicants Following Thermal Degradation of Perfluorinated Polymers: Evidence for a Critical Atmospheric Reaction. *Food Chem Toxicol*, 177–185, 1987.
11. Timperley, C. M., Highly-toxic fluorine compounds. in *Fluorine Chemistry at the Millennium*, 1st ed, edited by R.E. Banks, Elsevier, Oxford, 2000.
12. McKeen, L. W., *The Effect of Long Term Thermal Exposure on Plastics and Elastomers*, 2nd ed., Elsevier, Oxford, UK, 2021.
13. Data from: The Guide to Safe Handling of Fluoropolymers Resins, published by the Fluoropolymers Division of Plastics Industry Association, 5th ed, 2018, www.PlasticsIndustry.org - Original data was provided by Chemours Co (formerly DuPont) to the Plastics Ind. Association.
14. Gangal, S. V., Brothers, P. D., Perfluorinated Polymers, Tetrafluoroethylene–Perfluorovinyl Ether Copolymers, Wiley Online Library, June 2010, https://doi.org/10.1002/0471440264.pst418.
15. The Guide to Safe Handling of Fluoropolymers Resins, published by the Fluoropolymers Division of Plastics Industry Association (formerly Society of Plastics Industry), 3rd ed, June 1998.

16. Data Source: Lin S.C. Ignition resistance of fluoropolymers, polymer materials: science and engineering. Am. Chem. Soc., S Spring Meeting, New Orleans, April 6e10, 2008.

17. Ebnesajjad S., *Fluoroplastics Handbook: Non-melt processible fluoropolymers*, Vol. *1*, 2nd ed., Elsevier, Oxford, UK, 2015.

18. Data from: Voltalef PCTFE technical brochure. Arkema Corp.; May 10, 2004.www.voltalef.com.

19. Critchley, J. P., Knight, G. J., Wright, W. W.. *Heat Resistant Polymers*. Plenum Press, New York, 1983.

20. Zulfiqar, S., Zulfiqar, M., Rizvi, M., Munir, A., McNeili, I. C., Study of the Thermal Degradation of Polychlorotrifluoroethylene, Poly(vinylidene fluoride) and Copolymers of Chlorotrifluoroethylene And Vinylidene Fluoride. *Polymer Degradation and Stability*, *43*:.423–430, 1994.

21. Data from Solef® PVDF Design & Processing Guide, R 12/2017 | Version 2.7, Solvay Specialty Polymers, www.solvay.com, 2017.

22. Madorsky, S. L., Hart, V. E., Straus, S., Sedlak, V. A. Thermal Degradation of Tetrafluoroethylene and Hydrofluoroethylene Polymers in a Vacuum. *J Res Natl Bur Stand*, *51*:327–333, 1953.

17

The Effect of the Environment on Fluoropolymers

Sina Ebnesajjad

17.1 Introduction

One basic and important property of all fluoropolymers is resistance to chemicals. The first fluoropolymer, polytetrafluoroethylene (PTFE), was selected as material for the construction of valves and gaskets for the Manhattan Project because of its resistance to chemical attack by the highly reactive uranium hexafluoride [1]. Later it was tested and found to be the only polymer that could stand up to *most* organic and inorganic chemicals. The key to fluoropolymers' chemical resistance is the strength of the C–F bond.

A decrease in the fluorine content of the polymer molecule (see Table 17.1) generally results in the reduction of the chemical resistance of fluoropolymers [2].

In this chapter the chemical resistance of important commercial grades of fluoropolymers is discussed, if rather briefly. The best and perhaps only sensible approach for practitioners and users of fluoropolymer parts is to contact the manufacturers and suppliers of those plastics to obtain reliable chemical resistance data.

17.2 Polytetrafluoroethylene (PTFE)

The strength of C-F stems from the high electronegativity of fluorine which at 4 Pauling is the highest of all elements. The bond is highly polarized through the monomer units in a $-CF_2-CF_2-$, though the chain is neutral thanks to the structural symmetry. In the case of PTFE and other fluoropolymers containing chains consisting of $-CF_2-CF_2-$, the fluorine atoms act as a cover for the carbon backbone. A simplified analogy is an insulated wire containing a single conductor in which the insulation represents fluorine atoms and the conductor stands for the carbon backbone. The chemical properties of PTFE are not affected by fabrication conditions.

The TFE chains are completely linear and do not form side chains in contrast to polymers such as polyethylene. There are no defects in the polymer chain because of monomer head and tail differences. Conversely, polyvinylidene fluoride (CF_2-CH_2) contains defects because of such differences. One could say PTFE comes as close as possible to a perfect polymer with respect to chemical resistance.

Overall, PTFE has the highest chemical resistance properties among perfluorinated polymers, namely perfluorinated copolymers such as perfluoroalkoxy (PFA) and tetrafluoroethylene and hexafluoropropylene copolymers (FEP). Perfluorinated terpolymers approach and occasionally exceed the chemical resistance of PTFE. It does not absorb most chemicals or dissolve in them to any extensive degree under ambient conditions. PTFE resists environmental stress cracking. An overwhelming majority of PTFE (and fluoropolymer) applications take advantage of its resistance to chemicals.

There are, however, a few exceptions to the chemical resistance of PTFE. They include: highly oxidizing gases F_2, OF_2, and ClF_3; alkaline metals like sodium and potassium; and finely divided metal powder like aluminum and magnesium, which cause PTFE to combust at high temperatures [3]. PTFE is not

TABLE 17.1

Description of Commercial Fluoropolymers

Fluoropolymer	Abbreviation	Monomer	F/C Ratio	Comments on Chemical Resistance
Polytetrafluoroethylene	PTFE	-CF$_2$-CF$_2$-	2/1	Highest chemical resistance
Perfluoroalkoxy+ polymer**	PFA/MFA	-CF$_2$=CF-O-R$_f$-	2/1	Similar to PTFE
Hexafluoropropylene/ tetrafluoroethylene copolymer	FEP	-(CF$_2$-CF$_2$)$_n$-(CF$_3$-CF)$_m$ I CF$_2$	2/1	Similar to PTFE
Ethylene tetrafluoroethylene copolymer	ETFE	-CF$_2$-CF$_2$-CH$_2$-CH$_2$-	1/1	Good but <PTFE
Polyvinylidene fluoride	PVDF	-CH$_2$-CF$_2$-	1/1	Good but <PTFE
Polyvinyl fluoride	PVF	-CH$_2$-CHF-	1/2	Fine but <<PTFE
Ethylene chlorotrifluorothylene	ECTF	-CF$_2$-CClF-CH$_2$-CH$_2$-	3/4	Good <PTFE
Polychlorotrifluorothylene	PCTFE	-CF$_2$-CClF-	3/2	Good <PTFE

+ MFA is PFA made with perfluoro methyl vinyl ether;

** R$_f$ is a perfluoroalkyl group such as perfluoro propyl CF$_3$-CF$_2$-CF$_2$-. perfluoro ethyl CF$_3$-CF$_2$-. or perfluoromethyl CF$_3$-.

compatible with some chemicals such as concentrated sodium and potassium hydroxide at or near its melting point.

Another facet of polymer interaction with chemicals is permeation. Permeation requires a reagent to dissolve in PTFE. Even though a reagent may not react with a polymer it may permeate through the polymer structure. The extent and rate of permeation is dependent on the structure and properties of the plastic article as well as the type, size, and concentration of a permeant. The similarity of a liquid's structure to the polymer is one factor indicating possible permeation. For instance, PTFE is not a good barrier for fluorocarbon oils because of the similarity of its chemical composition to PTFE. Temperature and pressure usually expedite the permeation process [4].

Some chemicals interact with PTFE at elevated temperatures. For example carbon monoxide is absorbed by PTFE at temperatures exceeding 50°C over a year of exposure. The exposure impact time is drastically reduced as temperature increases. A more severe impact occurs after a few hours when the temperature is increased to >100°C. No chemical reactions occur [4].

Filled compounds of PTFE, generally, have inferior chemical resistance compared to neat PTFE. The loss of chemical resistance depends on the type of filler. PTFE compound manufacturers usually supply chemical resistance for the compounds.

17.3 Perfluorinated Copolymers of Tetrafluoroethylene

Generally speaking, the copolymers of TFE have similar chemical resistance characteristics as PTFE. There is one major difference between PTFE and its melt processible copolymers. Even the best processed PTFE parts contain small amounts of voids and pores which can result in liquid and gas permeation. In contrast, melt processible PFA, FEP, and MFA (tetrafluoroethylene perfluoromethyl vinyl ether) are virtually void free because of the ease of void closure during processing [5].

17.3.1 PFA and MFA

PFA resins have been tested against hundreds of organic and inorganic chemicals that are commonly used in the chemical processing industries [6]. Similar to PTFE, PFA is not affected when exposed to inorganic acids, bases, halogens, metal salt solutions, organic acids, and anhydrides. Aromatic and aliphatic hydrocarbons, alcohols, aldehydes, ketones, ethers, amines, esters, chlorinated compounds, and other polymer solvents do not impact PFA. Small amounts of some of these reagents especially amines are absorbed by PFA without chemical reaction. Use of MFA with amines is not recommended because of the extensive impact of those compounds [5].

Similar to PTFE, alkaline metals like sodium and potassium, as well as finely divided metal powder like aluminum and magnesium, cause PFA and FEP to degrade at elevated temperatures. Line PTFE, FEP, and PFA are susceptible to attacks by concentrated NaOH or KOH solutions at elevated temperatures.

17.3.2 FEP

Most chemicals and solvents do not affect FEP even at high temperatures and pressures. FEP is susceptible to reacting with fluorine, metallic sodium and potassium, and molten sodium hydroxide. Even at its maximum continuous temperature of 200°C FEP does not absorb acids or bases when exposed for up to one year. FEP absorbs less than 1% of organic solvents after extended is less than exposure at elevated temperatures. Even when absorbed, solvents and other reagents do not impact FEP chemically. After removal from the challenge fluids the absorbed material may be removed by heating [7].

FEP allows the permeation of a number of gases and vapors, though the rate and extent of permeation is significantly less than those of non-fluorinated thermoplastics. FEP is void free, thus only allowing permeation by diffusion through its molecules. More crystalline FEP has lower permeation rates. However, the variability of the crystallinity of FEP under most processing conditions is rather narrow. The low permeability of FEP renders this polymer suitable in the chemical, pharmaceutical, food and, beverage industries where it comes in contact with various chemicals and materials. By adjustment of the FEP film thickness, permeation can be reduced to virtually zero [5].

17.4 Ethylene Tetrafluoroethylene Copolymer

Commercial ETFEs are usually terpolymers (*modified*) in which a small amount of a third monomer is polymerized along with TFE and ethylene. An example is hexafluoropropylene which improves the mechanical properties and reduces the terpolymer's environmental stress cracking tendency. These terpolymers withstand inorganic chemicals and solvents that often cause the degradation of hydrocarbon based plastics [5].

The chemical resistance of ETFE performance resembles that of perfluorinated fluoropolymers. ETFE is also susceptible to those chemicals that attack PTFE, PFA, and FEP. ETFE is quite resistant to strong inorganic acids and bases and chlorine as well solvents and organic compounds, though with some exceptions. For example it is susceptible to amines and more generally organic bases. Highly oxidizing acids are incompatible with ETFE especially at elevated temperatures. For instance, a 50% concentration of chromic acid degrades ETFE after exposure for seven days at 125°C [4].

Water absorption of ETFE is extremely low (<0.03 wt%) which is an important characteristic. Humidity has no impact on ETFE, rendering it dimensionally stable. The mechanical and electrical properties of ETFE are independent of the environmental humidity. No changes in tensile properties were reported after 3,000 hours exposure to boiling water [5]. ETFE is also resistant to petroleum based products such as gasoline even at elevated temperatures, allowing its use in automotive applications [7].

17.5 Polyvinylidene Fluoride (PVDF)

Polyvinylidene fluoride is a versatile fluoropolymer. It can be processed by all melting techniques, as a dispersion or latex, and even blended with select polymers such as acrylics. It can be fabricated into virtually any shape for applications which do not quite require the extreme properties of perfluorinated fluoropolymers.

PVDF has a relatively low melting point compared to other fluoropolymers, resulting in a lower continuous use temperature [8]. It is resistant to a vast number of organic and inorganic chemicals. However, there are a number of exceptions, rendering PVDF susceptible to some chemicals. Some of these compounds are described in this section.

PVDF is not compatible with fuming sulfuric acid, concentrated nitric acid, concentrated alkalis, and bases with a pH of 12 or higher, such as 30% wt. ammonium hydroxide. PVDF swells in strong polar solvents including acetone and ethyl acetate, and partially dissolves in aprotic polar solvents (solvents that lack an acidic proton and are polar), such as dimethyl formamide, dimethyl acetamide, or dimethyl sulfoxide. PVDF is attacked by fluorine gas at room temperature. Exposure to atomic chlorine causes stress cracking in PVDF parts [9].

PVDF is not hygroscopic and adsorbs less than 0.05% of water at room temperature. It does not require drying before processing in contrast to many hydrocarbon-based plastics. PVDF is used in ultra pure water systems because it is superior to materials such as stainless steel or polyvinyl chloride. PVDF combines an excellent surface finish with low extractables to provide a high quality piping material for ultra pure water applications such as in semiconductor manufacturing [10].

17.6 Polychlorotrifluoroethylene (PCTFE)

PCTFE is a homopolymer of chlorotrifluoroethylene (CTFE). The monomer has one chlorine atom and three fluorine atoms. The substitution of the larger chlorine for fluorine leads to a massive change in its properties and applications. Some of those changes, compared to PTFE, include crystallinity reduction, a higher glass transition temperature, decrease in flexibility, and increase in mechanical strength. PCTFE melts at about 210°C but degrades before reaching its melting point, thus it has a continuous use temperature of 120°C [2].

PCTFE has excellent chemical resistance, though it does not match that of PTFE. PCTFE is resistant to attack by most chemicals and oxidizing agents. It is susceptible to attack by many organic solvents and swells in halogenated compounds, ethers, esters, and aromatic solvents, especially at elevated temperatures [8].

PCTFE has near-zero moisture absorption or permeation. This attribute has led to applications of PCTFE film as a moisture protective film or a coating in pharmaceutical blister packaging, due to its excellent water vapor barrier. In general PCTFE has extremely low gas permeability thus making it an outstanding barrier material against air, water, steam, fluids, and liquified gases.

17.7 Ethylene Chlorotrifluoroethylene Copolymer (ECTFE)

ECTFE is semicrystalline with a melting point of 240°C. An analog of ETFE, the conformation of ECTFE is an extended zigzag in which ethylene and CTFE alternate. Similar to ETFE, ECTFE terpolymers have better mechanical and abrasion and radiation resistance than PTFE and other perfluorinated polymers. ECTFE is resistant to a large number of inorganic and organic compounds.

ECTFE is resistant to: acids at high concentrations and temperatures, caustic media, oxidizing agents, and many solvents. Fuming sulfuric acid and concentrated hydrofluoric and hydrochloric acids discolor ECTFE when exposed at elevated temperatures. Alcohols such as methanol and isopropanol and concentrated potassium hydroxide, at elevated temperatures, all discolor ECTFE [11].

ECTFE is resistant to most chemicals except hot polar and chlorinated solvents. It does not stress crack in any solvents. ECTFE is attacked by amines, molten alkali metals, gaseous fluorine, and certain halogenated compounds such as CIF_3.

ETFE has better barrier properties against SO_2, Cl_2, HCl, and water than FEP and PVDF. ECTFE resists hydrolysis and UV aging and thus is durable outdoors.

17.8 Polyvinyl Fluoride (PVF)

PVF is only one fluorine atom apart from polyethylene, based on its chemical formula. Simply based on this formula, one may expect some differences of properties from polyethylene (PE) but not the extensive improvements that PVF exhibits over PE.

Alaaeddin et al. [12] call PVF "an exclusive polymer or special plastic employed in numerous industries and manufacturing processes since 1962. It is therefore an important type of polymer due to its excellent properties and high performance; e.g. chemical resistance, stain resistance, outdoor durability, high stability, and adherence". This statement aptly describes the many desirable attributes and properties of PVF [13].

PVF is used for surface protection because of its toughness, lightness, and high resistance to moisture, weathering, and ultraviolet radiation. PVF is inert to numerous inorganic acids/bases, staining agents, and a number of solvents. It stands up to cleaning and disinfecting agents used in operating room walls, aircraft cabin surfaces, and laminates protected with a top layer of transparent PVF film. PVF films do not readily absorb water and are highly resistant to degradation by hydrolysis [14].

REFERENCES

1. Lister, T., *The Discovery of Teflon®, Education in Chemistry*, Royal Soc. of Chemistry, London, UK, October 31, 2008.
2. Ebnesajjad S., *Fluoroplastics Handbook: Non-melt Processable Fluoropolymers*, Vol. *1*, 2nd ed., Elsevier, Oxford, UK, 2015.
3. An Introduction to Chemours™ Fluoropolymers, Chemours Co., publication C-11311, April 2018.
4. Ebnesajjad, S., Woishnis, W. A., *Chemical Resistance of Thermoplastics*, Vol. *1–3*, 4th ed., 2012.
5. Ebnesajjad S., *Fluoroplastics Handbook: Melt processable fluoropolymers*, Vol. 2, 2nd ed., Elsevier, Oxford, UK, 2015.
6. Teflon PFA Fluorocarbon Resins: Chemical Resistance, PIB #2 Bulletin, DuPont Co. 1972
7. Gangal, S. V., Brothers, P. D., *Perfluorinated Polymers, Perfluorinated Ethylene–Propylene Copolymers*, Wiley Online Library, June 2010, https://doi.org/10.1002/0471440264.pst417.pub2.
8. Ebnesajjad, S. *Introduction to Fluoropolymers - Materials, Technology and Applications*. 2st ed. Elsevier, Oxford, UK, 2020.
9. Dyneon™ PVDF Chemical Resistance Tables, Dyneon™ (a 3M Company), www.dyneon.com, 2004.
10. Application System Design - High Purity System Design, Rev. EDG–02/A, Asahi America, www.asahi-america.com, 2022.
11. Halar® ECTFE Design & Processing Guide, Solvay Specialty Polymers, SpecialtyPolymers.Americas@solvay.com, R 08/2016, version 2.7, 2016.
12. Alaaeddin, M. H., Sapuan, S. M., Zuhri, M. Y. M., Zainudin, E. S., Oqla, F. M., IOP Conf. Series: Materials Science and Engineering 538, 2019.
13. Ebnesajjad S., *Polyvinyl Fluoride: Technology and applications of PVF*. 1st ed, Elsevier, Oxford, UK, 2013.
14. Chemical Resistance of Tedlar® PVF Films, pub by DuPont Co., CDP, Rev. 0, January 2021, www.Tedlar.com.

18

The Effect of Radiation on Fluoropolymers

Jiri G. Drobny

18.1 Introduction

Radiant energy is one of the most abundant forms of energy available. Nature provides sunlight, the type of radiation that is essential for many forms of life and growth. There are some natural substances that generate yet another kind of radiation that can be destructive to life, but when harnessed it can provide other forms of energy or serve in medicine or industrial applications. These are radioactive substances that produce ionizing radiation by radioactive decay, which is the spontaneous breakdown of an atomic nucleus resulting in the release of energy and matter.

Scientific effort has created devices for generating radiant energy useful in a great variety of scientific, industrial, and medical applications. Cathode ray tubes emit impulses that activate the screens of computer monitors and televisions. X-rays are used not only as a diagnostic tool in medicine, but also as an analytical tool in the inspection of manufactured products such as tires and other composite structures. Microwaves are used not only in cooking or as a means of heating rubber or plastics, but also in a variety of electronic applications. Infrared radiation is used in heating, analytical chemistry, and electronics. Manmade ultraviolet radiation has been in use for decades in medical applications, analytical chemistry, and in a variety of industrial applications. Devices used to generate accelerated particles are not only valuable scientific tools but also important sources of ionizing radiation for industrial applications. Both ultraviolet (UV) and electron beam (EB) radiations are classified as electromagnetic radiation, along with infrared (IR) and microwave (MW) [1]. The differences between them in frequency and wavelength are shown in Table 18.1 [1, p. 2].

Polymeric substances, which are predominantly high molecular weight organic compounds, such as plastics and elastomers, respond to radiation in several ways. They may be gradually destroyed by UV radiation from sunshine when exposed for extended periods of time outdoors; and they may more or less change their properties. On the other hand, manmade UV radiation is actually used to produce polymers from monomers (low-molecular-weight building blocks for polymers) or from oligomers (essentially very low-molecular-weight polymers). In these reactions, almost always a liquid is converted into a solid more or less instantaneously. Ionizing radiation (gamma rays and high-energy electrons) is even more versatile: it is capable of converting monomeric and oligomeric liquids into solids but can also produce major changes in the properties of solid polymers [1, p. 2].

Industrial applications involving the large volume radiation processing of monomeric, oligomeric, and polymeric substances depend essentially on two electrically generated sources of radiation: accelerated electrons and photons from high-intensity UV lamps. The difference between these two is that accelerated electrons can penetrate matter and are stopped only by mass, whereas high-intensity UV light affects only the surface. Generally, the processing of monomers, oligomers, and polymers by the irradiation of UV light and an electron beam is referred to as "curing". This term encompasses chemical reactions including polymerization, cross-linking, surface modification, and grafting.

DOI: 10.1201/9781003204275-24

271

TABLE 18.1

Frequency and Wavelength of Various Types of Electromagnetic Radiation

Radiation	Wavelength, μm	Frequency, Hz
Infrared	$1-10^2$	$10^{15}-10^{12}$
Ultraviolet	10^2-1	$10^{17}-10^{15}$
Microwave	10^3-10^5	$10^{12}-10^{10}$
X-rays soft	$10^{-2}-10^{-3}$	$10^{17}-10^{16}$
X-rays hard	$10^{-4}-10^{-3}$	$10^{19}-10^{17}$
Gamma rays	$10^{-6}-10^{-5}$	$10^{-20}-10^{18}$

18.2 Effects of Ionizing Radiation on Fluoroplastics

As in the case of many other polymeric materials, an ionizing radiation electron beam (EB) and gamma radiation have a variety of effects on fluoropolymers. It may cross-link them, cause chain scission, or affect their surface. Quite often, these effects may occur simultaneously. The final result depends on the nature of the material, the dosage, dosage rate, and the energy of the radiation. Thus, there are many ways to exploit these processes technologically, such as cross-linking, reduction of molecular weight and surface modification alone, or grafting.

In general, perfluorinated polymers undergo chain scission, and those containing hydrogen atoms in their monomeric units tend to predominantly cross-link. In reality, however, the situation is not that simple. The reason for that is that the net effect is the result of two competing reactions, namely, cross-linking and chain scission. The structure of the monomeric units is the important factor here. Prorads (i.e., cross-linking promoters) are often used to enhance the cross-linking reaction.

18.2.1 Effects of Ionizing Radiation on Perfluoroplastics

PTFE is attacked and degraded by the irradiation of gamma rays, high-energy electron beams, or X-rays. The degradation of the polymer in air or oxygen occurs due to scission of the chain and is fairly rapid. Such scission results in molecular weight reduction [2]. When irradiated by an electron beam, the molecular weight is reduced by up to six orders of magnitude to produce micropowders [3]. However, there is evidence [4–6] that the irradiation of PTFE above its melting range (603–613 K) in a vacuum results in a significant improvement in tensile strength and elongation at 473 K and in an increase of tensile modulus at ambient (room) temperature. This clearly indicates cross-linking in the molten state, similar to effects caused by the irradiation of polyethylene. At temperatures above 623 K, thermal depolymerization is increasingly accelerated by irradiation and prevails over cross-linking at yet higher temperatures [7]. A fairly detailed discussion of the process is in [8].

FEP is degraded by radiation in a similar fashion to PTFE, namely, by chain scission and the resulting reduction in the molecular weight. The latter can be minimized by excluding oxygen. If FEP is lightly irradiated at elevated temperatures in the absence of oxygen, cross-linking offsets molecular breakdown [9, 10]. The degree to which radiation exposure affects the polymer depends on the amount of energy absorbed, regardless of the type of radiation. Changes in mechanical properties depend on total dosage but are independent of dose rate. The radiation tolerance of FEP in the presence or absence of oxygen is higher than that of PTFE by a factor of 10:1 [11]. However, if the polymer is irradiated above its glass transition temperature (80°C) [12], cross-linking predominates, and the result is an increase in viscosity. With doses above 26 kGy (2.6 Mrad), the ultimate elongation and resistance to deformation under load at elevated temperatures are improved, and the yield stress is increased. However, the improvements are offset by some loss in toughness [13].

PFA/MFA, like other perfluoroplastics, are not highly resistant to radiation [14]. Radiation resistance is improved in a vacuum, and strength and elongation are further increased after low dosage (up to 3 Mrad, or 30 kGy) when compared with FEP or PTFE. At 3 to 10 Mrad (30–100 kGy) it approaches the performance of PTFE and embrittles above 10 Mrad (100 kGy). After exposure to a dosage of 50 Mrad (500 kGy), PTFE, FEP, and PFA are all degraded [11].

18.2.2 Effects of Ionizing Radiation on Other Fluoroplastics

ETFE retains its tensile properties when exposed to low-level ionizing radiation, because the two competing processes, namely, chain scission and cross-linking, are occurring at an approximately equal rate so the net change in molecular weight is quite small. At higher levels of radiation, the tensile elongation of ETFE is severely affected and drops sharply when irradiated [11, p. 671]. The change in mechanical properties is much pronounced in the presence of air [15]. The radiation resistance of ETFE is superior to both PTFE and FEP as shown in Figure 18.1. For that reason, ETFE is used as cable insulation in nuclear energy installations [15].

Cross-linking of ETFE improves its high-temperature properties, such as cut-through by a hot soldering iron, and increases the continuous service temperature from 150°C (302°F) to 200°C (392°F) [16]. Further improvement is achieved by the use of prorads (radiation promoters) such as triallyl cyanurate (TAC) or triallyl isocyanurate (TAIC) in amounts up to 10% by weight [7].

PVDF and its copolymers undergo cross-linking when exposed to low-level radiation (up to 20 Mrad, or 200 kGy), as indicated by the increased gel fraction with an increased dose [17, 18]. Cross-linking in PVDF usually results in an increase in the tensile strength of the polymer and in a reduction of both the degree of crystallinity and melting point. The overall radiation resistance to nuclear radiation is very good. Its tensile strength is virtually unaffected after 1,000 Mrad (10,000 kGy) of gamma radiation in a vacuum, and its impact strength and elongation are slightly reduced as a result of cross-linking. The only adverse effect observed on PVDF after radiation is discoloration, which would already have occurred at relatively low doses [19]. It is attributed to the formation of double bonds due to dehydrofluorination [15, p. 60]. PVDF and its copolymers are significantly thermally destabilized by radiation doses above 100 kGy; therefore, for a sufficiently efficient cross-linking, polyunsaturated monomers such as TAC, TAIC, bis(maleimido-methyl) ether, and ethylene bis-maleimide should be used [20].

PVF becomes cross-linked as PVDF and forms a gel fraction when exposed to ionizing radiation. Irradiated PVF exhibits higher tensile strength, and resistance to etching increases with increasing dose; however, as in PVDF, the degree of crystallinity and melting point decrease [16]. As with PVDF, it is significantly destabilized by radiation doses above 100 kGy, also requiring prorads for a sufficient degree of cross-linking [20].

ECTFE behaves upon irradiation like ETFE, including improvement of cross-linking efficiency with the addition of prorads. Irradiation at room temperature followed by heat treatment at and above 435 K in nitrogen for 20 min. is reported to be most effective [15, p. 61].

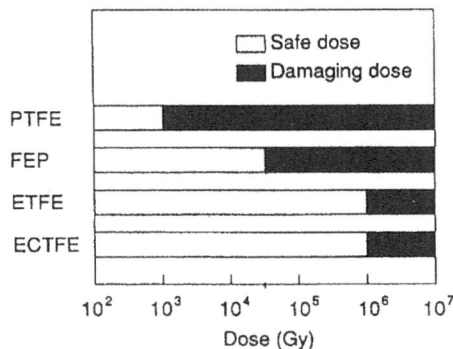

FIGURE 18.1 Radiation of common fluoropolymers used as cable insulation.
Source: Data from [15] and Schönbacher, H. and Stolarz-Izycka, A., *CERN Report 79-04*, CERN, Geneva, Switzerland (1979).

The reports on the effects of ionizing irradiation on *PCTFE* are not without controversy. One source claims that the resistance of PCTFE to ionizing radiation is superior to that of other fluoropolymers [15, p. 61], yet another work reports that poly(chlorotrifluoroethylene) degrades when exposed to ionizing radiation in a similar fashion as PTFE at ambient and elevated temperatures. Unlike PTFE, when irradiated above its crystalline melting point, it still exhibits chain scission [13]. For more on this see Section 5.2.7.2. In industrial practice, PCTFE is known to have a high enough resistance to ionizing radiation to be used in nuclear engineering components and uranium enrichment equipment [21].

Amorphous perfluoroplastics developed by DuPont (Teflon AF) and Asahi Glass (Cytop) contain bulky structures that are responsible for the absence of crystallinity. When Teflon AF was irradiated by low-energy X-rays, it was found that the inclusion of the dioxole monomer not only improves the optical properties but also increases the radiation tolerance of the homopolymer [22].

18.3 Effects of Ionizing Radiation on Fluorocarbon Elastomers

18.3.1 Effects of Ionizing Radiation on the FKM Type of Fluorocarbon Elastomers

The FKM type of fluoroelastomers, particularly the copolymers of HFP and VDF, are cross-linked by low-level gamma radiation (up to 20 Mrad or 200 kGy) in the same fashion as PVDF. The radiation degradation of a VDF-HFP copolymer was studied by Zhong and Sun [23]. Their finding was that there is a linear relationship between the dose and the weight loss of the polymer. Fluorocarbon elastomers based on the copolymers of TFE and propylene were found to be more resistant to gamma radiation than Viton™ elastomers [24]. The cross-linking of VDF-HFP, VDF-TFE, and tetrafluoroethylene-propylene (TFE/P) copolymers as well as of the terpolymer VDF-HFP-TFE by ionizing radiation (gamma or electron beam) can be enhanced by prorads, such as TAIC and trimethylolpropane trimethacrylate (TMPTM) [7, p. 339]. It appears that each fluorocarbon elastomer has the best cross-link yield with a specific prorad. In general, optimized compounds from fluorocarbon elastomers irradiated at optimum conditions attain considerably better thermal stability and mechanical properties than chemical curing systems [25–27]. A typical radiation dose for a sufficient cross-linking of most fluorocarbon elastomers is in the range 10 to 100 kGy.

18.3.2 Effects of Ionizing Radiation on Perfluoroelastomers

Perfluoroelastomers (ASTM designation FFKM) are essentially copolymers of two perfluorinated monomers, TFE and PMVE, with a cure site monomer (CSM), which is essential for cross-linking. Depending on the nature of the CSM and on the curing conditions, they are capable of service temperatures up to 300°C. Perfluoroelastomers can be cured by ionizing radiation (EB and gamma rays) without any additives. The advantage of radiation-cured FFKM is the absence of any additives, so the product is very pure. The disadvantage is the relatively low upper use temperature of the cured material, typically 150°C, which limits the material to special sealing applications only [28].

18.3.3 Effects of Ionizing Radiation on TFE/P Elastomers

TFE/P or FEPM fluoroelastomers (such as Aflas®) tend to cross-link when exposed to ionizing radiation [29] and the cross-linking reaction is further enhanced by the addition of allyl compounds such as allyl maleate into the rubber formulation [30]. Ito [31] found that the G values for cross-linking and chain scission in TFE/P to be 0.152 and 0.023 µmole • J^{-1} respectively.

18.4 Effects of Ionizing Radiation on Fluorosilicone Elastomers

In general, silicones based on the − Si−O−Si− chain with atoms attached to the Si atom, such as polydimethylsiloxanes, respond to irradiation by ionizing radiation and further respond by forming cross-links with negligible chain scissions [32]. The cross-link density increases linearly with a dose up to 160

Mrad (1,600 kGy) [33]. Standard (PMTFPS-based) fluorosilicone elastomers (FSRs) behave in a similar fashion [32, p. 129]. The most recent study of gamma ray irradiation of FSR [34] reported increasing cross-link density and hardness after irradiation, and the decrease of tensile strength and elongation at break with an increased radiation dose. The effects of the irradiation on the properties of the FSR under study were ascribed to the occurrence of degradation and cross-linking reactions. Overall the results proved that the degradation reaction dominated over the cross-linking reaction during the irradiation. One interesting application is the process of preparing blends of fluoroplastics, such as polyvinylidene fluoride with fluorosilicone elastomers, to obtain materials having a unique combination of flexibility at low temperatures and high strength [35].

18.5 Effects of UV Radiation on Fluoropolymers

Fluoropolymers as a group have intrinsically high resistance to degradation by UV radiation. The strength of the fluorine–carbon bond makes them resistant to pure photolysis. Moreover, they do not contain any light-absorbing chromophores either in their structure or as impurities. For example, no physical or chemical changes have been observed during 30 years of continuous exposure of PTFE in Florida [18]. The outdoor durability of a fluorinated coating is directly related to its fluorine content and is assessed by gloss retention [36].

The degradation of some fluoropolymers outdoors occurs very slowly and can be detected only by very sensitive analytical methods, such as X-ray photoelectron spectroscopy (XPS) [37] or electron spin resonance (ESR) spectroscopy [38].

To assess the propensity of a polymer to UV degradation accurately, it is important to pay attention to the wavelength of the UV light employed. For example, FEP polymers absorb UV radiation only at wavelengths below 180 nm, which makes them susceptible to degradation only in the space [15, p. 62]. Unlike FEP and other perfluoropolymers, degradation of PCTFE is greatly accelerated by UV light [39]. Copolymers of TFE and HFP have been reported to undergo scission and cross-linking when exposed to UV radiation [40].

REFERENCES

1. Drobny, J. G., *Radiation Technology for Polymers*, 3rd ed, CRC Press, Boca Raton, FL, p. 1 2021.
2. Blanchet, T. A., *Handbook of Thermoplastics* (Olabisi, O., Ed.), Marcel Dekker, New York, p. 987, 1997.
3. Lunkwitz, K., Brink, H. J., Handtke, D., Ferse, A., *Radiat Phys. Chem.*, *33*:523, 1989.
4. Tabata, Y., *Solid State Reactions in Radiation Chemistry*, Taniguchi Conference, Sapporo, Japan, 1992, Proceedings, p. 118.
5. Sun, J., Zhang, Y., Zheng, X., *Polymer*, 35:288, 1994.
6. Oshima, A., Tabata, Y., Kudoh, H., Seguchi, T., *Radiat. Phys. Chem.*, *45*(2):269, 1995.
7. Lyons, B. J., *Modern Fluoropolymers* (Scheirs, J., Ed.), John Wiley & Sons Ltd, Chichester, UK p. 341, 1997.
8. Lunkwitz K., Burger, W., Gessler, U., *J. Polym. Sci.*, *60*, 2017, 1996.
9. Eby, R. K., *J. Appl. Phys.*, *34*:2442, 1963.
10. Eby, R. K., Wilson, F. C., *J. Appl. Phys.* 33:2951, 1962.
11. Gangal, S. V., *Encyclopedia of Polymer Science and Engineering*, Vol. *16* (Mark, H. F. and Kroschwitz, J. I., Eds.), John Wiley & Sons, Ltd, Chichester, UK, p. 605, (1989).
12. Lovejoy, E. R., Bro, M., Bowers, G. H., *J. Appl. Polym. Sci.*, *9*:401, (1965).
13. Bowers, G. H., Lovejoy, E. R., *I&EC Product Research and Development*, Vol. 1, p. 89, American Chemical Society, Washington, D.C., June 1962.
14. *Teflon® PFA Fluorocarbon Resins: Response to Radiation*, APD#3, Bulletin, E. I. du Pont de Nemours, & Co., Wilmington, DE (1973).
15. Scheirs, J., *Modern Fluoropolymers* (Scheirs, J., Ed.), John Wiley & Sons Ltd, Chichester, UK, p. 59, (1997).
16. Clough, R. L., Gillen, K. T., Dole, M., in *Irradiation Effects on Polymers* (Clegg, D. W. and Collyer, A., Eds.), Elsevier Applied Science, London, p. 95, (1991).

17. Rosenberg, Y., Siegmann, A., Narkis, M., Shkolnik, S., *J. Appl. Polym. Sci.*, *45*:783, (1992).
18. Klier, I., Strachota, S., Vokal, A., *Radiat. Phys. Chem.*, *38*:457, (1991).
19. Kawano, Y., Soares, S., *Polym. Degrad. Stab.*, *35*:99, (1992).
20. Lyons, B. J., Weir, F. E., *Radiation Chemistry of Macromolecules*, (Dole, M., Ed.), Vol. *II*, Chapter 14, Academic Press, New York, p. 294, (1974).
21. *Professional Plastics- Overview of Fluoropolymers*, www.profesionalplastics.com/fluoropolymers, 2022
22. Jahan, M. S., Ermer, D. R., Cooke, D. W., *Radiat. Phys., Chem.*, *41*(3):481, (1993).
23. Zhong, X., Sun, J., *Polym. Degrad. Stab.*, *35*:99, (1992).
24. Ito, M., *Radiat. Phys. Chem.*, *47*:607, (1996).
25. Vokal, A., Pellanova, M., Cernoch, P., Klier, I., Kopecky, B., *Radioisotopy*, *29*(5–6):426, (1988).
26. McGinnis, V. D., *Encyclopedia of Polymer Science and Engineering*, Vol. *4* (Mark, H. F., and Kroschwitz, J. I., Eds.), John Wiley & Sons, New York, 1986, p. 438.
27. Banik, I., Bhowmick, A. K., Raghavan, S. V. Majali, A. B., Tikku, V. K., *Polymer Degradation and Stability*, *63*(3):413, (1999).
28. Marshall, J. B., *Modern Fluoropolymers* (Scheirs, J., Ed.), John Wiley & Sons Ltd, Chichester, UK, p. 352, (1997).
29. Yamamoto, T., Uchijima, K., Ito, Y., *Jpn Kokai*, JP *73*:37982, (1973).
30. Tabata, Y., Kojima, G., *Jpn Kokai*, JP *73*, 787465, (1973).
31. Ito, M., *Radiat. Phys. Chem.*, *31*:315, (1988).
32. Drobny, J. G., *Ionizing Radiation and Polymers*, Elsevier, Oxford, UK, p. 128, (2013).
33. Delides, C. G., Shepard I. W., *Radiat Phys Chem*, *10*:329, (1977).
34. Liu, Y., Zhu, C., Feng, S., *Materials Letters*, *78*:110–112, (2012).
35. Caporiccio, G., Mascia, L., US Patent 5,457,158 to Dow Corning Corporation (1995).
36. Scheirs, J., Fluoropolymer Coatings; New Developments, in *Polymeric Materials Encyclopedia* (Salamone, J., Ed.), CRC Press, Boca Raton, FL, pp. 2498–2507, (1996).
37. Sjostrom, C., Jernberg, P., Lala, D., *Mater. Struct.*, *24*:3, (1991).
38. Okamoto, S., Ohya-Nishiguchi, H., *J. Jpn. Soc. Colour Mater.*, *63*:392, (1990); *Chem. Abstr.* 114.
39. Wall, L. A., Straus, S., *J. Res. (Natl. Bur. Standards)*, *65*:227, (1961).
40. Bowers, G. H., Lovejoy, E. R., *Ind. Eng. Chem. Prod. Res. Dev.*, *1*:89, (1962).

Part VII

Safety and Sustainability

19

Safety Aspects of Fluoropolymers

Sina Ebnesajjad

This chapter includes brief information about the safe handling and processing of polytetrafluoroethylene (PTFE) and other fluoropolymers. It is *not* a complete guide for fluoropolymer safety, hygiene, and environmental topics. The discussions presented in this chapter are not intended to replace the information and data, including safety data sheets, provided by the suppliers and manufacturers of fluoropolymers or other authorities.

Two reliable resources of information are *The Guide to the Safe Handling of Fluoropolymer Resins* [1] and *Guide for the Safe Handling of Fluoropolymer Resins* [2]. This chapter makes frequent use of these resources.

19.1 Introduction

Some of the topics in this chapter have been the subject of a great deal of controversy during recent years. Some of the reasons for this are:

1. Extrapolation of issues with small fluorinated molecules conversion to polymers.
2. Conclusions based on unrealistic testing conditions and speculation.
3. Improvements in analytic technique detection limits down to parts per billion and even per trillion. Common thinking: if we can measure parts per billion of an impurity, why can't we eliminate the impurity?
4. Use of fluoropolymers beyond their temperature limits.
5. A phobia of "chemicals" and "chemistry" in general society.
6. Insufficient and inefficient communication by the industry.

While questions raised about the materials used to produce fluoropolymers are legitimate and require discussion and answers, fluoropolymers themselves are safe. Indeed, they are among the safest of all materials.

We have strived to put forward factual accounts for each topic based on available data within the scope of this book. Extensive coverage of the subjects of this chapter is not possible because of the large volume of available information. The reader, however, is advised strongly to develop a broader perspective on his or her own to ensure a comprehensive understanding of the topics of this chapter. Additional information may be obtained from published research articles, government rules and publications, and standards organizations [3–7].

Several areas related to the safety of fluoropolymers will be discussed in this chapter. The thermal properties of fluoropolymers have been discussed separately in Chapter 16. The subjects of recycling, reuse, and disposal methods will be discussed separately in Chapter 20 to ensure adequate coverage of new development in these areas.

19.2 Fluoropolymers: The Essential Plastic

The Plastics Industry Association [5] describes fluoropolymers as "fluorocarbon-based, high-performance plastics used in many unique and highly specialized applications where other materials simply cannot perform. They include PTFE, melt processible fluoropolymers and fluorinated elastomers. These polymers are used to protect and insulate wires in order to prevent overheating and potential fires."

Fluoropolymers have unmatched chemical and temperature resistance which is why they're used in products like cars and planes." Chapter 2 of this book has presented a number of examples of fluoropolymers. One key fact is that they are only specified for use in parts and equipment when they are *absolutely required*. These plastics play a key role in most industries including [8]:

- Chemical processing.
- Food and beverage.
- Automotive.
- Aerospace.
- Construction.
- Agriculture.
- Petroleum exploration and refining.
- Alternative energy.
- Semiconductor.
- Pharmaceutical and bio pharmaceutical.
- Sportswear and equipment.
- Apparel.
- Gas and liquid filtration.
- Environmental protection.
- Medical and surgical applications.

Fluoropolymers are an essential material in the enhancement of living standards throughout the world. However, these plastics *must be used* within the limits (especially of temperature) specified by the manufacturers and suppliers.

19.3 Polymerization Aids

Fluoropolymers are primarily polymerized in water. They used to be polymerized in the presence of two types of aids (Figure 19.1): alkaline salts of perfluoro-octanoic acid (PFOA) and perfluoro-octanesulfonic acid (PFOS). Because of their unique properties, these chemicals are persistent in the environment and, if

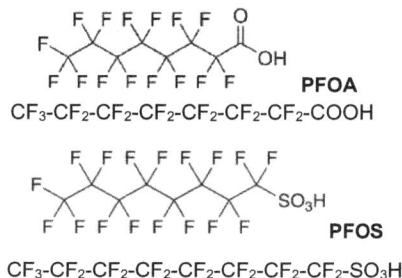

CF_3-CF_2-CF_2-CF_2-CF_2-CF_2-CF_2-COOH

CF_3-CF_2-CF_2-CF_2-CF_2-CF_2-CF_2-CF_2-SO_3H

FIGURE 19.1   Chemical structures of PFOA and PFOS.

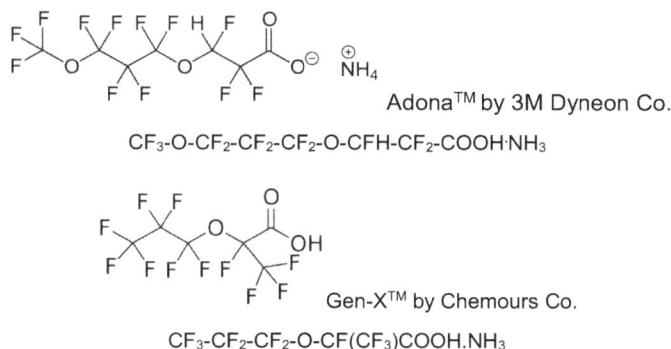

Adona™ by 3M Dyneon Co.

CF₃-O-CF₂-CF₂-CF₂-O-CFH-CF₂-COOH·NH₃

Gen-X™ by Chemours Co.

CF₃-CF₂-CF₂-O-CF(CF₃)COOH.NH₃

FIGURE 19.2 Chemical structures of Adona™ by 3M Dyneon and Gen X™ by Chemours Co.

found in the body, tend to accumulate over time. Their health effects are not completely understood, but it is known that exposure to PFOA and PFOS has negative impacts on human health.

PFOA and PFOS are, however, no longer manufactured in the USA, the EU, or Japan. When concerns arose about these polymerization aids, all eight major chemical manufacturers agreed to eliminate the use of PFOA and other "long-chain" fluorinated substances in their products under the US Environmental Protection Agency's (EPA) PFOA Stewardship Program. Significant New Use Rules (SNURs), which have the weight of EPA regulations, generally prevent companies from importing or manufacturing such perfluorinated chemicals without prior approval by the EPA.

19.3.1 Replacement of Polymerization Aids

To enable the removal of PFOA and PFOS, companies developed safer alternatives with shorter chain fluorinated substances containing ether bonds (-O-). Two examples of the new compounds Adona™ by 3M Dyneon [9] and Gen X™ by Chemours Co. [10] are shown in (Figure 19.2). The new aids perform the same tasks during polymerization without the same environmental or public health risks. The ether bonds render more easily the degradation of these compounds than PFOA and PFOS.

The replacement aids are considered safe to use according to governmental agencies like the Food and Drug Administration (FDA) and the EPA. Figure 19.3 shows a comparison of the relative risks of TEF, PTFE, and polymerization aids.

The most likely place for users to encounter TFE polymerization aids is during handling of PTFE dispersions. The reader is advised to contact the Chemours Company (https://www.chemours.com) for information about Gen X™ polymerization. 3M Dyneon should be contacted as well; consult the safety data sheet published by the National Library of Medicine (https://pubchem.ncbi.nlm.nih.gov/compound/52915299) or information about the safety and hygiene of Adona™.

19.4 Tetrafluoroethylene

This section describes some of the key characteristics of TFE because it is widely use in commercial fluoropolymers. By far TFE is the most reactive monomer among those used in fluoropolymers. It is a highly reactive compound and can autopolymerize exothermically in the presence of minor amounts of oxygen or metal powders and is a dangerous gas in that it can explode both in the presence and absence of oxygen.

Utmost care must be exercised in handling, transporting, storing, charging, and operating the polymerization reactor to avoid explosion, fire, environmental discharge, and personnel exposure. As Figure 19.4 illustrates, the nature of risk changes drastically after TFE is converted to PTFE, one of the safest and most inert among materials. Yet when overheated, in the presence or absence of oxygen, PTFE degrades into TFE (flammable) and a variety of small compounds, some of which are highly toxic.

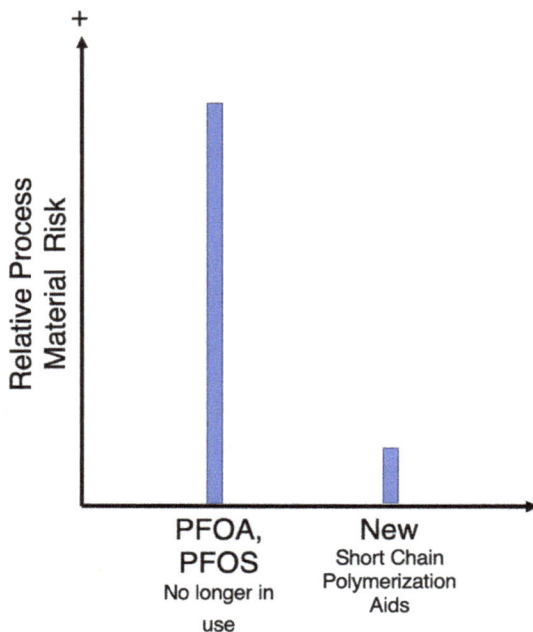

FIGURE 19.3   Relative risk of old and new polymerization aids, PTFE, and polymerization aids.

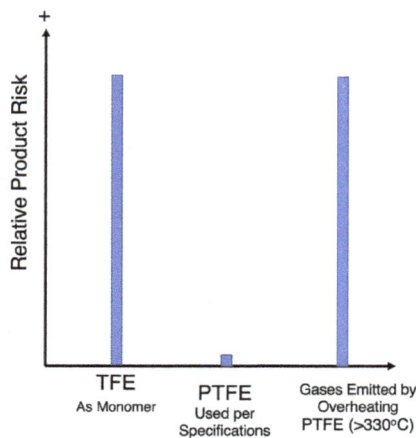

FIGURE 19.4   Relative risk of TFE, PTFE, and PTFE degradation products.

In commercial plants and semiworks, extensive measures including equipment and procedures have been put in place to mitigate the risks associated with TFE [11]. The *Guide* published by PlasticsEurope [2] is an outstanding source of detailed quantitative information about TFE and its safety.

The threshold limit value (TLV) of TFE according to the American Conference of Governmental Industrial Hygienists [12] averaged over an eight-hour period is two parts per million (ppm). PTFE resins usually do not contain TFE because the processing steps, including heat, during manufacturing remove as much of the residual TFE as possible. TFE monomer trapped within resin particles is likely to evolve during processing of the resins and so require adequate ventilation. It is advisable to open the containers of PTFE in a well-ventilated area (e.g., chemical hoods) to prevent exposure to any remaining trace residues of TFE.

Information about other monomers of commercial fluoropolymers can be obtained from the published safety data sheets, government publications, and other sources [1–4, 8, 12].

19.5 Toxicology of Fluoropolymers

Fluoropolymers have thermal, chemical, photochemical, hydrolytic, oxidative, and biological stability. They have relatively high thermal stability and can stand up to elevated temperatures without much degradation. PTFE is the best product in these areas [8, 13]. Generally, the reactivity of fluoropolymers decreases as the fluorine content of the polymer increases. This group of plastics has low toxicity and almost no toxicological activity. No fluoropolymers, including PTFE, have been known to cause skin sensitivity and irritation to humans.

Fluoropolymers are not classified as hazardous substances according to Globally Harmonized System (GHS) criteria (www.ccohs.ca/oshanswers/chemicals/ghs.html) [1]. Rather they have been assessed to show that they satisfy widely accepted assessment criteria to be considered as "polymers of low concern" (PLC) [14]. In their seminal paper, Henry et al. [14] indicate that fluoropolymers have these items of inherent toxicity:

- They have negligible residual monomer and oligomer content and low to no leachables.
- Fluoropolymers are practically insoluble in water and not subject to long-range transport.
- A molecular weight well over 100,000 Da (one atom of H has a molecular mass of 1 Da, so 1 Da = 1 g/mol) suggests fluoropolymers cannot cross the cell membrane.
- Fluoropolymers are not bioavailable or bio-accumulative, as evidenced by acute and sub-chronic systemic toxicity, irritation, sensitization, local toxicity on implantation, cytotoxicity, *in vitro* and *in vivo* genotoxicity, hemolysis, complement activation, and thrombogenicity studies on PTFE.
- Clinical studies of patients receiving permanently implanted PTFE cardiovascular medical devices demonstrate no chronic toxicity or carcinogenicity and no reproductive, developmental, or endocrine toxicity.
- Examples of PLC among fluoropolymers include PTFE, perfluoroalkoxy (PFA), fluorinated ethylene propylene polymer (FEP), and ethylene tetrafluoroethylene polymer (ETFE).

Since the publication of [14] other papers have appeared which express other points of view. For example, Lohmann et al. [15] recommend a broader study to include the entire life cycle of all fluoropolymers to determine whether they can be considered to be polymers of low concern: "Our recommendation is to move toward the use of fluoropolymers in closed-loop mass flows in the techno-sphere and in limited essential-use categories, unless manufacturers and users can eliminate PFAS [polyfluoroalkyl substances] emissions from all parts of the life cycle of fluoropolymers" [15].

Fluoropolymer compounds contain additives such as fillers, pigments, or surfactants to enhance processing or other characteristics. These additives are likely to give rise to hazards in the use of fluoropolymer resins. Compounds of fluoropolymers may have unknown toxicological properties. The reader is advised to consult the Safety Data Sheet from the manufacturer for health information on the specific product he or she intends to process.

19.6 Emissions During Processing

As discussed, although fluoropolymers are among the most thermally stable plastics, they will begin to generate toxic compounds at or above their processing temperatures (Table 19.1). The most important cautionary action that must be taken is to ensure ample ventilation around the equipment to capture and remove all evolved gases and particulates. The type and amount of degradation products depend on the types of additives, time, temperature, atmosphere (air, N_2, vacuum), and other variables.

TABLE 19.1

Melting Point and Maximum Processing and Continuous Use Temperature of Commercial
Fluoropolymers

Fluoropolymer	Melting Point, °C	Maximum Processing Temperature, °C	Maximum Continuous Use Temperature, °C
PTFE	342	715	260
PFA	305	380	260
MFA	215	360	249
FEP	260	360	200
ETFE	220–270	310	150
PCTFE	215	165	120
ECTFE	230	280–310	150
co-PVDF	115–117	232–249	100–150
PVDF	170	232	150
THV	120–230	171–310	70–130
EFEP	158–195	220–260	100–150

Four main types of products are formed during the decomposition of fluoropolymers. They include fluoroalkenes, hydrogen fluoride, oxidation products, and low-molecular-weight fluoropolymer particulates. The presence of other monomers or additives in the fluoropolymer resin changes the nature of the decomposition products.

This is why manufacturers recommend the use of localized exhaust ventilation during processing operations. The rate of generation rises as temperatures increase and usually results in degradation of the polymer to produce particulates and toxic gaseous by-products. In addition to general ventilation of the process area it is important to use localized ventilation at the feed and exit points of the extruder. An example of ventilation is exhibited in Figure 19.5 for the extrusion of fluoropolymers. The goal is the complete removal of emitted particulates and gaseous degradation by-products. Care must be taken to prevent any leaks of the gases and particulates to the process area. The *Guide to the Safe Handling of Fluoropolymer Resins* includes a number of designs and recommendations for effective exhaust systems.

In the case of PTFE, the evolved TFE may be a residual monomer trapped in the resin particles or a product of degradation of the resin. When the PTFE temperature increases to approximately 450°C in air, carbonyl fluoride and hydrogen fluoride are the main decomposition products. Carbonyl fluoride (COF_2) hydrolyzes in the presence of any form of water in air to carbon dioxide and hydrofluoric acid. Some hexafluoropropylene is generated at these elevated temperatures. Most concerning is the highly toxic perfluoroisobutylene (PFIB, $CF_2=C(CF_3)_2$) which is evolved at temperatures in excess of 475°C.

Other fluoropolymers decompose more or less similarly as documented by Baker and Kasprzak [1, 16]. The degradation onset of fluorinated copolymers is usually lower than those of homopolymers.

19.7 Polymer Fume Fever (PFF)

This section describes Polymer Fume Fever according to *The Guide to the Safe Handling of Fluoropolymer Resins* [1]. "The most common adverse effect associated with human exposure to fluoropolymer decomposition products is widely recognized as 'polymer fume fever' (PFF)". PFF is a temporary flu-like condition that lasts approximately 24 hours. PFF is similar to metal fume fever (foundry man's fever) [17] and other inhalation fevers. Symptoms include fever, chills and, sometimes, cough.

"The paper, 'Characterization of Early Pulmonary Inflammatory Response Associated with PTFE Fume Exposure', in Toxicology and Applied Pharmacology, Article No. 0208, Academic Press, May 1996] [18]" found that overheating PTFE, 420°C, evolves fumes containing ultra-fine particles that can be highly toxic to the lung, causing pulmonary edema (excessive fluid in cells in the lungs) with

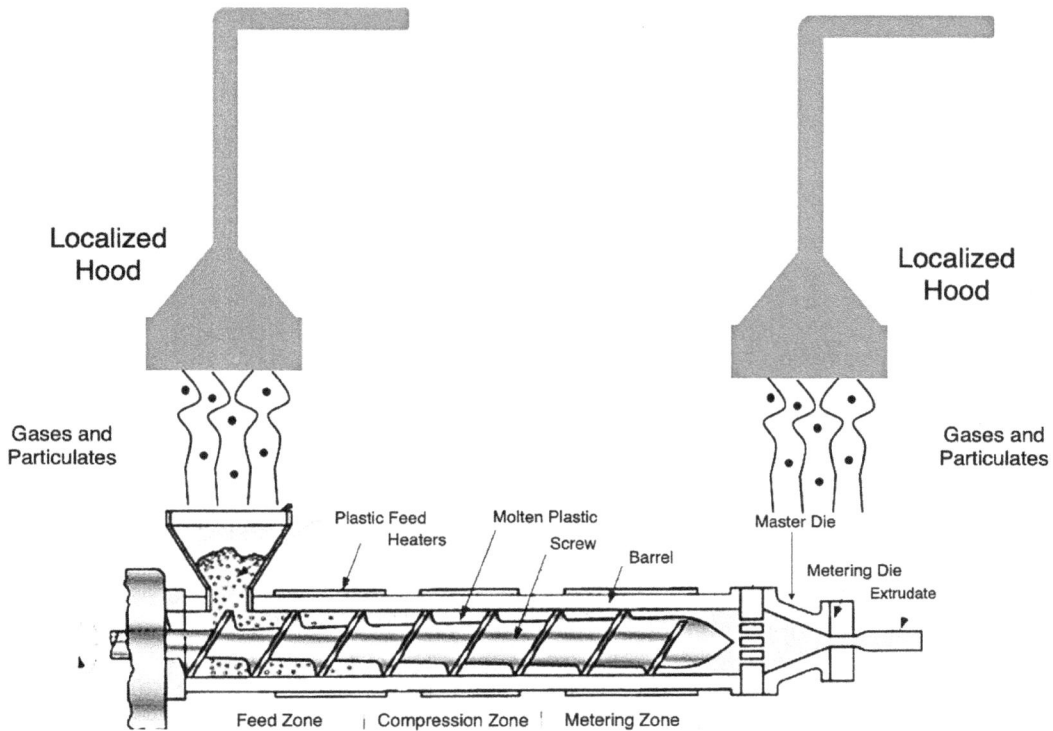

FIGURE 19.5 Example of ventilation during extrusion of fluoropolymers.

hemorrhagic inflammation (severe irritation of the tissue with release of blood from small blood vessels). In animal experiments in which these particles were removed from the air, signs and symptoms similar to those of polymer fume fever did not develop in the animals. Unfiltered air produced the expected polymer fume fever response.

"Inhalation of effluent products from overheated fluoropolymers or after smoking fluoropolymer-contaminated tobacco may also cause PFF. It is recommended that smoking and tobacco products be banned in work areas where fluoropolymer resins are processed. In addition, depending on the characteristics of the operation, local exhaust ventilation may be required."

19.8 Fluoropolymer Dispersions

Dispersions of fluoropolymers require special handling because in addition to resin they contain a number of compounds in their aqueous phase. The best sources for handling guidelines are the product-specific safety data sheets from suppliers. Those dispersions contain small amounts of polymerization aids which are carried through the finishing steps of the polymerization latexes. The polymerization aids are fluorinated examples which were presented in Figure 19.2. These compounds are released when dispersions are dried or heated. The primary exposure routes of polymerization aids are through the skin, ingestion, and inhalation.

The high vapor pressure of polymerization aids require the clean up of spills and overspray before they dry and allow the sublimation of those aids. Dispersion wastes should be disposed in closed containers. Their vapor pressures are high enough to cause exposure exceeds allowable levels under conditions of poor ventilation.

19.9 Hygiene and Personal Protective Equipment

There are three possible routes for exposure of personnel to hazardous materials: inhalation, ingestion, and dermal (skin). Besides engineering design controls of exposure, appropriate personal protective equipment (PPE) must also be worn. The major health hazard of handling fluoropolymers is the inhalation of decomposition products. While handling fluoropolymer dispersions, polymerization aids are released during the drying and heating. Sufficient ventilation must be provided to keep the exposures below the required/recommended limits. In the absence of adequate ventilation personnel must have suitable respiratory protection.

Manufacturers and suppliers of fluoropolymers provide safety data sheets and other guidance for the use of their products. It is important to have competent knowledge of the safety guidelines prior to handling and processing fluoropolymers. The *Guide to the Safe Handling of Fluoropolymer Resins* is a good place to start learning about the safety requirements of fluoropolymers.

REFERENCES

1. *The Guide to the Safe Handling of Fluoropolymer Resins*, published by the Fluoropolymers Division of Plastics Industry Association, 5th ed, 2018, www.PlasticsIndustry.org.
2. Guide for the Safe Handling of Fluoropolymer Resins, PlasticsEurope, www.plasticseurope.org.
3. Environmental Protection Agency (EPA), Office of Chemical Safety and Pollution Prevention (OCSPP), www.epa.gov/aboutepa/about-office-chemical-safety-and-pollution-prevention-ocspp, 2022.
4. Department of Labor, Occupational Health and Safety Administration, www.osha.gov, 2022.
5. Plastics Industry Association (Plastics), Fluoropolymers Division, https://thisisplastics.com/plastics-101/fluoropolymers-do-what-other-materials-cant/, 2022.
6. ASTM International, www.astm.org, 2022.
7. International Organization for Standardization, www.ISO.com, 2022.
8. Ebnesajjad, S., *Fluoroplastics Handbook: Non-Melt Processible Fluoropolymers*, vol *1*, 2nd ed, Elsevier, Oxford, UK, 2015.
9. 3M/DYNEON Company Progress Reports for 2009 Submission under the EPA 201012015.
10. PFOA Stewardship Program, www.epa.gov/opptintr/pfoa/pubs/stewardship/preports4.html#2010; 2010.
11. Guide for the Safe Handling of Tetrafluoroethylene, PlasticsEurope, Fluoropolymers Product Group, November 2017.
12. American Conference of Governmental Industrial Hygienists. 2010.TLVs and BEIs: Threshold limit values for chemical substances and physical agents and biological exposure indices. Cincinnati (OH): American Conference of Governmental Industrial Hygienists.
13. Ebnesajjad, S., *Fluoroplastics Handbook: Melt Processible Fluoropolymers*, vol *2*, 2nd ed, Elsevier, Oxford, UK, 2015.
14. Henry, B. J., Carlin, J. P., Hammerschmidt, J. A., Buck, R. C., Buxton, L. W., Fiedler, H., Seed, J., Hernandez, O., A Critical Review of the Application of Polymer of Low Concern and Regulatory Criteria to Fluoropolymers, *Integrated Environmental Assessment and Management*, *14*, (3):316–334, January 30, 2018.
15. Lohmann, R., Cousins, I. T., DeWitt, J. C., Glüge, J., Goldenman, G., Herzke, D., Lindstrom, A. B., Miller, M. F., Ng, C. A., Patton, S., Scheringer, M., Trier, X., Wang, Z., Are Fluoropolymers Really of Low Concern for Human and Environmental Health and Separate from Other PFAS? *Environ. Sci. Technol.* 54(20):12820–12828, 2020.
16. Baker, B. B., Kasprzak, D. J., Thermal Degradation of Commercial Fluoropolymer in Air. *Polym Degrad Stab*, *42*:181–188, 1994.
17. Harris, D. K., *Lancet*, *261*:1008, December, 1951.
18. Characterization of Early Pulmonary Inflammatory "Response Associated with PTFE Fume Exposure", Toxicology and Applied Pharmacology, Article No. 0208, Academic Press, May 1996.

20

Recycling, Reuse, and Disposal of Fluoropolymers

Sina Ebnesajjad

20.1 Introduction

One constant feature of fluoropolymers over their history has been their higher cost than most other hydrocarbon-based plastics. The reason for the difference is that H/C based plastics are derived from gaseous products of petroleum refining. In contrast, fluoropolymers require fluorine (F) which does not exist in organic form except in rare plants. Rather, fluorine must be extracted from a mineral ore by producing hydrofluoric acid (HF).

The cost of the ore fluorspar (CaF_2), the ensuing processes to obtain an organic fluorinated compound, the disposal of reaction by-products, and the significant capital required to render fluorinated monomers and polymers are more costly than hydrocarbons. In the following sections we describe the underlying reasons why there is a built-in cost incentive at every stage of manufacturing of fluoropolymers to recover the scrap and second grade material so as to recycle it for productive end uses.

A final point about recycling fluoropolymers: much more PTFE is manufactured and consumed than melt processible fluoropolymers, thus resulting in more second grade resin. More PTFE scrap is generated than for the melt processible grades because PTFE is not so processible and usually cannot be molded into the final shapes, thus resulting in machining and finishing generation scrap.

20.2 Fluorine Ore: Fluorspar

The main ore for extraction of fluorine is fluorspar, which is primarily comprised of calcium fluoride

(CaF_2). The origin of fluorine in all organic chemicals containing fluorine is hydrofluoric acid (HF), derived from acid-grade fluorspar, also know as acidspar. About 40% of fluorspar is used as metallurgical flux in the steel industry, some of which is recovered as synthetic fluorspar [1].

The highest grade of fluorspar (97% CaF_2) is reacted with sulfuric acid for the production of HF, which is the starting point of organic fluorinated compounds. Annually over 1 million metric tons of HF is consumed for the production of fluorocarbons, requiring over 2 million tons of fluorspar; the ratio of fluorspar to HF is 2:1 [2].

Tables 20.1 and 20.2 show the distribution of world reserves and production of fluorspar by country. Fluorspar reserves are all outside the United States, in countries including China, Mongolia, Mexico, and South Africa. United States' fluorspar imports from Mexico have declined significantly since 2011. China produces close to two-thirds of world consumption. It also has control over the supplies and price, thus a strong influence over the price of fluoropolymers, providing more incentive for recycling and reusing fluoropolymers.

TABLE 20.1

Approximate Distribution of World Reserves of Fluorspar by
Country [1]

Country	World Reserves, %
South Africa	16
China	15
Mexico	12
Mongolia	8
Spain	2
Kenya	2
USA	2
Iran	1
Rest of the World	42

TABLE 20.2

Approximate Production of Fluorspar by Country [1]

Country	World Production, %
China	66
Mexico	15
Mongolia	3
South Africa	3
Vietnam	3
Rest of the World	10

20.3 Melt Processible Fluoropolymers

This group of fluoropolymers was discussed in earlier chapters. Monomer production for this group requires complex reactions and special plant units. The melt processible fluoropolymers consumed in large quantities or with high value include perfluoroalkoxy (PFA), perfluorinated ethylene propylene copolymer (FEP), and ethylene tetrafluoroethylene copolymer (ETFE).

Typically, the fluoropolymer in a part is separated from metals, plastics, and other materials in the first step. For example, to recycle FEP-insulated wire, the FEP is stripped from the conductor (such as copper) and shredded into a coarse powder. The powder is melt-extruded, filtered, and pelletized usually using a single screw extruder. The pellets can be used in fabrication parts as long as the specifications are satisfied. Another application for FEP scrap is in compression molding which requires pulverizing the FEP particles into a fine powder material.

A number of companies specialize in recycling and reusing melt fluoropolymers in the USA and EU. One of the leading companies in this area is (https://frlusa.com), a division of Prime Materials Recovery, Inc. The Company is located in Connecticut, USA, but operates globally as a division of Prime Materials Recovery, Inc. They offer a range of standard and custom melt fluoropolymers. Some of their FEP resins are recognized by the Underwriters Laboratory for use at the 100% level in wire and cable insulation applications.

20.4 Polytetrafluoroethylene (PTFE)

To understand the roots of the higher cost of fluoropolymers one must look into the reactions required for the production of tetrafluoroethylene (TFE) which were described in Chapter 3. They are repeated here for the reader's convenience (shown below). Hydrofluoric acid is obtained by the reaction of sulfuric acid with feldspar (CaF_2). HF is reacted with chloroform to produce chlorodifluoromethane (CDFM), the precursor to TFE. Each reaction produces significant by-products including calcium sulfate, a range of chloromethanes, and large quantities of hydrochloric acid – all of which must be sold or if needed disposed. Additionally the pyrolysis of CDFM generates by-products, the most significant of which is hexafluoropropylene.

HF formation:

$$CaF_2 + H_2SO_4 \rightarrow 2HF + CaSO_4 \tag{20.1}$$

Chloroform formation:

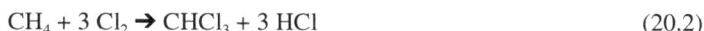

$$CH_4 + 3\,Cl_2 \rightarrow CHCl_3 + 3\,HCl \tag{20.2}$$

Chlorodifluoromethane (HCFC-22) preparation:

$$CHCl_3 + 2\,HF \rightarrow CHClF_2 + 2\,HCl \tag{20.3}$$

(SbF_3 catalyst)
TFE synthesis:

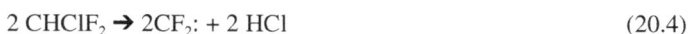

$$2\,CHClF_2 \rightarrow 2CF_2: + 2\,HCl \tag{20.4}$$

(*pyrolysis*)

$$2CF_2: \rightarrow CF_2{=}CF_2 \text{ (tetrafluoroethylene)} \tag{20.5}$$

The overall pyrolysis reaction is summarized as:

$$\text{``}2\,CHClF_2 \rightarrow CF_2{=}CF_2 + 2\,HCl \text{ (}an\ equilibrium\ reaction\text{)''} \tag{20.6}$$

Because of the corrosive and toxic nature of the chemicals involved in the production of TFE, significant capital and operating expenditures are required to generate a pure TFE monomer for polymerization. In comparison, ethylene, for polyethylene manufacturing, is one of many by-products of petroleum refining and petrochemical plants. There is high demand for every petroleum-based product that is produced during refining and petrochemical manufacturing.

Another factor is the difficulty of transportation of TFE. It simply cannot be transported over distances as hydrocarbon gases are. The reason is that TFE can explode, both in the presence and the absence of oxygen and with the force of black powder. In the absence of oxygen TFE explodes and generates carbon and tetrafluoromethane by a reaction called "deflagration". Consequently, the TFE polymerization plant must be in close proximity to the monomer plant. Even in TFE liquid storage and short distance movement it must be inhibited from auto-polymerization by adding a telogen such as di-limonene. The inhibitor is removed by distillation immediately prior to polymerization.

The polymerization and finishing of PTFE has to be done in high steel alloy equipment with extensive safety precautions. Both the capital and operating expenditure render PTFE and other fluoropolymers significantly more costly than hydrocarbon-based polymers with the exception of a few specialty ones.

The design, manufacturing, and operating of plants for TFE monomer and PTFE polymer production are akin to the management of hazardous materials. The costs and difficulties have made entry into PTFE and other fluoropolymers difficult. That has led to the existence of relatively few companies in the fluoropolymer business.

20.5 PTFE Scrap Sources for Recycling

PTFE recycling dates back to the 1960s, about a decade after its full commercialization. Resin manufacturers were looking for outlets for their second grade (off specification) products. These products usually deviated from the particle size or mechanical property specifications. Processors and fabricators tested these PTFE products and were able to produce useful parts for less demanding segments of the market. In a separate stream, resin manufacturers converted the "off spec" material to fine particle powders for use as additives.

There are a surprisingly large number of sources of recycled feed stuff. They come in two categories: PTFE resin-based and part scraps. Resin-based materials originate from polymer manufacturers whilst part scraps come from processors and fabricators. Sources of scrap from PTFE manufacturing plants are:

- Second quality PTFE (*uncontaminated*):
    - granular resins
    - fine powder resins.
- Other uncontaminated material:
    - reject polymerization reactor bead
    - heels from finishing equipment.
- Contaminated PTFE:
    - equipment failure
    - wet resin
    - *trench*: spills, wash, floor sweepings.

Sources of scrap PTFE processor and fabricator operations are:

- Processed PTFE *unsintered*:
    - defective preforms (e.g., cracked billets)
    - start-up and heel scrap paste extrusion.
- Processed PTFE *sintered*:
    - defective parts (e.g., cracked billets, ram extrusions)
    - start-up and heel ram extrusion.
- Small scrap (large volume):
    - edge trim (films and sheets)
    - machining scrap
        - dry and clean
        - contaminated with cutting oil
        - contaminated with aqueous cutting fluid
    - mixed scrap, spilled resin, floor sweepings, etc.

The various scrap PTFE types do not all have the same value for recycling and reusing. The best definition available has been offered by Reprolon Texas which is the largest recycler of PTFE in the world. Reprolon Texas has been recycling PTFE scrap, gathered globally, since 1992. Its purchasing and grading criteria for PTFE Scrap are (https://reprolon.com):

- **Grade 1 Scrap**: Includes clean items fabricated from virgin granular PTFE molding powders, unmodified by pigments or fillers. Typical examples are cores; cracked billets and preforms; rod, tube, and sheet ends; chucking stubs; out-of-tolerance and obsolete parts; heavy tape punching and trimmings; and exceptionally clean machine turnings uncontaminated by oil.
- **Grade 2 Scrap**: Includes items similar to Grade 1 Scrap; however, the products may be fabricated from partially or totally reprocessed PTFE, with some easily removable contamination allowable.
- **Grade 3 Scrap**: Includes etched, degraded virgin, or reprocessed PTFE. Items with deeply embedded contaminants such as rusty valve seats and gaskets, wire and cable jackets, oil contaminated machine turnings, and the like.
- **Unusable Scrap**: When it is extremely difficult to remove excessive amounts of oil and contaminants from some machine turnings and other types of scrap. Proper care of these contaminated

products should be exercised during their collection and storage. Any overly contaminated scrap, particularly oily pieces, will greatly reduce its value, and may downgrade the entire container as unusable. Freight charges on unusable scrap will be deducted from the value of the usable scrap.

There are no quantitative value systems for the various types of PTFE scrap. The author, solely based on his own experience in the industry, has ranked the value of different scrap types from 1 (low) to 10 (high) as seen in Tables 20.3–20.6.

TABLE 20.3

Relative Value of Scrap from PTFE Manufacturing Plant (High Value = 10, Low Value = 1)

Source	Relative Value
Second grade granular resins (uncontaminated)	10
Second grade fine powder resins (uncontaminated)	10
Heels from finishing equipment	4
Rejected reactor bead (uncontaminated)	9
Equipment failure (usually contaminated)	2
Wet resins (H_2O contaminated)	4
Trench: spills, wash, floor sweepings, etc. (contaminated)	1

TABLE 20.4

Value of Scrap PTFE Processor and Fabricator Operations (High Value = 10, No Value = 0): Unsintered and Uncontaminated

Source	Relative Value
Defective preforms (e.g., cracked billets)	9
Start-up scrap and heel of paste extrusion	4

TABLE 20.5

Value of Scrap PTFE Processor and Fabricator Operations (High Value = 10, No Value = 0): Sintered and Uncontaminated

Source	Relative Value
Defective parts (e.g., cracked billets, ram extrusions)	7
Start-up scrap and heel of ram extrusion	6

TABLE 20.6

Value of Scrap PTFE Processor and Fabricator Operations (High Value = 10, No Value = 0): Machining Waste

Source	Relative Value
Dry and mostly clean	4
Contaminated with cutting oil	2
Contaminated with aqueous cutting fluid	3
Mixed scrap, spilled resin, floor sweepings	1

20.6 Routes to the Reuse of Polytetrafluoroethylene

The term "recycling" can be confusing because it is used both in the context of its specific meaning and generally referring to reuse. There are many sources of PTFE scrap and second grade material recycled by the industry for useful end purposes. The recycling feed material can be classified into three general groups: virgin, recycled, and reprocessed. The definition of these three terms are:

1. Virgin PTFE is resin that has never been preformed or sintered.
2. Recycled PTFE consists of PTFE material made from feedstock without thermal or chemical treatment.
3. Reprocessed PTFE is resin that has been chemically and/or thermally treated.

There are three paths to recycling PTFE for reuse:

1. Virgin PTFE.
2. Physical processing:
   a. Recycling
   b. Reprocessing.
3. Radiation or thermal processing to reduce the molecular weight of PTFE (discussed in Chapter 18).

These methods are discussed next.

20.6.1 Virgin PTFE

The first question is why would virgin PTFE make its way to the recycle heap? The answer is PTFE manufacturing plants, like any others, produce a variety of unusable material which cannot be sold as first grade resin. The significant forms of unusable material are listed in Table 20.3–20.6, which do not all have the same values. The most valuable material is second grade resins that have been rejected because of not meeting specifications. Examples of off-spec reasons include particle size, apparent density, tensile strength, and other property deviations.

In such cases, the resin is classified as second grade which also means it cannot be certified for compliance with ASTM, ISO, or other standards. It is then sold into the recycle/reuse market. In some cases the second grade resin is used to make parts which are sold into less demanding applications. They can also become feed for fluorinated additive (micropowder) products. The resin is first treated by radiation (electron beam) or heat to reduce its molecular so that it can be ground into smaller particles. Micropowder particle size is an important property for most of its applications.

20.6.2 Physical Processing: Recycling

The first group of parts suitable for physical recycling include either unsintered parts/powder or those free of impurities and additives. They include parts fabricated from virgin granular PTFE molding powders, without pigments or fillers. Examples are cores; cracked billets and preforms; rod, tube, and sheet ends; chucking stubs; out-of-tolerance and obsolete parts; thick tape punching and trimmings; and exceptionally clean machine turnings. No contamination by oil or substances are allowed.

A second group of scrap parts include items similar to the first group. This group may include small amounts of products fabricated from partially or totally reprocessed PTFE. If there is contamination that could be easily removed, they can be used. For example light oil contamination can be removed by washing with warm water and soap followed by drying the wet parts.

This group of parts are ground into controlled particle sizes, which are classified. The previously unsintered powder may be used for billet, rod, and sheet molding. The previously sintered powders may be converted by ram (reciprocating) extrusion into rods and other cross-sections for machining. The small particles may also be used as feed for micropowder manufacturing [2].

20.6.3 Physical Processing: Reprocessing

Unfortunately most PTFE scrap is contaminated by organic and inorganic substances including oil, grease, pigments, and ordinary dirt. The only practical way to reuse this scrap group is to remove all the contaminants by a cleaning process. A typical process consists of:

1. Grind the PTFE scrap into a fine powder.
2. Wash the powder from Step 1 with an acid solution.
3. Separate the acid-washed powder from Step 2 followed by a clean water rinse.
4. Wash the powder from Step 3 by a base solution.
5. Separate the base-washed powder from Step 4 followed by a clean water rinse.
6. Dry the rinsed PTFE powder.
7. Bake the dry PTFE powder by sintering, followed by cooling.
8. Grind the PTFE slabs into a fine powder and classify by the particle size.

The powders made by this process are considered "presintered" and suitable for ram extrusion or compression molding under pressure at or above its melting point. In contrast, virgin PTFE is preformed into a shape and sintered freely under atmospheric pressure.

20.7 Disposal of Fluoropolymers

By now it is clear that fluoropolymers are very valuable and should be recycled if at all possible. Fluoropolymer scrap that is not useable (see p. 000) is disposed. To ensure the veracity of the statements about this critical subject we will quote from the authoritative source: *Guide to the Safe Handling of Fluoropolymer Resins* [3]. The reader is urged to contact the Plastics Industry Association (www.PlasticsIndustry.org) before implementing any statements made in this chapter, particularly with respect to the disposal of fluoropolymers.

Disposal Considerations

Fluoropolymer dispersions typically have the following waste disposal considerations. Consult your supplier for information on their particular product.

Preferred options for disposal are: (1) Separate solids from liquid by precipitation and decanting or filtering. Dispose of dry solids in a landfill that is permitted, licensed or registered to manage industrial solid waste. Consult your local regulations for discharge of liquid filtrate to a wastewater treatment system. (2) Incinerate only if incinerator is capable of scrubbing out hydrogen fluoride and other acidic combustion products.

RCRA/US EPA Waste Information[12]: Discarded product is not typically a hazardous waste under RCRA 40 CFR 261. Dispose in an authorized landfill site or incinerate under approved controlled conditions. This product may be incinerated above 800°C (1472°F) using a scrubber to remove hydrogen fluoride. Empty Container: Empty containers should be punctured or otherwise destroyed before disposal. Empty containers must not be used for home or personal uses. The processes are complex and costly. Any laps in thorough removal of HF results in corrosion at the incineration plant itself besides human exposure

All Disposal Methods: Treatment, storage, transportation, and disposal of this product and/or container must be in accordance with applicable federal, state/provincial, and local regulations.

20.8 Chemical Recycling ("Upcycling")

This development began in Europe and, specifically in Germany, is also referred to as "upcycling". There are a number issues that have arisen from the sustainability and regulatory standpoints. Often fluoropolymers are a small part of an apparatus or equipment along with metals, plastics, and other materials. It is impractical and economically not feasible to separate these components for recycling. Because of their extreme chemical resistance fluoropolymers are used as contact surfaces with a variety of chemicals. To date the majority of these parts have either been buried in the ground or incinerated. Many EU countries and other localities have banned and continue to ban the burial of PTFE waste.

There are a variety of problems with the burial of PTFE scrap especially dispersion, even in dry form. The contaminant on the parts/components, such as oils, other organics, and toxins, while in use, would be present underground, if buried. Impurities in PTFE including fillers, dies, and other additives are present in the resin and its compounds. Some of the most concerning agents include surfactants in the PTFE dispersions. A primarily bad actor was ammonium perfluorooctanoate (APFO and PFOA) phased out in the mid-2010s after their traces were detected in humans in all continents, even in polar bears. These surfactants have been found to be bio-cumulative and bio-persistent because of their resistance to degradation.

The replacement surfactants have shorter perfluoro alky chains and usually contain an ether bond in the alkyl chains which contribute to easier degradation; and the surfactants are less bio-cumulative. While the new surfactants are relatively safer than APFO and PFOA they are still considered toxic.

The incineration of fluoropolymer scraps is an option which is still in use. The process generates hydrofluoric acid (HF) and a wide variety of toxins that must be scrubbed and removed from the exhaust gases before atmospheric discharge.

In 2010 3M Dyneon launched a project to convert the end of life products (EOL) PTFE and other fluoropolymers to their monomers (upcycling) and other constituents. The project has been a collaboration among the University of Bayreuth, 3M, and other parties. The goal of the project was to convert fluoropolymers to monomers that could be repolymerized into fluoropolymers or used in other ways. The monomers mainly include tetrafluoroethylene (TFE) and hexafluoroethylene (HFP). Such a process would be an ideal way of recycling fluoropolymers. The concept was the short cycle pyrolysis of the scrap PTFE using heat from microwaves and other sources in a fluidized bed [4].

Dyneon 3M built a pilot plant with an annual capacity of 500 metric tons of perfluorinated polymers (PFP) that started operation in the second half of 2014 (Figure 20.1). The intention was to increase the operational intake to full capacity by 2015. Initially, the plant was commissioned to process fully fluorinated polymers such as PTFE, PFA, and FEP, but the second phase targeted polymer compounds containing fillers.

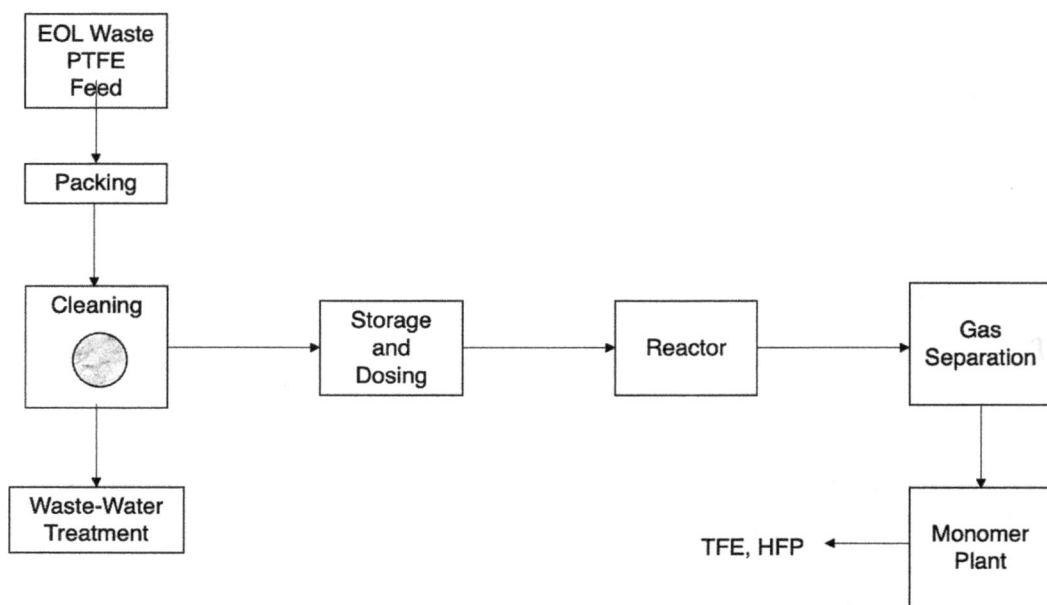

FIGURE 20.1 Schematic diagram of the 3M Dyneon upcycling process [5].

The new high-temperature recycling process includes a grinding stage, after which the PFPs, which are preferably end-of-life products, are decomposed into their monomers at temperatures above 600°C. These monomers are the same chemical components from which the polymers were produced. This process called "pyrolysis" primarily produces tetrafluoroethylene (TFE) and hexafluoropropylene (HFP) with a recovery rate of 90–95%. The resulting gas mixture is then passed to the Dyneon monomer plant and cleaned by distillation. After this step, TFE with a purity of 99.9999% is obtained and can be used to manufacture arbitrary new fluoropolymers with no loss in performance [6].

Notes

1 Resource Conservation and Recovery Act (RCRA).
2 Environmental Protection Agency (EPA).

REFERENCES

1. O'Driscoll, M. China Supply Shortages Hit Consumers as New Sources Emerge. *Aluminum International Today*; 2017.
2. Ebnesajjad, S. *Introduction to Fluoropolymers*, 2nd ed., Elsevier, Oxford, UK, 2021.
3. Guide to the Safe Handling of Fluoropolymer Resins, 5th edition, BP-101, pub by Plastics Industry Association, 2019, www.PlasticsIndustry.org.
4. Wißler, C., New Process for Recycling Fluoropolymers, pub in Chemanager On-line, www.chemanager-online.com, April 10, 2010.
5. Up-Cycling-Closing the loop, Protecting the environment takes more than just good intentions, pub by 3M™ Dyneon™ Fluoropolymers, www.dyneon.eu, 2016.
6. Closing the Recycling Loop-Up-Cycling of End-of-Life Fluoroplastics, published by 3M Company, 2018.

Appendix 1

Trade Names of Common Commercial Fluoropolymers and Films

Trade Name	Company	Web Page
3M™ Dyneon™ Fluoroelastomers	3M Dyneon	www.3m.com
3M™Dyneon™ THV Fluoroplastic	3M Dyneon	www.3m.com
Aclar® Film	Honeywell	www.honeywell.com
Aclon™	Honeywell	www.honeywell.com
AFLAS®	AGC Chemicals Americas	www.agcchem.com
Aflon®	AGC Chemicals Americas	www.agcchem.com
Algoflon®	Solvay	www.solvay.com
Chemraz® (FFKM parts)	Greene Tweed	www.gtweed.com
CYTOP®	AGC Chemicals Americas	www.agcchem.com
DAI-EL	Daikin America	daikin-america.com
DAI-EL THERMOPLASTIC	Daikin America	daikin-america.com
Dyneon™ THV Fluoroplastic	Dyneon	www.3m.com
Elaftor	HaloPolymer	www.halopolymer.com
Fluon®	AGC Chemicals Americas	www.agcchem.com
Fluoroplast	HaloPolymer	www.halopolymer.com
Fluoraz® (FEPM parts)	Greene Tweed	www.gtweed.com
Fluorplast	HaloPolymer	www.halopolymer.com
Ftorplast	HaloPolymer	www.halopolymer.com
Halar®	Solvay	www.solvay.com
HALEON™	HaloPolymer	www.halopolymer.com
Hylar®	Solvay	www.solvay.com
Hyflon®	Solvay	www.solvay.com
Kalrez® (FFKM parts)	DuPont	www.dupont.com
KF Polymer	Kureha Corporation	www.kureha.com
Kynar®	Arkema	www.arkema.com
Kynar Flex®	Arkema	www.arkema.com
LUMIFLON®	AGC Chemicals Americas	www.agcchem.com
Nafion™	Chemours	www.chemours.com
NEOFLON	Daikin America	daikin-america.com
POLYFLON	Daikin America	daikin-america.com
Polymist®	Solvay	www.solvay.com
Silastic™ (FSR)	Dow	www.dow.com
SKF	HaloPolymer	www.halopolymer.com
Solef®	Solvay	www.solvay.com
Tarflen®	Grupa Azoty, S.A.	www.grupaazoty.com
Tecnoflon®	Solvay	www.solvaycom
Tedlar® (films)	DuPont	www.dupont.com

Trade Name	Company	Web Page
Teflon™	Chemours	www.chemours.com
Tefzel™	Chemours	www.chemours.com
Viton™	Chemours	www.chemours.com
Voltalef®	Arkema	www.arkema.com
ZEFFLE	Daikin America	daikin-america.com
Zonyl™ Micropowders	Chemours	www.chemours.com

Appendix 2

PTFE Granular Resins Used for Compression Molding or Ram Extrusion or Both

This is a specification that covers the following six types of PTFE generally used for compression molding or ram extrusion, or both, as well as their classification and testing:

Type I: Resin used for general-purpose molding and ram extrusion.

Type II: Finely divided resin with an average particle size less than 100 micrometers.

Type III: Modified resins, either finely divided or free-flowing, typically used in applications requiring improved resistance to creep and stress-relaxation in end use.

Type IV: Free-flowing resins. Generally made by treatment of finely divided resin to produce free-flowing agglomerates.

Type V: Presintered resin. Resin that has been treated thermally at or above its melting point at atmospheric pressure without having been previously preformed.

Type VI: Resin, not presintered, but for ram extrusion only.

The classification based on ASTM D4894-15 covering details of requirements for tests on resins is provided in Table A1.1. A similar classification covering detailed requirements for tests on molded specimens is given in Table A1.2. Other details pertaining to sample preparation and test procedures are in the body of this ASTM standard.

TABLE A1.1

Detailed Requirements for Tests on Resins

Type	Grade	Bulk Density, g/L	Mean Particle Size,* μm	Water Content,** %	Melting Peak Initial, °C	Melting Peak Second, °C
I	1	700 ± 100	500 ± 150	0.04	*A*	327 ± 10
	2	675 ± 50	75	0.04	*A*	327 ± 10
II	—	—	<100	0.04	*A*	327 ± 10
III	1	400 ± 125	<100	0.04	*A*	327 ± 10
	2	850 ± 100	500 ± 150	0.04	*A*	327 ± 10
IV	1	650 ± 150	550 ± 225	0.04	*A*	327 ± 10
	2	>800	—	0.04	*A*	327 ± 10
	3	580 ± 80	200 ± 75	0.04	*A*	327 ± 10
V	—	635 ± 100	500 ± 250	0.04	327 ± 10	327 ± 10
VI	—	650 ± 150	800 ± 250	0.04	*A*	327 ± 10

* Diameter;

** Maximum. *Note*: A > 5°C above the second melting peak temperature.

TABLE A1.2

Detail Requirements for Tests on Molded Specimens

Type	Grade	Thermal Instability Index, Max	Standard Specific Gravity Min/Max	Tensile Strength, Min, MPa (psi)	Elongation at Break, Min %
I	1	50	213/2.18	13.8 (2000)	140
	2	50	213/2.18	17.2 (2500)	200
II	—	50	2.13/2.19	27.6 (4000)	300
III	1	50	2.14/2.22	28.0 (4060)	450
	2	50	2.14/2.18	20.7 (3000)	300
IV	1	50	2.13/2.19	25.5(3700)	275
	2	50	2.13/2.19	27.6 (4000)	300
	3	50	2.15/2.18	27.6 (4000)	200

Notes:
1. Type V and VI resins are not applicable by molding techniques included in this techniques.
2. Extrusions from the Type VI resin show different degrees of clarity from the other resins
3. Standard Specific Gravity is defined as the specific gravity of a specimen of PTFE material molded as described in this specification and sintered using the appropriate sintering schedule given in the specification.
4. Thermal Instability Index is a measure of the degree of the decrease of molecular weight of PTFE material, which has been treated for a prolonged period of time.

Appendix 3

PTFE Resins Produced from Dispersion

This is a specification that covers polytetrafluoroethylene (PTFE) resins prepared by coagulation of the aqueous PTFE dispersion. These resins are homopolymers of tetrafluoroethylene or modified homopolymers containing no more than 1% by weight of other fluoromonomers. The materials covered herein do not include mixtures of PTFE with additives such as colors, fillers, or plasticizers, nor do they include reprocessed or reground resin or any fabricated articles because the properties of such materials have irreversibly changed when they were fibrillated or sintered.

The classification based on ASTM D4895-15 covers PTFE Type I and Type II: resin produced from dispersion, each type of resin having the same requirements for bulk density, particle size, water content, melting peak temperature, tensile quality, and elongation. Each type of resin is divided into grades in accordance with standard specific gravity (SSG), thermal instability index (TII), and stretching void index (SVI). Grades are divided into classes according to extrusion pressure. Details about grades and classes are given in Tables A3.1 and A3.2.

Other details pertaining to sample preparation and test procedures are in the body of this ASTM standard.

TABLE A3.1

Detail Requirements for All Types, Grades, and Classes

Type	Bulk Density, g/L	Particle Size, Average Diameter, μm	Water Content, Max. %	Melting Peak, °C Initial	Melting Peak, °C Second	Tensile Strength, MPa	Elongation at Break, %
I	550 ±150	500 ±200	0.04	[a]	327 ±10	19	200
II	550 ±150	1050 ±350	0.04	[a]	327 ± 10	19	200

[a] Greater than 5.0°C above the second melting peak temperature.

TABLE A3.2

Detail Requirements for All Types,[A] Grades, and Classes

Type	Grade	Class	Standard Specific Gravity		Extrusion Pressure, MPa	Thermal Instability Index, Max.	Stretching Void Index, Max.
			Min.	Max.			
I	1	A	2.14	2.18	5 to <15[B]	50	NA[C]
		B	2.14	2.18	15 to <55[D]	50	NA[C]
		C	2.14	2.18	15 to <75[E]	50	NA[C]
	2	A	2.17	2.25	5 to <15[B]	50	NA[C]
		B	2.17	2.25	15 to <55[D]	50	NA[C]
		C	2.17	2.25	15 to <75[E]	50	NA[C]
	3	C	2.15	2.19	15 to <75[E]	15	200
		D	2.15	2.19	15 to <65[E]	15	100
		E	2.15	2.19	15 to <65[E]	50	200
	4	B	2.14	2.16	15 to <55[D]	15	50
II	1	A	2.14	2.25	5 to <15[B]	50	NA[C]

Notes:

[A] The types, grades, and classes are not the same as those in previous editions of Specification D4895.

[B] Tested at the reduction ratio 100:1.

[C] Not applicable.

[D] Tested at the reduction ratio 400:1.

[E] Tested at the reduction ratio 1600:1.

The tests listed in Tables A3.1 and A3.2, as they apply, are sufficient to establish conformity of a material to this specification.

Appendix 4

Applications of Fluoropolymer Films

Practically, all fluoropolymers can be converted into films, either by casting and evaporating from aqueous systems or from a solution or by melt processing techniques, predominantly by extrusion. These films can consist of a single polymer or of a blend of two or more polymers or can be composed of several layers forming a laminate. Such a laminate may also include a fabric or a nonwoven layer. Table A4.1 includes most currently known industrial applications of commercial fluoropolymer films.

TABLE A4.1

Examples of Applications of Fluoropolymer Films by Industry and/or Activity

Industry/Activity	Type of Films Used	Examples of Applications
Aerospace	Skived PTFE films; cast PTFE films; MFA films; PCTFE films; ECTFE films; PVF films	Mold release films for the manufacture of composite structures; EL lamps; aircraft cabin interiors
Automotive	Skived PTFE films; cast PTFE films; MFA films; PCTFE films; THV films; PVF films	Mold release films for the manufacture of composite structures; EL lamps for automotive applications; automotive interiors, layer in safety glass; truck and trailer siding
Chemical processing industry	Skived PTFE films; cast PTFE films; FEP films; MFA films; PCTFE films; ECTFE films; PVDF films; THV films	Inner cores of chemically resistant hoses, barrier films; sample bags and containers; transparent sight glasses; tank liners and roll covers; pump diaphragms; lining in chlorine cells and water treatment; chemical storage bags, flowmeters, gas sampling bags
Construction	ETFE films; PCTFE films; ECTFE films; THV films; PVF films	Architectural roofing designs; greenhouses; EL lamps in buildings; architectural and protective coating; canopies, skylights; commercial sidings and trim; conformable building panels
Consumer products	Cast PTFE films; PVF films	Laminate sheets for home baking; labels; flexible signs; laminates for silk-screening graphics
Defense	Skived PTFE films; cast PTFE films; PCTFE films; PVF films	Mold release films for the manufacture of composite structures; EL lamps for military applications; package films for corrosion-resistant parts; special laminates
Electrical and electronic	Skived PTFE films; cast PTFE films; unsintered PTFE films; FEP films; PFA films; MFA films; PCTFE films; ECTFE films; PVDF films; THV films; Teflon™ AF films; PVF films	Bonding films for microwave circuit boards; insulations, flexible printed circuits; flat cables; computer LCD panels; dissipative laminates; electret condenser microphone; semiconductor technology; harness wraps
Energy and environment	Unsintered PTFE films; MFA films; ETFE films; ECTFE films; PVDF films; THV films; Teflon™ AF films; PVF films	Batteries (lithium ion, zinc-air); fuel cell components; solar collector applications; anti-graffiti protection; deep UV-resistant films and components; solar energy applications; membranes in fuel cells; protection of photovoltaic panels

(Continued)

TABLE A4.1 (*Continued*)

Examples of Applications of Fluoropolymer Films by Industry and/or Activity

Industry/Activity	Type of Films Used	Examples of Applications
Engineering	Skived PTFE films; cast PTFE films; unsintered PTFE films; PFA films; MFA films; ETFE films; PCTFE films; PVDF films; PVF films	Gas and liquid barrier in compensators, expansion joints, and bellows; plumbers sealing tape; release films in manufacture of composite structures; EL lamps in business equipment; backlit signs; shielding of UV radiation; billboards, highway signs; highway sound barriers
Food processing	Skived PTFE films: cast PTFE films	Release film and composite sheets for food processing, baking, and grilling
Medical	Cast PTFE films	Surgical gowns; towels, protective gowns; packaging of disinfection materials packages of drugs and medical devices
Packaging	Unsintered PTFE films; PCTFE films	Pharmaceutical packaging and laminations; packaging of corrosion-sensitive components
Safety and protection	Skived PTFE films; cast PTFE films; MFA films	Protective clothing; UV-protective films
Other	Skived PTFE films; cast PTFE films; unsintered PTFE films: PVF films	Substrate (backing) of heat-resistant pressure sensitive tapes, billboards, other graphics; interiors of transit vehicles and passenger trains

Note: EL = electroluminescent; FRP = fiberglass-reinforced plastic; LCD = liquid crystal display; UV = ultraviolet.

Source: Data from Drobny, J.G., *Applications of Fluoropolymer Films – Properties, Processing, and Products*, Chapter 21, Elsevier, 2020.

Appendix 5

Tedlar® Film Designation Guide

The identification codes consist of a combination of upper case letters and numbers, such as **TTR10SG3**

T stands for *Tedlar® symbol*

TR stands for *end use property or color*

10 refers to the *nominal thickness**

S refers to *surface*

G is for *gloss or other properties*

3 indicates the *type of PVF film*

Details regarding the above letters and numbers are shown in the table below.

End Use Property and Color	Surface	Type	Gloss and/or Other Properties
FM Flame modified	**S** Strippable, nonadherable release films	**1** High shrinkage compatible with curing polyester resins	**G** Glossy
MR High temperature release	**A** One side adherable for use with adhesives	**2** High tensile, excellent fold and flex endurance	**M** Medium gloss
PC Epoxy board release	**B** Both sides adherable	**3** Standard tensile and elongation	**L** Low gloss
ST Special transparent		**4** High elongation, good formability, good heat sealing properties	**S** Satin gloss
TR Transparent		**5** Excellent formability for release and lamination of engineering plastics and metals	**E** Meets aircraft specifications
UT Ultraviolet screening film, transparent			
UW Ultraviolet screening film, translucent white			
WH White			

* Nominal thickness or approximate gauge thickness times 10; for example 10: Nominal 100 gauge = Nominal 1 mil.
Source: TEDLAR® Polyvinyl Fluoride Film, Film Designation Guide, H-49723, DuPont, www.dupont.com, 2019.

Appendix 6

Heat Sealing Temperatures for Common Fluoroplastics

Material	Temperature Range, °F (°C)
ECTFE	475–500 (245–260)
ETFE	530–600 (275–315)
FEP	540–700 (285–370)
PCTFE	450–500 (230–260)
PFA	600–750 (315–400)
PVDF	400–425 (205–220)
PVF	400–425 (205–220)

Bibliography

Books

Introduction to Fluoropolymers, Second Edition (Ebnesajjad, S., Editor), Elsevier, Oxford, UK, 2021.

Drobny, J.G., *Application of Fluoropolymer Films, Properties, Processing, and Products*, Elsevier, Oxford, UK, 2020.

Ebnesajjad S., *Expanded PTFE – Technology, Manufacturing and Applications*, Plastics Design Library. Elsevier, Oxford, UK, 2017.

Drobny J.G., *Handbook of Fluoroelastomers, The Definitive User's Guide and Data Book*, 2nd ed., Plastics Design Library, Elsevier, Oxford, UK, 2016.

Ebnesajjad S, *Fluoroplastics, Vol. 2, Melt Processible Fluoropolymers, The Definitive User's Guide and Data Book*, 2nd ed., Plastics Design Library, Elsevier, Oxford, UK, 2016.

Ebnesajjad S, *Fluoroplastics, Vol. 1, Non-Melt Processible Fluoropolymers, The Definitive User's Guide and Data Book*, 2nd ed., Plastics Design Library, Elsevier, Oxford, UK, 2015.

Handbook of Fluoropolymer Science and Technology (Smith, D.W., Iacono, S.T., Iyer, S.S., Eds.), John Wiley & Sons, Hoboken, NJ, 2014.

Drobny J.G., *Handbook of Thermoplastic Elastomers*, 2nd ed., Elsevier, Oxford, UK, 2014.

Introduction to Fluoropolymers – Materials, Technology, and Applications (Ebnesajjad S., Ed.) Plastic Design Library, Elsevier, Oxford, UK, 2013.

Ebnesajjad S. *Polyvinyl Fluoride – Technology and Applications of PVF*. Elsevier, Oxford, UK, 2012.

Ebnesajjad S., Morgan R. A., *Fluoropolymer Additives*, Elsevier, Oxford, UK, 2012.

Grot, W., *Fluorinated Ionomers*, 2nd ed., Plastics Design Library. Elsevier Oxford, UK, 2011.

Drobny J.G., *Technology of Fluoropolymers*, 2nd ed., Boca Raton, FL, CRC Press, 2009.

Moore A.L., *Handbook of Fluoroelastomers, The Definitive User's Guide and Data Book*, Plastics Design Library, William Andrew Publishing, Norwich, NY, 2005.

Ebnesajjad S., Khaladkar, P.R., *Fluoropolymers Applications in Chemical Processing Industries, The Definitive User's Guide and Data Book*, Plastics Design Library, William Andrew Publishing, Norwich, NY, 2005.

Améduri B., Boutevin B., *Well-Architectured Fluoropolymers: Synthesis, Properties and Applications*. Amsterdam, Elsevier BV, 2004.

Fluoropolymers 2, Properties (Hougham, G., Cassidy P., Johns, K., and Davidson T., Eds.), Kluver Academic/Plenum Publishers, New York, 1999.

Fluoropolymers 1, Synthesis (Hougham, G., Cassidy, P., Johns, K., and Davidson, T., Eds.) Kluver Academic/Plenum Publishers, New York, 1999.

Modern Fluoropolymers (Scheirs, J, Ed.). John Wiley & Sons Ltd; Chichester, UK, 1997.

Fluoropolymers (Wall L.A., Ed.), Wiley-Interscience, New York, 1972.

Advances in Fluorine Chemistry (Stacey M., Tatlow, J.C., Sharpe A.G., Eds.), Butterworth & Co., Ltd., Kent, UK, 1963.

Reviews

Teng, H., Overview of the Development of the Fluoropolymer Industry, *Appl. Sci.* 2012, *2*, p. 496.

Drobny J.G., *Fluoroplastics*, Volume *16*, Number 4, Review Report 184, Rapra Technology Ltd, Shawbury, Shrewsbury, Shropshire, UK. 2006.

Souzy, R. Améduri, B., Functional Polymers for Fuel Cells Membranes. *Progr. Polym. Sci. 30* (2005) p. 644.

Scheirs J, *Fluoropolymers: Technology, Markets and Trends*. Rapra Industry Report Series, Rapra Technology Ltd., Shawbury, Shrewsbury, Shropshire, 2001.

Magazines

The *Journal of Fluorine Chemistry* is published monthly by Elsevier Science (Amsterdam) and contains original papers and short communications describing both pure and applied research on the chemistry and applications of fluorine and of compounds where fluorine exercises significant effects. The journal covers inorganic, organic, organometallic, and physical chemistry and also includes papers on biochemistry and environmental industrial chemistry.

Glossary of Terms

A

Abrasion resistance   Wear rate or abrasion rate measured by a number of methods, such as *Taber Abrasion Test* (ASTM D3389).

Adherend   A part to be covered by an adhesive and then joined into an adhesive joint.

Adhesive   A material, usually polymeric, capable of forming permanent or temporary surface bonds with another material as it is or after processing, such as curing. The main classes of adhesives include hot melt, pressure sensitive contact, UV cured, and EB cured.

Adhesive bond strength   The strength of a bond formed by joining of two materials using an adhesive. Bond strength may be measured by peeling or shearing the two adherends (see above) using extensiometry.

Amorphous polymer   A polymer having a non-crystalline or amorphous supramolecular structure or morphology. Amorphous polymers may have some molecular order but usually are substantially less ordered than crystalline polymers and consequently have inferior mechanical properties.

Annealing   A process in which a material, such as plastic, glass, or metal, are heated and then cooled slowly. In plastics and metals, it is used to reduce stresses formed during fabrication.

APA   Advanced Polymer Architecture is a process technology producing fluorocarbon elastomers with optimized structure (polymer branching) and an innovative cure-site monomer. The products exhibit significantly improved processing characteristics, rapid cure, and good physical properties and low compression set without post-curing.

ASTM   American Society for Testing and Materials is a nonprofit organization with the purpose of developing standards on characteristics and performance of materials, products, systems, and services and promoting the related knowledge. Now ASTM International.

Average particle size   The average diameter of solid particles as determined by various test methods.

B

Bar   A metric (SI) unit of pressure, equal to 1.0×10^6 dyne/cm^2, or 1.0×10^5 pascals (Pa). It has the dimension of unit of force per unit of area and is used to denote the pressure of gases, vapors, and liquids.

Biaxial orientation   Orientation in which the material is drawn in two directions, usually perpendicular to one another. It can be either sequential or simultaneous. Commonly used for films and sheets.

Blow Molding   The process of forming hollow articles by expanding a hot plastic element against the internal surfaces of a mold. In its simplest form the process comprises extruding a tube (parison) downward between opened halves of a mold, closing the mold and injecting air to expand the tube, pinched on the bottom.

C

Calander   A processing equipment used to form a thermoplastic material into a sheet or film. It consists of two or more steel (often heated) rolls with an adjustable Gap between them.

Coalesce   To combine particles into one body or to grow together.

Coefficient of Friction   A number expressing the amount of frictional effect usually expressed in two ways: static and dynamic.

Cold Flow (creep)   Tendency of a material to flow slowly under load and or/over time.

Comonomer   A monomer reacting with a different monomer in a polymerization reaction, the result of which is a copolymer.

Contact angle   A measure of the ability of a liquid to wet solid surfaces. It expresses the relationship between the surface tension of a liquid and the surface energy of the surface on which the liquid rests. As the surface energy decreases, the contact angle increases.

Corona treatment   A method to render inert polymers more receptive to wetting by solvents, adhesives, coatings, and inks using high voltage discharge. The corona discharge oxidizes the surface thus making it more polar.

Cross-linking   A reaction during which chemical links are formed between polymeric chains. The process can be carried out by chemical agents (e.g., organic peroxides), reactive sites on the polymeric chains, or by high energy radiation (e.g., electron beam).

Cryogenic   Refers to very low temperatures, below about –150°C (–238°F).

Crystalline Melting Point   A temperature at which the crystalline portion of the polymer melts.

Crystallinity   A state of molecular structure attributed to the existence of solid crystals with a definite geometric form.

Cure   A process of changing the properties of a polymer by a chemical reaction (condensation, polymerization, or addition). In elastomers it means mainly cross-linking or vulcanization.

D

Degradation   Loss of or undesirable change in polymer properties as a result of aging, chemical reactions, wear, use, exposure, etc. The properties include color, size, and strength.

Density   The mass of any substance (gas, liquid, or solid) per unit volume at specified temperature and pressure.

Dielectric constant   The ratio of the capacitance assembly of two electrodes separated by a plastic insulating material to its capacitance when the electrodes are separated by air only.

Dielectric heating   The heating of polymeric materials by dielectric loss (see below) in a high-frequency electrostatic field.

Dielectric loss   A loss of energy evidenced by the rise in heat of a dielectric placed in an alternating electric field. It is usually observed as a frequency-dependent conductivity.

Dielectric loss factor   The product of the dielectric constant and the tangent of the dielectric loss angle for a material.

Dielectric loss tangent   The difference between 90° and the dielectric phase angle for a material.

Dielectric strength   Ability of a material to resist the passage of electric current. It is expressed in volts per thickness required to break down through the thickness of a dielectric (insulation) material and create a puncture. ASTM D149 is the standard used to measure dielectric strength of plastic insulation materials.

Differential scanning calorimetry (DSC)   The method to measure the heat flow to a sample as a function of temperature. It is used to measure specific heats, glass transition temperatures, melting points, melting profiles, degree of crystallinity, and degree of cure, purity, and more.

Dynamic vulcanization   A process to produce a *thermoplastic vulcanizate* (TPV), which is a type of thermoplastic elastomer consisting of a soft, elastic phase dispersed in a hard thermoplastic matrix. It is essentially mixing a raw (unvulcanized) elastomer, a hard thermoplastic, and vulcanization ingredients in an internal mixer at high shear conditions. The mixing temperature must be high enough to melt the plastic and to cause vulcanization (cross-linking). (See also *TPE* and *FTPE*).

E

Elasticity   The ability of a material to quickly recover its original dimensions after removal of the load that has caused the deformation.

Elastomer   A polymeric substance with elastic properties. Such material can be stretched repeatedly at room temperature to at least twice its original length and upon immediate release of the stress will return with force to its approximate original length.

ETFE   Copolymer of ethylene and tetrafluoroethylene noted for an exceptional chemical resistance, toughness, and abrasion resistance.

Electron beam cure   A process using electron beam radiation to promote reactions in a polymeric materials leading to cross-linking, polymerization, modification, and degradation.

Electron beam (EB) radiation   Ionizing radiation propagated by electrons accelerated by very high voltage (typically kilovolts to megavolts). This radiation is used frequently for processing of polymeric materials. (See *Electron Beam Cure*).

F

FEP   Fluorinated ethylene propylene having excellent nonstick and non-wetting properties.

Film formation   A process in which a film is formed after solvent or water evaporates or due to a chemical reaction.

Friction, dynamic   Resistance to continued motion between two surfaces, also known as *sliding friction*.

Friction, static   Resistance to initial motion between two surfaces.

FTPE   *Fluorinated thermoplastic elastomer* is a polymeric material exhibiting elastic behavior similar to cross-linked (vulcanized) rubber but can be processed by conventional thermoplastics methods without curing (cross-linking). This allows flash from molding and other scrap as well as post-consumer waste to be recovered and reused. FTPEs are essentially phase-separated systems. Usually, one phase is hard and solid at ambient temperature and the other one is soft and elastic. The hard phase forms the *physical cross-links*, which are thermoreversible. Essentially, there are two main types of FTPEs, namely *block copolymers*, and fluorinated *thermoplastic vulcanizates* (FTPVs, see below); several other types are also phase separated systems but prepared by different production methods.

FTPV   *Fluorinated thermoplastic vulcanizate* is a thermoplastic vulcanizate produced from two fluorinated components, one fluoroplastic and one fluoroelastomer, using *dynamic vulcanization*. The product is composed of two phases, one is soft and elastic and the other one is hard. The vulcanized (cross-linked) phase is dispersed in the thermoplastic phase.

Fuel cell   An electrochemical energy conversion device. It produces electricity from various external quantities of fuel (on the anode side) and oxidant (on the cathode side). These react in the presence of an electrolyte. Fuel cells are different from batteries in that they consume the reactant, which must be replenished, while batteries store electrical energy chemically in a closed system.

Fusion   A process in which a continuous film or a solid body is formed by melting and a flowing (coalesce) of polymer particles.

G

Glass transition temperature (T_g)   A point below which an amorphous polymer behaves as glass does. It is very strong and rigid, but brittle. Above this temperature it exhibits leathery or rubbery behavior.

H

Heat buildup   Heat generated within a polymeric material due to its viscoelasticity (hysteresis) and friction. It occurs during processing (mainly friction and kneading) and in service (mainly repeated cycling).

HFP   Hexafluoropropylene, a monomer used for the production of FEP and other copolymers, such as THV, and of fluorinated elastomers.

Hysteresis   Incomplete recovery of strain during unloading cycle due to energy consumption. This energy is converted from mechanical to frictional energy (heat).

I

Ionizing radiation   Any electromagnetic or particulate radiation, which in its passage through matter is capable of producing ions directly or indirectly. Examples are electron beams, X-rays, and gamma radiation.

Ionomer resins   Modified polymers obtained by heating and pressing certain polymers containing carboxylic groups in the presence of metallic ions.

L

Laminate   A product made by bonding together one or more layers of material or materials. It is frequently assembled by simultaneous application of heat and pressure. A laminate may consist of coated fabrics, metals, or films or it may be different combinations of these.

Latex   A stable dispersion of a polymeric substance (most frequently of an elastomer) in an essentially aqueous medium.

Limiting oxygen index   LOI is defined as the required minimum percentage of oxygen in a mixture with nitrogen, which allows a flame to be sustained by an organic material such as a polymer.

M

Melt-processible polymer   A polymer that melts when heated to its melting point and forms a molten material with a definite viscosity value at or somewhat above its melting temperature. Such a melt can be pumped and should flow when subjected to a shear rate using commercial processing equipment such as extruders and molding machines.

MFA   A copolymer of TFE and perfluoro (methyl vinyl ether) with properties similar to PFA; it has an approximately 20°C lower melting temperature than PFA.

Micron (micrometer)   A unit of length equal to 1×10^{-6} m. The *micrometre* (international spelling as used by the International Bureau of Weights and Measures; SI symbol: μm) or *micrometer* (American spelling), also commonly known by the previous name *micron*, is an SI derived unit of length.

Modified PTFE   Copolymer of TFE and of a small amount (less than 1 %) of other perfluorinated monomers (e. G. Perfluoroalkoxy monomer) exhibiting considerably improved physical properties, moldability, and much lower microporosity.

Monomer   A relatively simple compound, usually containing carbon and of low molecular weight, which can react to form a polymer by combination with itself or with other similar molecules or compounds.

N

Nanometer   A unit of length equal to 1×10^{-9} m. Often used to denote the wavelength of radiation especially UV and the visible spectral region and size of very small particles. Its symbol is nm.

Newtonian fluid   A term to describe an ideal fluid in which shear stress is proportional to shear rate with *viscosity* being the proportionality coefficient. In this case viscosity is independent of shear rate in contrast to non-ideal fluids, where viscosity is a function of shear rate. The latter represents non-Newtonian fluids that include paints and polymer melts.

O

Orientation   A process of drawing or stretching of thermoplastic films and fibers in order to orient polymer in the direction of stretching. While fibers are drawn in one direction, films may be drawn in another direction (uniaxially, either longitudinally or transversely) or two directions (biaxially). Oriented films and fibers have enhanced properties in the direction of stretching. The film will shrink in the direction of stretching when reheated without tension.

Orientation Index (OI)   A measure of the degree of orientation in the machine (longitudinal) direction versus that in the cross-machine (transverse) direction.

Ozone   Molecule consisting of three atoms of oxygen, i.e., O_3.

P

Perfluorinated resin   A polymer consisting of monomers where all main chain carbons are combined with fluorine atoms only (PTFE, FEP, PFA, and MFA).

Permeability   The capacity of a material to allow another substance to pass through it; or the quantity of a specific gas or other substance, which passes through under specific conditions.

PFA   Copolymer of TFE with perfluoro (propyl vinyl ether), an engineering thermoplastic characterized by excellent thermal stability, release properties, low friction, and toughness. Its performance is comparable to PTFE with the difference that it is melt-processible.

Piezoelectricity   The ability of some materials (notably crystals and certain ceramics) to generate an electric potential in response to applied mechanical stress. This may take the form of a separation of electric charge across the crystal lattice. If the material is not short-circuited, the applied charge induces a voltage across the material. The effect finds useful applications such as the production and detection of sound, generation of high voltages, electronic frequency generation, microbalances, and ultrafine focusing of optical assemblies.

PMVE   Perfluoro (methyl vinyl ether) a monomer used for the production of MFA.

Polymer fume fever   An illness characterized by temporary flu-like symptoms caused by inhaling the products released during the decomposition of fluoropolymers, mainly PTFE. Tobacco smoke enhances the severity of this condition.

Postcure   A second cure at high temperatures enhancing some properties and or removing decomposition products of the primary reaction.

PPVE   Perfluoro (propyl vinyl ether) monomer used for the production of PFA.

Prorad   Radiation promoter, a compound promoting or enhancing the cross-linking reaction by high energy (ionizing) radiation.

Pyroelectricity   The ability of certain materials to generate an electrical potential when they are heated or cooled. As a result of this change in temperature, positive and negative charges move to opposite ends through migration (i.e., the material becomes polarized) and hence an electrical potential is established.

R

Radiation dose   Amount of ionizing radiation energy absorbed by a material during irradiation. The unit of radiation dose is a grey (Gy), defined as 1J per kilogram. In practical application a larger unit, namely kGy (10^3 Gy) is used. The previously used unit, no longer official since 1986, was megarad (Mrad), equal to 10 kGy. (See *Ionizing radiation*).

Reduction Rate   RR is defined as the ratio of the cross-section of the polymer before extrusion to that after extrusion. It is an important characteristic applicable to polymer and paste extrusion.

Rheology   A science that studies and characterizes the flow of polymers, resins, gums, and other materials.

S

Sintering   A process in which particles are heated, softened, and coalesced, thus forming a continuous film or a solid body, used typically in the processing of PTFE.

Specific gravity   The ratio of the density of a substance to the density of a standard, usually water for a liquid or solid, and air for a gas. Specific gravity has no unit because of that.

Substrate   Any surface to be coated or bonded by an adhesive.

Surface energy   The energy associated with the intermolecular forces at the interface between two media. The *surface energy* per unit area equals the surface tension, called also the *free surface energy*. Solid materials can be divided into two categories: high or low surface energy. High energy solids include metals and inorganic compounds, with typical values of 200 to 500 mN/m. Low energy materials are generally organic materials, including polymers with values below 100 mN/m. (See also *Surface tension*).

Surface resistance   The surface resistance between two electrodes in contact with a material is the ratio of the voltage applied to the electrodes to that portion of the current between them, which flows through the surface layers.

Surface tension   The cohesive force at a liquid surface measured as a force per unit length along the surface or the work, which has to be done to extend the area of the surface by a unit area, e.g., by a square centimeter. It is an important factor in the wetting of solids by liquids and in the formation of adhesive bonds.

Surfactant   A widely used contraction of *surface active agent*, a compound that alters the surface tension of a liquid in which it is dissolved.

T

Terpolymer   The product of simultaneous polymerization of three different monomers, or of the grafting of one monomer to the copolymer of two monomers.

TFE   Tetrafluoroethylene, a perfluorinated monomer used as a feedstock for the production of PTFE and as a comonomer for the production of a variety of other fluoropolymers.

Thermoforming   A process of forming a plastic film or sheet into a three-dimensional shape by clamping it, heating it, and then applying a differential pressure to make the film or sheet conform to the shape of the mold.

Thermogravimetric analysis (TGA)   A widely used method to determine weight change upon heating, such as decomposition, and the amount of volatile components, including moisture.

THV   A terpolymer of TFE, HFP, and VDF. The trade name is 3M™ Dyneon™ THV Fluoroplastic.

TPE Thermoplastic elastomer is a polymeric material exhibiting elastic behavior similar to cross-linked (vulcanized) rubber but can be processed by conventional thermoplastic methods without curing (cross-linking). This allows flash from molding and other scrap as well as post-consumer waste to be recovered and reused. TPEs are essentially phase-separated systems. Usually, one phase is hard and solid at the ambient temperature and the other one is soft and elastic. The hard phase forms the *physical cross-links*, which are thermoreversible. Essentially, there are two main types of TPEs, namely *block copolymers*, and *thermoplastic vulcanizates* (TPVs, see below).

TPV Thermoplastic vulcanizate is a thermoplastic elastomer produced from two components, one thermoplastic and one elastomer, using *dynamic vulcanization*. The product is composed of two phases, one is soft and elastic and other one is hard. The vulcanized (cross-linked) phase is dispersed in the thermoplastic phase.

U

Ultraviolet (UV) radiation Electromagnetic radiation in the 40 to 400 nm wavelength region. UV radiation causes polymer degradation and other chemical reactions, including polymerization and cross-linking of monomeric and oligomeric systems.

V

Viscoelasticity The tendency of polymers to respond to stress as if they were a combination of elastic solids and viscous fluids.

Viscosity The property of resistance of flow exhibited within the body of a material. Units of viscosity are Pascal (traditional) and Pa.s (SI). Conversion: 10 P = 1 Pa•s or 1 cP = 1 mPa•s.

Viscosifying agent A substance used to increase the viscosity of a liquid mainly by swelling.

Volume resistivity The electrical resistance between opposite sides of a cube.

W

Wainscot Interior paneling in general and, more specifically, paneling that covers only the lower portion of an interior wall or partition. It has a decorative or protective function.

Wetting The spreading out (and sometimes absorption) of a fluid onto (or into) a surface. In adhesive bonding, the wetting occurs when the surface tension of the liquid adhesive is lower than the critical surface tension of the substrate. Good surface wetting is essential for high strength adhesive bond. Wetting can be increased by preparation of the part surface prior to adhesive bonding.

Y

Yield deformation The strain at which the elastic behavior begins, while the plastic is being strained. Deformation beyond the yield deformation is not reversible.

Index

Pages in *italics* refer to figures and **bold** refer to tables

R

For Product Safety Concerns and Information please contact our EU
representative GPSR@taylorandfrancis.com
Taylor & Francis Verlag GmbH, Kaufingerstraße 24, 80331 München, Germany